Monografie Matematyczne
Instytut Matematyczny Polskiej Akademii Nauk (IMPAN)

Volume 65
(New Series)

Founded in 1932 by
S. Banach, B. Knaster, K. Kuratowski,
S. Mazurkiewicz, W. Sierpiński, H. Steinhaus

Volumes 31–62 of the series
Monografie Matematyczne were published by
PWN – Polish Scientific Publishers, Warsaw

Lucian Bădescu

Projective Geometry and Formal Geometry

Springer Basel AG

Author:

Lucian Bădescu
Dipartimento di Matematica
Università degli Studi di Genova
Via Dodecaneso 35
16146 Genova
Italy
e-mail: badescu@dima.unige.it

2000 Mathematics Subject Classification 14B20, 14E25, 14J99, 14L30, 14N05, 32A20, 32D15

A CIP catalogue record for this book is available from the Library of Congress, Washington D.C., USA

Bibliographic information published by Die Deutsche Bibliothek
Die Deutsche Bibliothek lists this publication in the Deutsche Nationalbibliografie;
detailed bibliographic data is available in the Internet at <http://dnb.ddb.de>.

ISBN 978-3-0348-9626-9 ISBN 978-3-0348-7936-1 (eBook)
DOI 10.1007/978-3-0348-7936-1

Originally published by Birkhäuser Verlag, Basel, Switzerland in 2004
Softcover reprint of the hardcover 1st edition 2004
Cover design: Micha Lotrovsky, CH-4106 Therwil, Switzerland
Printed on acid-free paper produced of chlorine-free pulp. TCF∞

ISBN 978-3-0348-9626-9
9 8 7 6 5 4 3 2 1

www.birkhauser-science.com

For Mihaela, Miruna and Ciprian

Preface

The aim of this book is to introduce the reader to some modern methods of projective geometry involving, in one way or another, certain techniques of formal geometry. Some of these methods are illustrated in the first part through the proofs of a number of results of projective geometry (of a rather classical flavor and having an obvious interest in themselves). In three of these proofs the crucial idea is to consider the first infinitesimal neighbourhood of a certain closed subvariety Y of an ambient projective variety X. In other words, the schemes with nilpotents can play a very important role, sometimes even when one deals with classical problems of projective geometry. When we want to study a certain closed embedding $Y \subset X$ of algebraic varieties, quite often we are naturally led to consider all higher order infinitesimal neighbourhoods of Y in X. In other words, one is led to work with the formal completion $X_{/Y}$ of X along Y, which is a formal scheme, introduced into algebraic geometry by Zariski and Grothendieck in the 1950s. The formal scheme $X_{/Y}$ is very well suited to study the given embedding $Y \subset X$, and is an analogue of the concept of tubular neighbourhood of a submanifold of a complex manifold.

The second part starts with a systematic and rather self-contained presentation of the basic concepts of formal geometry that are necessary to prove a number of basic results concerning the extension of formal-rational functions to rational functions. As we mentioned before, this theory is due to Zariski and Grothendieck. It was subsequently developed by Hironaka, Matsumura, Hartshorne, Chow, Faltings and others. It provides powerful tools for proving deep results in algebraic geometry, such as the famous Zariski connectivity theorem. In fact, the second part will make clear the deep relationship between formal geometry and connectivity results in projective geometry (such as the Fulton–Hansen connectivity theorem). The formal geometry is closely related with a number of questions in complex analysis and complex-analytic geometry (considered by Severi, Remmert, Van de Ven, Barth, Rossi, Chow and others), such as the extension problem of meromorphic functions or (more generally) of certain analytic entities. This relationship is also explained in the second part of the book.

Besides, we have tried to apply several results of formal geometry (especially concerning the extension of formal-rational functions to rational functions) to projective geometry, in particular to the geometry of homogeneous spaces. These aspects are also closely related with various kinds of Grothendieck–Lefschetz theorems in projective geometry. We hope to convince the reader that the interplay between projective and formal geometry is very important and fruitful.

The book is self-contained modulo the basic concepts and results of algebraic geometry and cohomology (for instance Hartshorne's textbook [73] is in principle enough as background material). It contains many recent results which have not appeared in any other textbook. The second part could be considered in particular as an introduction to formal schemes. The book addresses persons working in algebraic geometry and in complex geometry, as well as all mathematicians working in other fields, but interested in questions related to the ones touched upon here.

Acknowledgment

Part 1 is based on a Ph.D. course I gave at the University of Milan (Italy) in the Spring 1999. I am grateful to Antonio Lanteri for inviting me at the University of Milan to give this "corso di dottorato INDAM", to Leonede De Michele for supporting this idea, and to the Istituto Nazionale di Alta Matematica (Roma) for financial support. I enjoyed the pleasant and stimulating atmosphere of the whole group of algebraic geometers of Milan: Alberto Alzati, Marina Bertolini, Elisabetta Colombo, Antonio Lanteri, Marino Palleschi, Cristina Turrini, and Alfonso Tortora. I thank them all. My thanks also go to Francesco Russo and his family for their friendship and warm hospitality I enjoyed during my stay in Milan (Rho).

I also want to thank the members of the Algebraic Geometry Seminar organized at the Institute of Mathematics of the Romanian Academy (Bucharest) in collaboration with the Department of Mathematics of the University of Bucharest (Cristian Anghel, Vasile Brînzănescu, Nicolae Buruiană, Iustin Coandă, Gabriel Chiriacescu, Paltin Ionescu and Cristian Voica) for allowing me to present most of the material of the second part in the Fall of 2001 and for their criticism and remarks; in particular, that was an excellent occasion for me to revise the presentation and to correct some errors. I am especially grateful to Paltin Ionescu for reading carefully large parts of various preliminary versions and for suggesting a number of improvements of the presentation. The final version of the book was written at the University of Genova.

Finally, I want to thank Piotr Pragacz for his interest in this book and for suggesting to publish it in the Series "Monografie Matematyczne" edited by the Institute of Mathematics of the Polish Academy of Sciences and Birkhäuser Verlag.

Contents

Introduction

This book is devoted to some aspects concerning the interplay between projective geometry and formal geometry. It consists of two parts. The main aim of Part 1 is to introduce the reader to some modern aspects and methods of projective geometry. The material presented here has been conceived to touch three important themes: the connection between projective geometry and deformations of quasi-homogeneous singularities, cohomological dimension and connectedness results, and applications of formal geometry to projective geometry. Part 1 is based on a series of lectures the author gave in the Spring of 1999 at the Dipartimento di Matematica, Università degli Studi di Milano (see [13]).

As far as the first theme is concerned, Chapter 1 starts with the classical problem of classifying the extensions in $\mathbb{P}^{n+1}$ of a given closed subvariety Y in $\mathbb{P}^n$. This problem was very popular among Italian algebraic geometers in the first part of the last century. A renewed interest in it started in the 1970s. As a consequence of the modern methods used, the classical picture has been completed and better understood and many new results have been proved. In particular, the whole story has been put in a new perspective. It is beyond our scope to discuss all its aspects and results here. We rather restrict ourselves to presenting a remarkable result due to Zak–L'vovsky (Theorem 1.2), also because its proof presented here is relevant for the philosophy of this book. This proof relies on another fundamental result due to Mori–Sumihiro–Wahl (Theorem 1.3) which is also of interest in itself. Chapters 1–3 provide the proofs of these two results and a number of applications, comments and examples. The condition involved in the Zak–L'vovsky theorem (the surjectivity of the Zak map) is better understood in the case of curves in terms of the so-called Gaussian maps (see Chapter 4). The study of Gaussian maps was initiated in 1987 by J. Wahl with the main motivation of understanding the geometry of curves lying on $K3$-surfaces. In Chapter 5 we present a result of Schlessinger which relates the deformation theory of the vertex of the affine cone C_Y over a smooth projectively normal closed subvariety Y in $\mathbb{P}^n$ with the projective geometry of Y. In particular, it becomes transparent that the surjectivity of the Zak map is naturally interpreted in terms of deformations of the vertex of the affine cone C_Y over $Y \subset \mathbb{P}^n$.

The second theme is presented in Chapter 7. First we prove a special case of a result of Hartshorne–Lichenbaum (conjectured by Grothendieck) which asserts that the cohomological dimension of a quasi-projective variety U of dimension n is $\leq n-1$ if and only if U is not a projective variety. This result is then applied to prove a generalization of the Fulton–Hansen connectedness theorem to weighted projective spaces. Then some applications of this connectedness result are given. We will come back to Fulton–Hansen's result in the second part of the book to reinterpret it in the framework of formal functions and then to prove a significant improvement of it.

In Chapters 6 and 8 we present two results whose proofs involve, in an essential way, the first infinitesimal neighbourhood of a closed subvariety $X \subset \mathbb{P}^n$. The first result, due to Van de Ven, characterizes the linear subspaces as the only irreducible smooth closed subvarieties of $\mathbb{P}^n$ for which the normal sequence splits. The second, due to Ellingsrud–Gruson–Peskine–Strømme, gives a necessary and sufficient condition for a curve Y lying on a complete intersection surface X in $\mathbb{P}^n$ to be the scheme-theoretic intersection of X with a hypersurface of $\mathbb{P}^n$ (see Theorem 8.4). The methods of proving this latter result also yield a geometric proof of a result of Barth–Larsen which asserts that $\mathrm{Pic}(X) = \mathbb{Z}$ for every smooth closed subvariety X of $\mathbb{P}^n$ of dimension $\geq \frac{n+2}{2}$. We hope that the method of using the first infinitesimal neighbourhood will convince the reader less familiar with formal methods that formal geometry deserves to be studied and applied to projective geometry; in particular, this provides an excellent motivation for the second part of the book. At the end of Chapter 8, two appendices are added. In Appendix A we derive a special case of Lefschetz's theorem for the Picard group from the well-known topological Lefschetz theorem on hyperplane sections (see e.g. [105]). In Appendix B we review some elementary facts about cyclic covers.

A very preliminary version of Part 2 is our survey [14]. The main sources of Part 2 are Grothendieck–Dieudonné [66], Hironaka–Matsumura [81], Hartshorne [75], Faltings [47], [48], Gieseker [58], Bădescu [11], [15], Bădescu–Schneider [17], [18] and Bădescu–Beltrametti–Ionescu [19]. Fix an irreducible projective variety X over an algebraically closed field k. Let $K(X)$ be the field of rational functions of X, and let $K(X_{/Y})$ denote the ring of formal-rational functions of X along a closed subscheme Y of X. Sometimes we shall call the formal-rational functions simply formal functions. If X is normal and Y is connected of positive dimension, then $K(X_{/Y})$ is a field. The field $K(X)$ can be considered in a natural way as a subfield of $K(X_{/Y})$, and the extension problem of formal-rational functions asks when can any formal-rational function on X along Y be extended to a rational function on X, i.e. when do we have $K(X) = K(X_{/Y})$? If the latter equality holds, according with the terminology of Hironaka and Matsumura [81], we say that Y is $G3$ in X.

The formal functions have been introduced and studied systematically in algebraic geometry by Zariski in [151]. In particular, Zariski used them to prove his famous connectedness theorem. Later on this theory was considerably extended and deepened by Grothendieck in the framework of formal schemes, by providing powerful tools (such as Grothendieck's existence theorem, see [66] III, Theorem (5.1.4)) to study related questions (see [67], [68]).

There is an analogous problem in the complex-analytic setting. Given a pair (X, Y) consisting of a complex irreducible projective variety X and a closed subvariety Y of X, let U be a connected open neighbourhood of Y in X (in the complex topology), and denote by $\mathcal{M}(U)$ the $\mathbb{C}$-algebra of all meromorphic functions on U. We get the field extensions $K(X) \subseteq \mathcal{M}(U) \subseteq K(X_{/Y})$, and therefore the extension problem for meromorphic functions can be solved if it can be solved for formal functions, i.e. the equality $K(X) = \mathcal{M}(U)$ is a consequence of the equality $K(X) = K(X_{/Y})$.

Extension of analytic or formal objects like functions or subvarieties is a rather classical subject. The earliest reference we are aware of is a paper of Severi [137], in which he proved that any meromorphic function defined in a small complex connected neighbourhood of a smooth hypersurface in $\mathbb{P}^n(\mathbb{C})$ $(n \geq 2)$ can be extended to a meromorphic function on $\mathbb{P}^n(\mathbb{C})$. After some partial extension of Severi' s result by Remmert and Van de Ven [123], Barth [21] generalized the above-mentioned result of Severi to arbitrary closed positive-dimensional connected subvarieties of $\mathbb{P}^n(\mathbb{C})$. On the other hand, Hironaka [78] extended Severi's result to arbitrary smooth projective varieties X and smooth hypersurfaces Y in X with ample normal bundle. His approach is algebraic and he actually considers the extension problem of formal-rational functions along Y. Further important contributions to the extension problem of meromorphic functions or analytic objects are due to Griffiths [63], Chow [35], Rossi [124], and others.

The second part (which, as we said before, is a natural continuation of part one) contains six chapters. In Chapter 9 we present the basic definitions and concepts as well as some of the fundamental results (due to Grothendieck, Hironaka, Hironaka–Matsumura, Hartshorne, Faltings, and others) concerning the problem of extending formal-rational functions to rational functions. This (rather long) chapter can also be considered as an introduction to formal schemes and formal functions.

Chapter 10 starts by explaining the relationship between formal geometry and the so-called Grothendieck–Lefschetz conditions $\mathrm{Lef}(X, Y)$ and $\mathrm{Leff}(X, Y)$ for a pair consisting of a closed subvariety Y of a projective ambient variety X. These conditions are crucial to prove algebro-geometrically various types of Lefschetz theorems (see [68]). Moreover it turns out that, under some additional mild hypotheses, one has

$$\mathrm{Leff}(X, Y) \implies Y\ G3 \text{ in } X \implies \mathrm{Lef}(X, Y).$$

In Chapter 10 one also relates, via Serre's GAGA [135], the problem of extending formal-rational functions with the analogous problem in Complex Analysis of extending meromorphic functions defined in a small complex neighbourhood of a closed subvariety.

Motivated by the strong connection of formal geometry with connectedness theorems in algebraic geometry given by Theorem 9.21 and Corollary 9.22, in Chapter 11 we interpret the Fulton–Hansen connectedness theorem in terms of formal-rational functions. Then we strengthen this result to Theorem 11.2. One of the relevant consequences of this strengthening asserts that the diagonal embedding Δ_Y of the product $Y \times Y$ is $G3$ in $Y \times Y$ for every closed irreducible subvariety Y of $\mathbb{P}^n$ such that $\dim(Y) > \frac{n}{2}$ and $n \geq 3$. In other words, for every closed irreducible subvariety Y as above, the diagonal embedding $\Delta_Y \subset Y \times Y$ enjoys an important positivity property. Some further applications of Theorem 11.2 are given. We also include Zak's original proof of the finiteness of the Gauss map which used formal functions. At the end of this chapter we prove an impossibility cohomological criterion for a smooth projective variety Y of dimension ≥ 2 to occur as the zero locus of a global section of an ample vector bundle on a smooth projective variety X with prescribed normal bundle (Theorem 11.25).

In Chapter 12 one proves two criteria for extending formal-rational functions. The first one (Theorem 12.6) is a refinement of a result of Hartshorne (see [75], [74], or also Theorem 9.28). This result will be applied in the next chapter to homogeneous spaces. The second result (Theorem 12.9) asserts that the $G3$ condition holds when Y is an effective Cartier divisor on X such that $X \setminus Y$ is a semi-affine variety in the sense of Goodman–Landman [61]. In Appendix C following Chapter 12 one proves a result of Hartshorne concerning the ampleness of the normal bundle of a curve lying in an abelian variety and an ampleness criterion of Gieseker for vector bundles on curves.

Chapter 13 deals with formal functions on homogeneous spaces and presents some fundamental results due to Chow, Faltings, and others. In particular, applying the main result of the previous chapter, we get more elementary proofs of some results of Faltings [47] concerning formal functions on homogeneous spaces. Various applications are given, e.g. concerning the s-geometrically non-degenerate closed subvarieties of an abelian variety. The concept of s-geometrically non-degenerate subvariety of an abelian variety is due to Ran–Debarre [122], [42], and some of its main properties are presented in Appendix D following Debarre [42].

In Chapter 14 the projective manifolds carrying quasi-lines are studied (following [19]). A quasi-line Y of a smooth projective complex variety X of dimension $n \geq 2$ is a smooth rational curve $Y \subset X$ whose normal bundle is isomorphic to the normal bundle of a line in $\mathbb{P}^n$. One proves (among other things) that the quasi-lines of X are precisely those smooth curves $Y \subset X$ with ample normal bundle for which the canonical restriction map $\mathrm{Pic}(X) \to \mathrm{Pic}(Y(1))$ has finite cokernel (see Theorem 14.2). One shows by examples that there are pairs (X, Y), (X', Y'), with Y quasi-line in X and Y' quasi-line in X' (with X and X' projective manifolds of dimension $n \geq 3$), such that (X, Y) and (X', Y') are not formally equivalent (i.e. $X_{/Y} \not\cong X'_{/Y'}$). The projective manifolds carrying quasi-lines are special cases of rationally connected manifolds (see [94], or also [93]). Some of their properties are also included (e.g. they are stable under small projective deformations).

Throughout this book the standard terminology and notation of algebraic geometry (see e.g. [73] or [138]and [139]) are used, unless otherwise specified.

Part I

Special Chapters of Projective Geometry

Chapter 1

Extensions of Projective Varieties

We shall fix throughout an algebraically closed ground field k (of arbitrary characteristic, unless otherwise specified).

Let Y be a smooth connected closed subvariety of dimension ≥ 1 of the n-dimensional projective space $\mathbb{P}^n$ over k.

Definition 1.1. A closed irreducible subvariety X of the projective space $\mathbb{P}^{n+1}$ of dimension $n+1$ is said to be an *extension* of Y (in $\mathbb{P}^{n+1}$) if the following two conditions are satisfied:

1. $\dim(X) = \dim(Y) + 1$.
2. There exists a linear embedding $i : \mathbb{P}^n \hookrightarrow \mathbb{P}^{n+1}$ such that $Y = X \cap H$, where $H := i(\mathbb{P}^n)$ and the intersection is taken in the scheme-theoretical sense.

Example 1.1. Fix $Y \subset \mathbb{P}^n$ as above and a linear embedding $i\colon P^n \hookrightarrow \mathbb{P}^{n+1}$, and set $H := i(\mathbb{P}^n)$. Pick an arbitrary point $x \in \mathbb{P}^{n+1} \setminus H$, and let us denote by X the projective cone in $\mathbb{P}^{n+1}$ over Y of vertex x. Clearly, X is an extension of Y in $\mathbb{P}^{n+1}$. These extensions will be called *trivial extensions*.

One of the important problems of classical projective geometry is to classify all extensions in $\mathbb{P}^{n+1}$ of a given closed subvariety $Y \subset \mathbb{P}^n$. We shall prove a remarkable result due to Zak–L'vovsky in connection with this problem. In order to state it we need to fix some notation.

For every algebraic variety Z we shall denote by Ω^1_Z the sheaf of differential forms of degree 1 on Z (over k). Then we shall define the tangent sheaf T_Z of Z as the dual $(\Omega^1_Z)^* = \underline{\mathrm{Hom}}_Z(\Omega^1_Z, \mathcal{O}_Z)$ of Ω^1_Z. It follows that T_Z is a reflexive sheaf on Z (as the dual of a coherent sheaf). If Z is smooth then T_Z is locally free, i.e. is a vector bundle on Z. Moreover, if Z is a closed subscheme of a scheme W of ideal sheaf $\mathcal{I}$, then the $\mathcal{O}_Z$-module $\mathcal{I}/\mathcal{I}^2$ is called the conormal sheaf of Z in W. The normal sheaf $N_{Z|W}$ of Z in W is by definition the dual $(\mathcal{I}/\mathcal{I}^2)^* = \underline{\mathrm{Hom}}_Z(\mathcal{I}/\mathcal{I}^2, \mathcal{O}_Z)$ of $\mathcal{I}/\mathcal{I}^2$. It follows that $N_{Z|W}$ is a reflexive sheaf on Z. If Z is smooth and W is smooth along Z (i.e. at each point of Z) then $N_{Z|W}$ is locally free, i.e. $N_{Z|W}$ is a vector bundle on Z.

Coming back to our situation, let

$$0 \longrightarrow T_Y \longrightarrow T_{\mathbb{P}^n}|Y \xrightarrow{\;a\;} N_{Y|\mathbb{P}^n} \longrightarrow 0$$

be the normal sequence of Y in $\mathbb{P}^n$. Consider also the Euler sequence of $\mathbb{P}^n$ restricted to Y:

$$0 \longrightarrow \mathcal{O}_Y \longrightarrow (n+1)\mathcal{O}_Y(1) \xrightarrow{\;b\;} T_{\mathbb{P}^n}|Y \longrightarrow 0,$$

where $\mathcal{O}_Y(1)$ is the sheaf of hyperplane sections of Y (with respect to the embedding $Y \hookrightarrow \mathbb{P}^n$), and $(n+1)\mathcal{O}_Y(1)$ denotes the direct sum of $n+1$ copies of $\mathcal{O}_Y(1)$. In particular, we get the surjective maps

$$a(-1) : T_{\mathbb{P}^n}(-1)|Y \to N_{Y|\mathbb{P}^n}(-1) \text{ and } b(-1) : (n+1)\mathcal{O}_Y \to T_{\mathbb{P}^n}(-1)|Y,$$

and therefore the surjective map of vector bundles

$$c := a(-1) \circ b(-1) : (n+1)\mathcal{O}_Y \to N_{Y|\mathbb{P}^n}(-1).$$

Passing to global sections we get the map of k-vector spaces

$$z := H^0(c) : H^0(Y, (n+1)\mathcal{O}_Y) \to H^0(Y, N_{Y|\mathbb{P}^n}(-1)), \tag{1.1}$$

which we call *the Zak map* of Y in $\mathbb{P}^n$.

Now, we can state the main result of this chapter:

Theorem 1.2 (Zak–L'vovsky [149], [103]). *In the above situation, assume furthermore that Y is of codimension $r \geq 2$ and non-degenerate in $\mathbb{P}^n$, and that the Zak map* (1.1) *is surjective. Then every extension of Y in $\mathbb{P}^{n+1}$ is trivial.*

Theorem 1.2 is valid in arbitrary characteristic, even under more general hypotheses (see e.g. [9]; for earlier related results see [10]). The proof we shall give below is one of the proofs of [9] and is valid only in characteristic zero. This proof is based on the following fundamental result proved by Wahl in [146] and inspired by a paper of Mori and Sumihiro [108] (this result will be proved in the next chapter):

Theorem 1.3 (Mori–Sumihiro–Wahl [146]). *Let (X, L) be a normal polarized variety (i.e. a normal projective variety X endowed with an ample line bundle L) of dimension ≥ 2. Assume that the characteristic of k is zero and that $H^0(X, T_X \otimes L^{-1}) \neq 0$. Then there exists an effective divisor E in the complete linear system $|L|$ such that X is isomorphic to the projective cone over the polarized scheme (E, L_E), where $L_E := L|E$. In other words, $X \cong \operatorname{Proj}(A[T])$, where $A := \bigoplus_{i=0}^{\infty} H^0(E, L_E^i)$, with T an indeterminate over A, and the gradation of $A[T]$ is given by $\deg(aT^m) = m + \deg(a)$ whenever $a \in A$ is a homogeneous element. Moreover, $L \cong \mathcal{O}_{\operatorname{Proj}(A[T])}(1)$.*

We shall also make use of the following two elementary results (which will be proved later):

Proposition 1.4 (Schlessinger [129]). *Let X be a normal variety of dimension ≥ 2 over k, Y a closed subvariety of X of codimension ≥ 2, and F a coherent reflexive sheaf. Then the restriction map $H^0(X, F) \to H^0(X \setminus Y, F)$ is an isomorphism of k-vector spaces.*

Proposition 1.5 (Bertini–Serre). *Let E be a vector bundle of rank r on an algebraic variety X over k. Assume that V is a finite-dimensional k-vector subspace of $H^0(X, E)$ which generates E (this means that for every $x \in X$ the $\mathcal{O}_{X,x}$-module E_x is generated by V). Then there is a non-empty Zariski open subset V_0 of V such that $\operatorname{codim}_X(Z(s)) \geq \min\{r, \dim(X)+1\}$ for every $s \in V_0$, where $Z(s)$ denotes the zero locus of s, and $\operatorname{codim}_X(Z(s)) > \dim(X)$ means $Z(s) = \varnothing$.*

Proof of Theorem 1.2. The proof which follows is taken from [9] and works only in characteristic zero (because it makes use of Theorem 1.3 which is in general false in positive characteristic). However, Theorem 1.2 is valid in arbitrary characteristic (see [9] for another proof which is characteristic free).

Claim. (Mumford [109]) Under the hypotheses of Theorem 1.2, for every $i \geq 2$ one has $H^0(Y, N_{Y|\mathbb{P}^n}(-i)) = 0$.

Indeed, since $N_{Y|\mathbb{P}^n}(-i-1) \subseteq N_{Y|\mathbb{P}^n}(-i)$ for every i (via the multiplication by a global equation of a hyperplane in $\mathbb{P}^n$), it will be sufficient to prove the statement for $i = 2$. Assume that there exists a non-zero section $s \in H^0(Y, N_{Y|\mathbb{P}^n}(-2))$. Since Y is non-degenerate in $\mathbb{P}^n$ the k-linear map of vector spaces

$$H^0(P^n, \mathcal{O}_{\mathbb{P}^n}(1)) \to H^0(Y, N_{Y|\mathbb{P}^n}(-1)), \quad \text{given by } h \to hs,$$

is injective. Moreover, the surjectivity of the Zak map (1.1) implies that the second space is of dimension $\leq n+1$. Since the first space is of dimension $n+1$ this map is an isomorphism. In particular, every global section of $N_{Y|\mathbb{P}^n}(-1)$ is of the form hs, with $h \in H^0(P^n, \mathcal{O}_{\mathbb{P}^n}(1))$, whence the zero locus of every global section of $N_{Y|\mathbb{P}^n}(-1)$ contains the support of a non-zero divisor of Y.

On the other hand, the surjective map $c \colon (n+1)\mathcal{O}_Y \to N_{Y|\mathbb{P}^n}(-1)$ considered above shows that the vector bundle $N_{Y|\mathbb{P}^n}(-1)$ of rank $r = \operatorname{codim}_{\mathbb{P}^n}(Y) \geq 2$ is generated by its global sections. Then, by Proposition 1.5, the zero locus of a general section of $H^0(Y, N_{Y|\mathbb{P}^n}(-1))$ should be of codimension $\geq r \geq 2$, a contradiction. The claim is proved.

Let X be an arbitrary extension of Y in $\mathbb{P}^{n+1}$. The hypotheses imply that Y is a Cartier divisor on X. Since Y is smooth, X is smooth at each point of Y. In other words, Y is contained in the smooth locus $V := \operatorname{Reg}(X)$ of X. Moreover, since Y is a hyperplane section of X, $\operatorname{Sing}(X) = X \setminus V$ is a finite (possibly empty) set of points.

On the other hand, the equality $Y = X \cap H$ (scheme-theoretically) means that Y is the proper intersection of X with H. Then a simple general property of proper intersections implies that

$$N_{X|\mathbb{P}^{n+1}} \otimes \mathcal{O}_Y \cong N_{Y|\mathbb{P}^n}. \tag{1.2}$$

Let $f\colon X' \to X$ be the normalization of X (in its field of rational functions). Clearly, $f|f^{-1}(V) : f^{-1}(V) \to V$ is an isomorphism. In other words, V can be identified with a Zariski open subset of X', denoted again by V; in particular, Y is contained both in X and in X' as an ample Cartier divisor ($\mathcal{O}_{X'}(Y) = f^*(\mathcal{O}_X(Y))$ is ample because Y is ample on X and ampleness is preserved under inverse images of finite morphisms). Note also that $\mathcal{O}_{X'}(Y)$ is generated by its global sections because it is the inverse image of the very ample line bundle $\mathcal{O}_X(Y) = \mathcal{O}_X(1)$.

Set $N_{X'} := f^*(N_{X|\mathbb{P}^{n+1}})^{**}$ (bidual) and $N_{X'}(i) := N_{X'} \otimes \mathcal{O}_{X'}(iY)$ for all $i \in \mathbb{Z}$. Clearly, $N_{X'}|V \cong N_{X|\mathbb{P}^{n+1}}|V$. Now, using (1.2) and all these observations, for every $i \geq 1$ we get the exact sequence

$$0 \longrightarrow N_{X'}(-i-1) \xrightarrow{h'} N_{X'}(-i) \longrightarrow N_{Y|\mathbb{P}^n}(-i) \longrightarrow 0,$$

which yields the exact sequence

$$0 \to H^0(X', N_{X'}(-i-1)) \to H^0(X', N_{X'}(-i)) \to H^0(Y, N_{Y|\mathbb{P}^n}(-i)). \tag{1.3}$$

Here $h' \in H^0(X', \mathcal{O}_{X'}(Y))$ is a global equation of Y in X'. By the above claim the last space is zero for every $i \geq 2$, therefore the first map (between the H^0's) is an isomorphism for all $i \geq 2$.

On the other hand, since $\mathcal{O}_{X'}(1)$ is ample, by a general simple statement, $H^0(X', N_{X'}(-i)) = 0$ for every $i \gg 0$. Therefore by induction on i we get that $H^0(X', N_{X'}(-i)) = 0$ for all $i \geq 2$. Then the exact sequence (1.3) (with $i = 1$), the claim and the surjectivity of the Zak map yield

$$\dim_k H^0(X', N_{X'}(-1)) \leq \dim_k H^0(Y, N_{Y|\mathbb{P}^n}(-1)) \leq n+1. \tag{1.4}$$

On the other hand, since X' is normal, $V \subseteq X'$, $\operatorname{codim}_{X'}(X' \setminus V) \geq 2$ and $N_{X'}$ is reflexive, by Proposition 1.4 we get

$$\dim_k H^0(X', N_{X'}(-1)) = \dim_k H^0(V, N_{X|\mathbb{P}^{n+1}}(-1)). \tag{1.5}$$

Then (1.4) and (1.5) yield

$$\dim_k H^0(V, N_{X|\mathbb{P}^{n+1}}(-1)) \leq n+1. \tag{1.6}$$

Now look at the commutative diagram

$$
\begin{array}{ccccccccc}
 & & 0 & & 0 & & & & \\
 & & \downarrow & & \downarrow & & & & \\
 & & \mathcal{O}_V & \xrightarrow{\text{id}} & \mathcal{O}_V & & & & \\
 & & \downarrow & & \downarrow & & & & \\
0 & \longrightarrow & F & \longrightarrow & (n+2)\mathcal{O}_V(1) & \longrightarrow & N_{X|\mathbb{P}^{n+1}}|V & \longrightarrow & 0 \\
 & & \downarrow & & \downarrow & & \downarrow{\scriptstyle \text{id}} & & \\
0 & \longrightarrow & T_V & \longrightarrow & T_{\mathbb{P}^{n+1}}|V & \longrightarrow & N_{X|\mathbb{P}^{n+1}}|V & \longrightarrow & 0 \\
 & & \downarrow & & \downarrow & & & & \\
 & & 0 & & 0 & & & &
\end{array}
$$

in which the last row is the normal sequence of X in $\mathbb{P}^{n+1}$ restricted to $V = \operatorname{Reg}(X)$, the second column is the Euler sequence of $\mathbb{P}^{n+1}$ restricted to V, and $F := \operatorname{Ker}((n+2)\mathcal{O}_V(1) \to N_{X|\mathbb{P}^{n+1}}|V)$. By Proposition 1.4 we have

$$\dim_k H^0(V, (n+2)\mathcal{O}_V) = \dim_k H^0(X', (n+2)\mathcal{O}_{X'}) = n+2.$$

The top long row (after tensored by $\mathcal{O}_V(-1)$) yields the exact sequence

$$0 \to H^0(V, F(-1)) \to H^0(V, (n+2)\mathcal{O}_V) \to H^0(V, N_{X|\mathbb{P}^{n+1}}(-1)),$$

where for every coherent $\mathcal{O}_{X'}$-module G we put $G(-1) := G \otimes \mathcal{O}_{X'}(-Y)$. Therefore the last equalities together with the inequality (1.6) yield $H^0(V, F(-1)) \neq 0$. Then from the first column of the above diagram, taking into account that

$$H^0(V, \mathcal{O}_V(-1)) = H^0(X', \mathcal{O}_{X'}(-1)) = 0,$$

it follows that $H^0(V, T_V(-1)) \neq 0$. Recalling that $T_{X'} = (\Omega^1_{X'})^*$, X' is normal and $\operatorname{codim}_{X'}(X' \setminus V) \geq 2$, this last fact translates — via Proposition 1.4 — into

$$H^0(X', T_{X'}(-1)) \neq 0. \tag{1.7}$$

Now, (1.7) allows one to apply Theorem 1.3 to the normal polarized variety $(X', \mathcal{O}_{X'}(Y))$ (in which $\mathcal{O}_{X'}(Y)$ is not only ample, but also generated by its global sections) to deduce that X' is isomorphic to the projective cone over $(E, \mathcal{O}_{X'}(Y)\otimes \mathcal{O}_E)$ for some $E \in |\mathcal{O}_{X'}(Y)|$. This implies that X' is in fact isomorphic to the projective cone over $(Y, \mathcal{O}_Y(Y))$. Finally, a simple standard argument shows that this implies that X must be the cone over Y. □

Proof of Proposition 1.4. We shall make use of the following well-known general facts (see [65]):

a) Let X be a scheme, Y a closed subsecheme of X, and set $U := X \setminus Y$. Let F be a coherent $\mathcal{O}_X$-module. Then for every $i \geq 0$ one can define the cohomology spaces $H^i_Y(X, F)$ with support in Y, such that there is a canonical exact sequence (called the exact sequence of local cohomology)

$$0 \to H^0_Y(X,F) \to H^0(X,F) \to H^0(U,F) \to H^1_Y(X,F) \to \cdots$$

$$\cdots \to H^q_Y(X,F) \to H^q(X,F) \to H^q(U,F) \to H^{q+1}_Y(X,F) \to \cdots,$$

where the maps $H^q(X,F) \to H^q(U,F)$ are the natural restriction maps.

b) Assume now that $X = \mathrm{Spec}(A)$ is affine and $Y = V(I)$ is given by the ideal I of the commutative Noetherian ring A. Let M be a finitely generated A-module, and let $F := \tilde{M}$ be the coherent sheaf on X associated to M. Let $f_1, \ldots, f_p \in A$ be arbitrary elements of I. Then $f_1, \ldots, f_p$ is said to be an *M-sequence* if f_1 is not a zero divisor in M (i.e. $f_1 m = 0$, with $m \in M$, implies $m = 0$), and f_{i+1} is not a zero divisor in $M/(f_1M + \cdots + f_iM)$ for all $i = 1, \ldots, p-1$. The maximal non-negative integer p such that there is an M-sequence $f_1, \ldots, f_p \in I$ is called the I-depth of the A-module M (denoted by I-depth(M)). One can prove that the following equality holds:

$$I - \mathrm{depth}(M) = \inf_{p \in V(I)} \{\mathrm{depth}(M_p)\},$$

where $V(I) := \{p \in \mathrm{Spec}(A) \mid I \subseteq p\}$, and $\mathrm{depth}(M_p) := pA_p\text{-depth}(M_p)$.

c) Let $r \geq 0$ be a non-negative integer. Then the following two statements are equivalent:

i) I-depth$(M) \geq r$, i.e. pA_p-depth$(M_p) \geq r$ for all $p \in V(I)$.

ii) $H^i_Y(X, F) = 0$ for all $i < r$, where $X = \mathrm{Spec}(X)$, $Y = V(I)$, and $F = \tilde{M}$.

In particular, if I-depth$(M) \geq 2$, then the restriction map $H^0(X,F) \to H^0(U,F)$ is an isomorphism, where $X = \mathrm{Spec}(A)$, $Y = V(I)$ and $F = \tilde{M}$.

d) If A is a normal ring (i.e. a Noetherian domain which is integrally closed in its fraction field) and if I is an ideal of A of height ≥ 2 (which by definition means that every minimal prime ideal of A containing I has height ≥ 2), then I-depth$(A) \geq 2$. This follows from a well-known criterion of normality due to Serre (see [133], IV-44, Théorème 11, or also [1]).

Now we can proceed to the proof of Proposition 1.4. The conclusion of our proposition is local, so we may assume $X = \mathrm{Spec}(A)$ affine, $Y = V(I)$, with I an ideal of A, and $F = \tilde{M}$, with M an A-module of the form $M = \mathrm{Hom}_A(N, A)$, with N a finitely generated A-module (because M is reflexive). Since A is normal of dimension ≥ 2 and $\mathrm{codim}_X(Y) \geq 2$, the criterion of normality of Serre quoted above (see [133], IV-44, Théorème 11) implies that pA_p-depth$(A) \geq 2$ for every $p \in Y = V(I)$. By what we have said above it follows that I-depth$(A) \geq 2$, i.e. there is an A-sequence $f_1, f_2 \in I$. Using the properties recalled above, the conclusion of the proposition is a consequence of the following:

Claim. f_1, f_2 is an M-sequence, and in particular, I-depth$(M) \geq 2$.

To prove the claim observe that since f_1 is not a zero divisor in A we have the exact sequence

$$0 \longrightarrow A \xrightarrow{\ f_1\ } A \longrightarrow B := A/f_1A \longrightarrow 0,$$

in which the map f_1 is the multiplication by f_1. Since the functor Hom is left exact we get the exact sequence

$$0 \longrightarrow \mathrm{Hom}_A(N, A) \xrightarrow{f_1} \mathrm{Hom}_A(N, A) \longrightarrow \mathrm{Hom}_A(N, B).$$

Since $M = \mathrm{Hom}_A(N, A)$, this shows first that f_1 is not a zero-divisor in M, and second, that

$$M/f_1M \subseteq \mathrm{Hom}_A(N, B) \cong \mathrm{Hom}_B(\overline{N}, B),$$

where $\overline{N} := N/f_1N$ (it is immediate to see that $\overline{N}$ becomes a B-module and that there is the above identification of B-modules).

Now we can apply the same argument to $B = A/f_1A$ and to the non-zero divisor $f' := f_2 \mod f_1A$ in B to prove that f_2 is not a zero divisor in $\mathrm{Hom}_B(\overline{N}, B)$, whence, *a fortiori*, not a zero divisor in the B-submodule M/f_1M. The claim (and thereby Proposition 1.4) is proved. □

Proof of Proposition 1.5. (See [51], Appendix B9 for a more general formulation.) Since V generates E, the canonical evaluation map $\varphi\colon X \times V \to E$ defined by $\varphi(x, s) := s(x)$, is a surjective smooth morphism such that every fiber of φ is a k-vector subspace of V of dimension $v - r$, where $v := \dim_k(V)$. Let C be the zero section of the canonical projection $\pi\colon E \to X$, so that $C \cong X$, and in particular, $\dim(C) = \dim(X) =: d$. It follows that $\varphi^{-1}(C) = \{(x, s) \mid s(x) = 0\}$ is a closed irreducible subset of $X \times V$ of dimension $d+v-r$. Indeed, only the irreducibility of $\varphi^{-1}(C)$ has to be checked. Since φ is smooth with irreducible fibers, the restriction $\varphi' := \varphi|\varphi^{-1}(C) : \varphi^{-1}(C) \to C$ is a smooth morphism with irreducible fibers. In particular, φ' is an open morphism (because it is flat). Then the irreducibility of $\varphi^{-1}(C)$ is a consequence of the following general simple fact: if $\varphi'\colon U' \to U$ is an open surjective morphism of algebraic varieties with all fibers irreducible and U irreducible, then U' is also irreducible (exercise).

Let $p\colon X \times V \to V$ be the second projection of $X \times V$, and let us denote by Y the closure of $p(\varphi^{-1}(C))$ in V. Then there are two cases to be considered:

i) $Y \neq V$. (This is always the case for example when $r > d$.) Then, setting $V_0 := V \setminus Y \neq \varnothing$, for every $s \in V_0$ we get

$$p^{-1}(s) \cap \varphi^{-1}(C) \cong \{x \in X \mid s(x) = 0\} = \varnothing,$$

i.e. $Z(s) = \varnothing$, whence $\operatorname{codim}_X(Z(s)) \geq d+1 \geq \min\{r, d+1\}$.

ii) $Y = V$. Then we get the dominant morphism $q := p|\varphi^{-1}(C) : \varphi^{-1}(C) \to V$ of irreducible varieties. By the theorem of the dimension of fibers (see e.g. [138], Theorem 7, page 76) there is a non-empty open subset V_0 of V which is contained in $q(\varphi^{-1}(C)) = p(\varphi^{-1}(C))$ such that

$$\begin{aligned}\dim(q^{-1}(s)) &= \dim(p^{-1}(s) \cap \varphi^{-1}(C)) = \dim(\varphi^{-1}(C)) - \dim(V)\\ &= (d + v - r) - v = d - r, \quad \forall s \in V_0.\end{aligned}$$

In other words, $\operatorname{codim}_X(Z(s)) = r \leq d$ for all $s \in V_0$.

Proposition 1.5 is proved. □

Remark 1.6. The proof of Proposition 1.5 actually shows that if $Z(s) \neq \varnothing$ for s general in V, then $r \leq \dim(X)$ and $\operatorname{codim}_X(Z(s)) = r$.

Now we want to prove a generalization Theorem 1.2 of Zak–L'vovsky. First we shall need a definition.

Definition 1.7. Let Y be a smooth connected closed subvariety of $\mathbb{P}^n$ of dimension $d \geq 1$. Let $m \geq 1$ be an integer. A closed irreducible subvariety X of $\mathbb{P}^{n+m}$ of dimension $d+m$ is called an *extension of* Y *in* $\mathbb{P}^{n+m}$ if there is a linear embedding $i : \mathbb{P}^n \hookrightarrow \mathbb{P}^{n+m}$ such that $Y = X \cap H$ (in the scheme-theoretic sense), where $H := i(\mathbb{P}^n)$. If there exists an extension X of Y in $\mathbb{P}^{n+m}$ which is not a cone, we sometimes also say that Y *extends non-trivially* m *steps*.

Theorem 1.8. *Let* Y *be a smooth connected non-degenerate closed subvariety of codimension* ≥ 2 *of* $\mathbb{P}^n$. *Assume* $\operatorname{char}(k) = 0$, $\dim(Y) \geq 1$, *and*

$$H^0(Y, N_{Y|\mathbb{P}^n}(-2)) = 0.$$

Set $\dim_k H^0(Y, N_{Y|\mathbb{P}^n}(-1)) =: n + r + 1$ *(by Proposition* 3.1 *below,* $r \geq 0$*). If* $m \geq r+1$ *then every extension of* Y *in* $\mathbb{P}^{n+m}$ *is a cone over a closed subvariety of* $\mathbb{P}^{n+m-1}$.

Remark 1.9. By the claim of the proof of Theorem 1.2, the surjectivity of the Zak map (1.1) implies that $H^0(Y, N_{Y|\mathbb{P}^n}(-2)) = 0$. Therefore Theorem 1.8 is a generalization of Theorem 1.2 in characteristic zero.

Proof. We shall proceed in a similar way as in the last part of the proof of Theorem 1.2. Since $\dim(X) = \dim(Y) + m$ and $Y = X \cap H$ (scheme-theoretically), $N_{Y|X} \cong m\mathcal{O}_Y(1)$ and $N_{X|\mathbb{P}^{n+m}}|Y \cong N_{Y|H}$. (Notice that since X is smooth at each point of Y, $N_{X|\mathbb{P}^{n+m}}$ is a vector bundle along Y.) Let $f\colon X' \to X$ be the normalization of X. Then, exactly as in the proof of Theorem 1.2, $Y \subset X'$ and $N_{Y|X'} \cong m\mathcal{O}_Y(1)$. Moreover, $N_{X'} := f^*(N_{X|\mathbb{P}^{n+m}})^{**}$ is a vector bundle along Y such that $N_{X'}|Y \cong N_{Y|H}$. Since f is a finite morphism, $\mathcal{O}_{X'}(1) := f^*(\mathcal{O}_X(1))$ is an ample line bundle on X', generated by its global sections. For every coherent sheaf F on X' and for every integer i, set $F(i) := F \otimes \mathcal{O}_{X'}(i)$. For an integer $i > 0$ consider the exact sequence ($p \geq 0$) coming from these isomorphisms,

$$0 \to \mathcal{I}^{p+1} \otimes N_{X'}(-i) \to \mathcal{I}^p \otimes N_{X'}(-i) \to \mathcal{I}^p/\mathcal{I}^{p+1} \otimes N_{X'}(-i) \to 0, \qquad (1.8)$$

where $\mathcal{I}$ is the ideal of Y in X. Then we have

$$\begin{aligned} \mathcal{I}^p/\mathcal{I}^{p+1} \otimes N_{X'}(-i) &\cong S^p(m\mathcal{O}_Y(-1)) \otimes N_{Y|H}(-i) \\ &\cong \tbinom{m-1+p}{p} N_{Y|H}(-i-p). \end{aligned} \qquad (1.9)$$

The hypothesis that $H^0(Y, N_{Y|\mathbb{P}^n}(-2)) = 0$ implies that $H^0(Y, N_{Y|\mathbb{P}^n}(-i)) = 0$ for every $i \geq 2$. Then by ((1.9) we get $H^0(Y, \mathcal{I}^p/\mathcal{I}^{p+1} \otimes N_{X'}(-1)) = 0$ for all $p \geq 1$. Therefore (1.8) implies that the canonical maps

$$H^0(X', \mathcal{I}^{p+1} \otimes N_{X'}(-1)) \to H^0(X', \mathcal{I}^p \otimes N_{X'}(-1))$$

are isomorphisms for every $p \geq 1$. This yields

$$H^0(X', \mathcal{I} \otimes N_{X'}(-1)) \cong H^0(X', \mathcal{I}^p \otimes N_{X'}(-1)), \quad \forall p \geq 1. \qquad (1.10)$$

On the other hand we have the following:

Claim. Let $Y \neq \varnothing$ be a closed subvariety of the normal projective variety X' defined by the ideal sheaf $\mathcal{I}$. Assume that Y meets every hypersurface of X'. Let F be a coherent torsion free $\mathcal{O}_{X'}$-module such that the canonical maps $H^0(X', \mathcal{I}^p F) \to H^0(X', F)$ are isomorphisms for every $p \geq 1$. Then $H^0(X', F) = 0$.

To prove the claim, let $s \in H^0(X', F)$ be a section of F and pick an arbitrary point $y \in Y$. Then the hypothesis implies that $s_y \in \mathcal{I}_y^p F_y$ for all $p \geq 1$. Since F_y is an $\mathcal{O}_{X',y}$-module of finite type and $\mathcal{I}_y$ is contained in the maximal ideal of the local ring $\mathcal{O}_{X',y}$, by a well-known theorem of Krull (see [133]) we get

$$\bigcap_{p=1}^{\infty} \mathcal{I}_y^p F_y = 0.$$

It follows that $s_y = 0$. Since y was an arbitrary point of Y, s vanishes along Y, whence $s|V = 0$ for a certain open neighbourhood V of Y in X. By hypothesis $\operatorname{codim}_{X'}(X' \setminus V) \geq 2$. On the other hand, since F is torsion free the canonical map

$F \to F^{**}$ into the bidual is injective, whence s is also a section of the reflexive sheaf F^{**}. By Proposition 1.4 the restriction map $H^0(X', F^{**}) \to H^0(V, F^{**})$ is an isomorphism. Since $s|V = 0$, we infer that $s = 0$ as a section of F^{**}, whence also as a section of F because $F \subseteq F^{**}$. This proves the claim.

Coming back to the proof of Theorem 1.8, in the claim we take $F = \mathcal{I} \otimes N_{X'}(-1)$. Since $N_{X'}$ is reflexive F is torsion free. Moreover, by (1.10), F satisfies the hypotheses of the claim. Therefore by the claim we get

$$H^0(X', \mathcal{I} \otimes N_{X'}(-1)) = 0. \tag{1.11}$$

Taking $i = 1$ and $p = 0$ in (1.8) and using (1.11) we infer that the restriction map $H^0(X', N_{X'}(-1)) \to H^0(Y, N_{Y|H}(-1))$ is injective, whence by hypotheses we get

$$\dim H^0(X', N_{X'}(-1)) \leq n + r + 1 \leq n + m. \tag{1.12}$$

At this point, if we set $V := \operatorname{Reg}(X)$, we have $Y \subset V \subset X'$. Then a diagram completely similar to the diagram considered in the proof of Theorem 1.2 (with $(n+2)\mathcal{O}_V(1)$ replaced by $(n+m+1)\mathcal{O}_V(1)$), and the inequality (1.12) yield $H^1(V, T_V(-1)) \neq 0$. Since X' is normal and $\operatorname{codim}_{X'}(X' \setminus V) \geq 2$ we get

$$H^0(X', T_{X'}(-1)) \cong H^0(V, T_V(-1)) \neq 0.$$

Now the conclusion follows from Theorem 1.3. □

Remark 1.10. Under the hypotheses of Theorem 1.8, the closed subvariety Y of $\mathbb{P}^n$ cannot be extended non-trivially more than r steps. In this sense the hypothesis that $H^0(Y, N_{Y|\mathbb{P}^n}(-2)) = 0$ is essential. For, let Y be a smooth complete intersection in $\mathbb{P}^n$ of multi-degree $(d_1, \ldots, d_r)$, such that $2 \leq r \leq n-1$ and $d_i \geq 2$, $i = 1, \ldots, r$. Then $H^0(Y, N_{Y|\mathbb{P}^n}(-2)) \neq 0$, and clearly Y can be extended non-trivially in $\mathbb{P}^{n+m}$ for every $m \geq 1$. We also note that by a result of Barth (see [22]), every smooth closed subvariety Y of $\mathbb{P}^n$ of dimension ≥ 1 which can be extended smoothly in $\mathbb{P}^{n+m}$ for all $m \geq 1$ is necessarily a complete intersection.

Examples of closed subvarieties $Y \subset \mathbb{P}^n$ satisfying $H^0(Y, N_{Y|\mathbb{P}^n}(-2)) = 0$ are given by the following:

Proposition 1.11 (Wahl [147]). *Let Y be a smooth projectively normal closed subvariety of $\mathbb{P}^n$ of dimension ≥ 1. Let I be the saturated homogeneous ideal defining Y in the polynomial k-algebra $P := k[T_0, T_1, \ldots, T_n]$, and denote by $A := P/I$ the homogeneous coordinate k-algebra of Y in $\mathbb{P}^n$. Assume furthermore that there is an exact sequence*

$$uP(-3) \to sP(-2) \to I \to 0, \quad \textit{with } s, t \geq 1, \tag{1.13}$$

where $P(i)$ is the graded P-module such that $P(i)_j = P_{i+j}$, $\forall i, j \in \mathbb{Z}$, and $aP(i)$ is the direct sum of $a \geq 1$ copies of $P(i)$. (This means that the ideal I is generated by s independent homogeneous polynomials of degree 2, and the relations among them are generated by independent linear ones.) Then $H^0(Y, N_{Y|\mathbb{P}^n}(-2)) = 0$.

We shall prove this result in Chapter 5 after we shall interpret the vanishing of $H^0(Y, N_{Y|\mathbb{P}^n}(-2))$ in terms of the deformation theory of the vertex of the affine cone over Y in $\mathbb{P}^n$. The next result produces examples satisfying the hypotheses of Proposition 1.11.

Theorem 1.12 (Mumford–Green [111], [62]). *Let L be a line bundle of degree $\geq 2g+3$ on a smooth projective curve Y of genus g. Let $Y \subset P := \mathbb{P}(H^0(Y, L)^*)$ be the linearly normal projective embedding of Y given by the complete linear system $|L|$. Then Y is projectively normal in P and there is an exact sequence of the form* (1.13).

Remark 1.13. In fact, Mumford proved in [111] that there is a surjective map $sP(-2) \to I \to 0$ (if $\deg(L) \geq 2g+3$) and subsequently M. Green refined Mumford's result in [62] in the above form (see [62] for a more general result, or also [98] or [100] for a simpler proof of Green's result).

Chapter 2

Proof of Theorem 1.3

We shall prove Theorem 1.3 under the additional hypothesis that L is generated by its global sections. In fact we used Theorem 1.3 in the proofs of Theorems 1.2 and 1.8 only under this additional hypothesis.

Set $R := \bigoplus_{i=0}^{\infty} H^0(X, L^i)$. Since L is ample and X projective, there exists a canonical isomorphism $X \cong \mathrm{Proj}(R)$ such that $L \cong \mathcal{O}_{\mathrm{Proj}(R)}(1)$. Moreover, since X is normal, R is a normal finitely generated k-algebra (see [66], III). Consider the normal affine cone $\mathrm{Spec}(R)$ over the polarized variety (X, L), and let $m_R := \bigoplus_{i=1}^{\infty} H^0(X, L^i)$ be the irrelevant maximal ideal of R which corresponds to the vertex of $\mathrm{Spec}(R)$. Set $U := \mathrm{Spec}(R) \setminus \{m_R\}$. Then there is a canonical morphism $\pi\colon U \to X$. Since L is generated by its global sections, π is the projection of a locally trivial $\mathbb{G}_m$-bundle, where $\mathbb{G}_m = k \setminus \{0\}$ is the multiplicative group of the ground field k. The next lemma and its proof take care of the structure of π more closely.

Lemma 2.1. *Under the hypotheses of Theorem* 1.3, *assume furthermore that L is generated by its global sections. Then there is a canonical exact sequence*

$$0 \to \mathcal{O}_U \to T_U \to \pi^*(T_X) \to 0.$$

Proof. There is a general canonical exact sequence associated to the morphism $\pi\colon U \to X$ (this exact sequence makes actually sense for every morphism $f\colon V \to W$ of algebraic varieties over k, see [73], Proposition 8.11, page 176)

$$\pi^*(\Omega^1_{X|k}) \to \Omega^1_{U|k} \to \Omega^1_{U|X} \to 0,$$

which upon dualizing gives the exact sequence ([73], Proposition 8.11, page 176)

$$0 \longrightarrow T_{U|X} \longrightarrow T_U \xrightarrow{\ \alpha\ } \pi^*(T_X). \tag{2.1}$$

Here $T_{U|X}$ is the relative tangent sheaf of $\pi\colon U \to X$. To prove Lemma 2.1 it will be sufficient to check the following two facts:

i) $T_{U|X} \cong \mathcal{O}_U$, and

ii) The map α is surjective.

Let us first fix some notation. For every $s \in R_1 \setminus \{0\} = H^0(X, L) \setminus \{0\}$, let R_s be the ring of fractions of R with denominators in the multiplicative subset $\{1, s, s^2, \dots, s^n, \dots\}$. Since s is homogeneous, R_s becomes a graded k-algebra by setting $\deg(\frac{t}{s^n}) = \deg(t) - n$ whenever $t \in S$ is homogeneous. Then it makes sense to consider the subring $R_{(s)}$ of R_s whose elements are the fractions $\frac{t}{s^n} \in R_s$ such that $t \in R_n = H^0(X, L^n)$ is a homogeneous element of degree n, i.e. $\deg(\frac{t}{s^n}) = 0$. Clearly, $R_{(s)}$ is a k-subalgebra of R_s. Moreover, since $s \in R_1$, the inclusion $R_{(s)} \subset R_s$ is identified with the inclusion of $R_{(s)}$ in the $R_{(s)}$-algebra $R_{(s)}[T, T^{-1}]$ of Laurent polynomials in the indeterminate T. This is done by sending $T \to s$ and $T^{-1} \to \frac{1}{s}$, as is easily checked. Since L is generated by its global sections,

$$X = \mathrm{Proj}(R) = \bigcup_{s \in R_1 \setminus \{0\}} D_+(s), \text{ where } D_+(s) := \mathrm{Spec}(R_{(s)}),$$

$$U = \bigcup_{s \in R_1 \setminus \{0\}} D(s), \text{ where } D(s) := \mathrm{Spec}(R_s).$$

Then $\pi^{-1}(D_+(s)) = D(s)$ for all $s \in R_1 \setminus \{0\}$ (and in particular, π is an affine morphism). Moreover, the restriction $\pi|D(s) : D(s) \to D_+(s)$ corresponds to the inclusion $R_{(s)} \subset R_s$, and since there is an isomorphism of $R_{(s)}$-algebras $R_s \cong R_{(s)}[T, T^{-1}]$, we see that $D(s) \cong D_+(s) \times \mathbb{G}_m$. This gives explicitly the local structure of the (locally trivial) $\mathbb{G}_m$-bundle $\pi \colon U \to X$.

Let A be a commutative ring, B a commutative A-algebra, and M a B-module. Denote by $\mathrm{Der}_A(B, M)$ the set of all A-derivations $D \colon B \to M$. One can define the sum $D + D'$ of two derivations $D, D' \in \mathrm{Der}_A(B, M)$ and the multiplication bD (with $b \in B$) in an obvious way and one easily checks that $D + D', aD \in \mathrm{Der}_A(B, M)$. In other words, $\mathrm{Der}_A(B, M)$ becomes a B-module in a natural way. If $M = B$, we shall also denote $\mathrm{Der}_A(B) := \mathrm{Der}_A(B, B)$.

The restriction of the exact sequence (2.1) to the affine open subset $D(s)$ corresponds to the obvious exact sequence

$$0 \longrightarrow \mathrm{Der}_{R_{(s)}}(R_s) \longrightarrow \mathrm{Der}_k(R_s) \xrightarrow{\alpha_s} \mathrm{Der}_k(R_{(s)}, R_s), \tag{2.2}$$

where α_s associates to any k-derivation $D \in \mathrm{Der}_k(R_s)$ the restriction $D|R_{(s)} \in \mathrm{Der}_k(R_{(s)}, R_s)$.

Now, it is clear that the map α is surjective (i.e. that condition ii) above holds). Indeed, since this verification is local, it is sufficient to check that the map α_s (of (2.2)) is surjective for every $s \in R_1 \setminus \{0\}$. In fact, if $\delta \in \mathrm{Der}_k(R_{(s)}, R_s)$ then we can define $D \in \mathrm{Der}_k(R_s)$ by $D|R_{(s)} = \delta$ and $D(T) = u$ for an arbitrary $u \in R_s$ (via the above isomorphism of $R_{(s)}$-algebras $R_s \cong R_{(s)}[T, T^{-1}]$). Then, of course, $D(\frac{1}{T}) = -\frac{u}{T^2}$. This proves condition ii) above.

It remains to prove condition i) above. To this end, consider the map $D\colon R \to R$ defined by $D(r) := \deg(r)r$, for every homogeneous element $r \in R$. It is obvious that D is a k-derivation such that $D(R_n) \subseteq R_n$ for every $n \geq 0$. Since $\mathrm{char}(k) = 0$, it is also clear that D is surjective. Then by the universal property of $\Omega^1_{R|k}$, there is a unique homomorphism $w \in \mathrm{Hom}_R(\Omega^1_{R|k}, R)$ of R-modules which, composed with the canonical derivation $d\colon R \to \Omega^1_{R|k}$, coincides to D. Since D is surjective, w is also surjective. Passing to sheaves on $\mathrm{Spec}(R)$ and restricting to U we get the surjective map of $\mathcal{O}_U$-modules $w : \Omega^1_{U|k} \to \mathcal{O}_U$.

If we denote by $\beta : \Omega^1_{U|k} \to \Omega^1_{U|X}$ the canonical surjection, we claim that there exists a unique map of $\mathcal{O}_U$-modules $w' : \Omega^1_{U|X} \to \mathcal{O}_U$ such that $w' \circ \beta = w$. To check this, observe that $D|R_{(s)} = 0$ for every $s \in R_1 \setminus \{0\}$, whence $D \in \mathrm{Der}_{R_{(s)}}(R_s)$, so that the existence of w' comes from the universal property of $\Omega^1_{R_s|R_{(s)}}$ taking into account that $U = \bigcup_{s \in R_1\setminus\{0\}} D(s))$. The surjectivity of w implies the surjectivity of $w' : \Omega^1_{U|X} \to \mathcal{O}_U$. But since $\pi\colon U \to X$ is a smooth morphism of relative dimension 1, $\Omega^1_{U|X}$ is an invertible $\mathcal{O}_U$-module. Therefore w' is a surjective map between two invertible $\mathcal{O}_U$-modules, whence w' is necessarily an isomorphism.

This proves condition i) above, and thereby Lemma 2.1. □

Lemma 2.2. *Under the hypotheses of Lemma 2.1, for every coherent $\mathcal{O}_X$-module F and for every $p \geq 0$ there is a natural isomorphism of graded R-modules*

$$H^p(U, \pi^*(F)) \cong \bigoplus_{i \in \mathbb{Z}} H^p(X, F \otimes L^i).$$

Proof. In the proof of Lemma 2.1 we already observed that the morphism π is affine ($\pi^{-1}(D_+(s)) = D(s)$ for every $s \in R_1 \setminus \{0\}$). In particular, for every coherent $\mathcal{O}_X$-module F,

$$H^p(U, \pi^*(F)) \cong H^p(X, \pi_*(\pi^*(F))), \quad \forall p \geq 0.$$

To conclude the proof of Lemma 2.2, it will be sufficient to show that for every coherent $\mathcal{O}_X$-module F one has a canonical identification

$$\pi_*(\pi^*(F)) \cong \bigoplus_{i \in \mathbb{Z}} F \otimes L^i. \tag{2.3}$$

To check this, since R is a finitely generated k-algebra such that $X = \mathrm{Proj}(R) = \bigcup_{s \in R_1\setminus\{0\}} D_+(s)$, the sheaf F is associated to a finitely generated graded R-module M. Then by the proof of Lemma 2.1, $R_s \cong R_{(s)}[T, T^{-1}]$, whence we get canonical isomorphisms of $R_{(s)}$-modules

$$M_s \cong M \otimes R_s \cong M \otimes R_{(s)}[T, T^{-1}] \cong \bigoplus_{i \in \mathbb{Z}} M_{(s)} T^i.$$

Since this happens for every $s \in R_1 \setminus \{0\}$, these local isomorphisms patch together to give the isomorphism (2.3). □

The tangent sheaf $T_{\mathrm{Spec}(R)}$ of the cone $\mathrm{Spec}(R)$ corresponds to the graded R-module

$$T_R = \bigoplus_{i \in \mathbb{Z}} T_R(i),$$

where $T_R := \mathrm{Der}_k(R)$, and the piece $T_R(i)$ of weight i is given by

$$T_R(i) := \{D \in \mathrm{Der}_k(R) \mid D(R_n) \subseteq R_{n+i} \ \ \forall n \geq 0\}.$$

Now, the next step in the proof of Theorem 1.3 is the following:

Lemma 2.3. *Under the above hypotheses, the k-vector space $H^0(X, T_X \otimes L^{-i})$ is canonically identified to the k-vector space $T_R(-i)$, for all $i \geq 1$.*

Proof. Since R is normal of dimension ≥ 2 and $U = \mathrm{Spec}(R) \setminus \{m_R\}$, we can apply Proposition 1.4 to deduce a canonical isomorphism of graded R-modules $T_R \cong H^0(U, T_U)$. Then the exact sequence of Lemma 2.1 yields the cohomology exact sequence of R-modules

$$0 \to H^0(U, \mathcal{O}_U) \to H^0(U, T_U) \to H^0(U, \pi^*(T_X)) \to H^1(U, \mathcal{O}_U). \tag{2.4}$$

By Lemma 2.2, we have for every $p \geq 0$:

$$H^p(U, \mathcal{O}_U) \cong \bigoplus_{i \in \mathbb{Z}} H^p(X, L^i) \ \text{ and } \ H^0(U, \pi^*(T_X)) \cong \bigoplus_{i \in \mathbb{Z}} H^0(X, T_X \otimes L^i).$$

Now, by Mumford's vanishing theorem (see [110]), $H^p(X, L^i) = 0$, for every $i < 0$ and $p \leq 1$ (because $\dim(X) \geq 2$ and $\mathrm{char}(k) = 0$; note that this vanishing for $p = 0$ is trivial). Therefore the above identification and the exact sequence (2.4) yield

$$T_R(i) \cong H^0(X, T_X \otimes L^i), \ \ \forall i < 0.$$

Lemma 2.3 is proved. □

In view of Lemma 2.3, Theorem 1.3 is a consequence of the following:

Proposition 2.4. *Let (X, L) be a normal polarized variety of dimension ≥ 2 over k, with $\mathrm{char}(k) = 0$. Assume that there is a non-zero k-derivation $D\colon R \to R$ of weight -1, i.e. $D(R_i) \subseteq R_{i-1}$ for all $i \geq 0$, where $R = \bigoplus_{i=0}^{\infty} H^0(X, L^i)$. If L is generated by its global sections, then the conclusion of Theorem 1.3 holds true.*

To prove Proposition 2.4 we need another two lemmae.

Lemma 2.5 (Zariski). *Let R be a graded k-algebra such that* $\mathrm{char}(k) = 0$ *and there is a k-derivation $D\colon R \to R$ of weight -1 and an element $t \in R_1$ with the property that $D(t) = 1$. If $A := \{r \in R \mid D(r) = 0\}$, then A is a graded k-subalgebra of R, t is transcendental over A and $R = A[t]$. Moreover, $D = \frac{\partial}{\partial t}$ on $A[t]$.*

Proof. The fact that A is a graded k-subalgebra of R is immediate. Let $A[T]$ be the polynomial A-algebra in the indeterminate T. Grade $A[T]$ by $\deg(aT^m) = \deg(a) + m$ for every $a \in A$ homogeneous and $m \geq 0$. Consider the homomorphism of graded k-algebras $\varphi\colon A[T] \to R$ such that $\varphi|A = \mathrm{id}_A$ and $\varphi(T) = t$. Then it is immediate to check that the diagram

$$\begin{array}{ccc} A[T] & \xrightarrow{\ \varphi\ } & R \\ {\scriptstyle \frac{\partial}{\partial T}}\big\downarrow & & \big\downarrow{\scriptstyle D} \\ A[T] & \xrightarrow[\ \varphi\ ]{} & R \end{array}$$

is commutative. So, it will be enough to show that φ is bijective.

Injectivity of φ: Assume that there exists a non-zero polynomial $f(T) = \sum_{i=0}^{n} a_i T^i \in A[T]$ such that $\varphi(f(T)) = 0$. We may assume that $f(T)$ is homogeneous in $A[T]$ and of minimal degree. Thus $a_i \in A_{n-i} \subseteq R_{n-i}$. Then $\frac{\partial f(T)}{\partial T}$ is a polynomial of smaller degree and still in $\mathrm{Ker}(\varphi)$. Thus $\frac{\partial f(T)}{\partial T} = 0$, and since $\mathrm{char}(k) = 0$, $f \in A$. Since $\varphi(f) = 0$ it follows that $f = 0$, a contradiction.

Surjectivity of φ: We proceed by induction on n, the case $n = 0$ being clear. If $r \in R_n$ $(n > 0)$ then $D(r) \in R_{n-1}$, whence $D(r) = \sum_{i=0}^{n-1} a_i t^i$, with $a_i \in A_{n-1-i}$, $\forall i$ (by induction hypothesis). Thus

$$D(r - \sum_{i=0}^{n-1} \frac{a_i}{i+1} t^{i+1}) = 0,$$

i.e. $r - \sum_{i=0}^{n-1} \frac{a_i}{i+1} t^{i+1} \in A$. Lemma 2.5 is proved. □

Lemma 2.6. *Let (X, L) be a normal polarized variety over k such that L is generated by its global sections and* $\mathrm{char}(k) = 0$. *Assume that there is a non-zero k-derivation $D\colon R \to R$ of weight -1, where $R = \sum_{i=0}^{\infty} H^0(X, L^i)$. Then there exists an element $t \in R_1 = H^0(X, L)$ such that $D(t) = 1$.*

Proof. Let $t_0, \dots, t_n$ be a basis of the k-vector space $H^0(X, L)$. Since L is generated by its global sections we get a morphism $u\colon X \to \mathbb{P}^n$ such that $u^*(\mathcal{O}_{\mathbb{P}^n}(1)) \cong L$, and since L is also ample, u is a finite morphism (otherwise there would exist a curve C such that $u(C)$ is a point, in which case $L|C$ cannot be ample). If for some i, $D(t_i) \neq 0$, then we can take $t = \alpha^{-1} t_i$, with $\alpha := D(t_i) \in k \setminus \{0\}$, and get $D(t) = 1$.

Assume therefore that $D(t_i) = 0$ for every $i = 0, 1, \dots, n$. We shall show that this leads to a contradiction. Consider the map of graded k-algebras

$$k[T_0, \dots, T_n] \cong \bigoplus_{i=0}^{\infty} H^0(\mathbb{P}^n, \mathcal{O}_{\mathbb{P}^n}(i)) \to R = \bigoplus_{i=0}^{\infty} H^0(X, L^i),$$

which sends T_j to t_j, $j = 0, 1, \dots, n$. Since u is finite the morphism of affine varieties $u^{-1}(D_+(T_0)) \to D_+(T_0)$ is finite, whence the map of k-algebras

$$k[T_0, \dots, T_n]_{(T_0)} \to R_{(t_0)}$$

is finite. In particular, every homogeneous element $t \in R$ satisfies a non-trivial algebraic equation

$$a_0 t^m + a_1 t^{m-1} + \ldots + a_m = 0, \quad \text{with } a_i \in k[t_0, \dots, t_n],\ a_0 \neq 0,\ m \geq 1.$$

We may assume m minimal with this property. Since the derivation D vanishes on $k[t_0, \dots, t_m]$ (by our assumption), we get by derivating

$$(m a_0 t^{m-1} + (m-1) a_1 t^{m-2} + \ldots + a_{m-1}) D(t) = 0.$$

If $D(t) \neq 0$, since R is a domain, we get

$$m a_0 t^{m-1} + (m-1) a_1 t^{m-2} + \ldots + a_{m-1} = 0.$$

Recalling also that $\operatorname{char}(k) = 0$, this contradicts the minimality of m. Therefore $D(t) = 0$ for every homogeneous $t \in R$, whence $D = 0$, a contradiction. □

Lemmae 2.5 and 2.6 imply the following:

Corollary 2.7. *In the hypotheses of Lemma* 2.6, *there exists* $t \in R_1$ *such that* $D(t) = 1$, $R = A[t]$, *and* t *is transcendental over* $A := \{r \in R \mid D(r) = 0\}$.

Proof of Proposition 2.4. By Corollary 2.7, $R = A[t]$, with $t \in H^0(X, L)$ such that $D(t) = 1$ and t is transcendental over $A = \{r \in R \mid D(r) = 0\}$. Since R is normal it follows that A is also normal. Set $E := \operatorname{div}_X(t) \in |L|$, i.e. $L = \mathcal{O}_X(E)$, and consider the canonical exact sequences ($n \geq 0$)

$$0 \longrightarrow L^{n-1} \xrightarrow{\ t\ } L^n \longrightarrow \mathcal{O}_E(nE) = L_E^n \longrightarrow 0.$$

Taking cohomology we get

$$0 \to R \to R \to \bigoplus_{n=0}^{\infty} H^0(E, L_E^n) \to \bigoplus_{n=0}^{\infty} H^1(X, L^{n-1}). \tag{2.5}$$

We claim that $H^1(X, L^{n-1}) = 0$, $\forall n \in \mathbb{Z}$. Assuming this claim, the proof of Proposition 2.4 is finished, because from the exact sequence (2.5) it follows that

$$\bigoplus_{n=0}^{\infty} H^0(E, L_E^n) \cong R/tR = A[t]/tA[t] \cong A.$$

It remains therefore to prove the claim. Since A is normal, m_A-depth$(A) \geq 2$, whence m_R-depth$(R)= m_R$-depth$(A[t]) \geq 3$. If $U := \mathrm{Spec}(R)\setminus\{m_R\}$, we can write the local cohomology exact sequence

$$0 = H^1(\mathrm{Spec}(R), \mathcal{O}_{\mathrm{Spec}(R)}) \to H^1(U, \mathcal{O}_U) \to \tag{2.6}$$

$$\to H^2_{m_R}(\mathrm{Spec}(R), \mathcal{O}_{\mathrm{Spec}(R)}) \to H^2(\mathrm{Spec}(R), \mathcal{O}_{\mathrm{Spec}(R)}) = 0.$$

By Lemma 2.2,

$$H^1(U, \mathcal{O}_U) \cong \bigoplus_{i=-\infty}^{\infty} H^1(X, L^i). \tag{2.7}$$

Since m_R-depth$(R) \geq 3$, $H^2_{m_R}(\mathrm{Spec}(R), \mathcal{O}_{\mathrm{Spec}(R)}) = 0$ (see property c) in the proof of Proposition 1.4), whence from (2.6) and (2.7) we get

$$H^1(X, L^i) = H^1(X, \mathcal{O}_X(iE)) = 0 \ \text{ for all } \ i \in \mathbb{Z}.$$

In this way Proposition 2.4 and thereby Theorem 1.3 (under the extra hypothesis that L is generated by its global sections) are completely proved. □

Remark 2.8. The proof of Theorem 1.3, without the hypothesis that L is generated by its global sections, follows the same main ideas, but technically is more involved (see [146]).

Chapter 3

Counterexamples and Further Consequences

First we show by counterexamples that Theorem 1.3 is in general false in positive characteristic.

A Counterexample in Characteristic 2

This example is given in [146]. Let $X \subset \mathbb{P}^{2n}$ $(n \geq 2)$ be the hyperquadric in $\mathbb{P}^{2n}$ of equation

$$f = T_0^2 + T_1T_{n+1} + T_2T_{n+2} + \ldots + T_nT_{2n}$$

over an algebraically closed field k of characteristic 2, and set $L := \mathcal{O}_X(1) = \mathcal{O}_{\mathbb{P}^{2n}}(1)|X$. Then X is a smooth hyperquadric in $\mathbb{P}^{2n}$, and in particular, X cannot be isomorphic to a cone over a polarized variety (E, L_E). Set

$$R = \bigoplus_{i=0}^{\infty} H^0(X, \mathcal{O}_X(i)) \cong k[T_0, \ldots, T_{2n}]/(f)$$

(being a smooth hypersurface, X is projectively normal in $\mathbb{P}^{2n}$).

Observe that because $\operatorname{char}(k) = 2$, all the derivatives of the polynomial f vanish at the point $P = [1, 0, \ldots, 0]$ (and only at this point). In other words, all the projective tangent spaces at X have the point P in common, i.e. X is a strange variety in the terminology of [73], page 311. Moreover, the derivative

$$\frac{\partial}{\partial T_0} : k[T_0, \ldots, T_{2n}] \to k[T_0, \ldots, T_{2n}]$$

has the property that $\frac{\partial f}{\partial T_0} = 0$. Therefore

$$\frac{\partial(gf)}{\partial T_0} = f\frac{\partial g}{\partial T_0}, \quad \forall g \in k[T_0, \ldots, T_n], \text{ i.e. } \frac{\partial}{\partial T_0}((f)) \subseteq (f).$$

This implies that $\frac{\partial}{\partial T_0}$ yields a non-zero k-derivation $D\colon R \to R$ of weight -1.

On the other hand, since X is a hypersurface in $\mathbb{P}^{2n}$ $(n \geq 2)$, $H^p(X, L^{-i}) = 0$ for all $p = 0, 1$ and for all $i \in \mathbb{Z}$. Therefore the arguments in the proof of Lemma 2.3 can be carried out to prove that $H^0(X, T_X(-1)) \neq 0$ (due to these vanishings, we don't have to appeal to Mumford's vanishing theorem).

This shows that (X, L) is a smooth polarized variety of dimension ≥ 3 (which is not a cone), such that $H^0(X, T_X \otimes L^{-1}) \neq 0$.

A Counterexample in Characteristic 3

Let k be an algebraically closed field of characteristic 3. In $\mathbb{P}^3$ consider the surface X of equation

$$f = T_0^3 + T_1T_2^2 + T_2T_3^2 + T_3T_1^2.$$

We have $\frac{\partial f}{\partial T_0} = 0$, $\frac{\partial f}{\partial T_1} = T_2^2 + 2T_1T_3 = T_2^2 - T_1T_3$, $\frac{\partial f}{\partial T_2} = T_3^2 - T_1T_2$, $\frac{\partial f}{\partial T_3} = T_1^2 - T_2T_3$. The closed subvariety in $\mathbb{P}^3$ of equations

$$T_1^2 - T_2T_3 = T_2^2 - T_1T_3 = T_3^2 - T_1T_2 = 0$$

is (at least set-theoretically) the line

$$L := \{[\lambda, \mu, \mu, \mu] \in \mathbb{P}^3 \mid [\lambda, \mu] \in \mathbb{P}^1\},$$

as one can easily see. Thus $\operatorname{Sing}(X)$ is the point $A = [0, 1, 1, 1]$ and all other points of $L \setminus \{A\}$ are strange points of X (i.e. points in the intersection of all projective tangent spaces at all smooth points of X). In particular, X is a normal surface. Observe that $B := [0, 0, 0, 1] \in X$, but the line AB has in common with X only the points A and B. In particular, X cannot be a cone.

On the other hand, as in the previous example, since $\frac{\partial f}{\partial T_0} = 0$, we have $\frac{\partial (fg)}{\partial T_0} = f\frac{\partial g}{\partial T_0}$, whence the derivation

$$\frac{\partial}{\partial T_0} : k[T_0, T_1, T_2, T_3] \to k[T_0, T_1, T_2, T_3]$$

yields a non-zero k-derivation $D\colon R \to R$ of weight -1, where

$$R = \bigoplus_{i=0}^{\infty} H^0(X, \mathcal{O}_X(i)) \cong k[T_0, T_1, T_2, T_3]/(f).$$

As above we get $H^0(X, T_X \otimes L^{-1}) \neq 0$, where $L := \mathcal{O}_X(1)$. In other words, the normal polarized surface (X, L) (which is not a cone) is a counterexample to Theorem 1.3 in characteristic 3.

Here is another consequence of Theorem 1.3.

Proposition 3.1. *Let Y be a smooth, connected, non-degenerate closed subvariety of codimension ≥ 2 of $\mathbb{P}^n$. Assume $\dim(Y) \geq 1$ and $\operatorname{char}(k) = 0$. Then the Zak map $z\colon H^0(Y,(n+1)\mathcal{O}_Y) \to H^0(Y, N_{Y|\mathbb{P}^n}(-1))$ is injective. In particular, $\dim_k H^0(Y, N_{Y|\mathbb{P}^n}(-1)) \geq n+1$, with equality if and only if the map z is surjective.*

Proof. Consider the following commutative diagram with exact rows and columns:

$$
\begin{array}{ccccccccc}
 & & 0 & & 0 & & & & \\
 & & \downarrow & & \downarrow & & & & \\
 & & \mathcal{O}_Y(-1) & \xrightarrow{\text{id}} & \mathcal{O}_Y(-1) & & & & \\
 & & \downarrow & & \downarrow & & & & \\
0 & \longrightarrow & F(-1) & \longrightarrow & (n+1)\mathcal{O}_Y & \longrightarrow & N_{Y|\mathbb{P}^n}(-1) & \longrightarrow & 0 \\
 & & \downarrow & & \downarrow & & \downarrow{\scriptstyle \text{id}} & & \\
0 & \longrightarrow & T_Y(-1) & \longrightarrow & T_{\mathbb{P}^n}(-1)|Y & \longrightarrow & N_{Y|\mathbb{P}^n}(-1) & \longrightarrow & 0 \\
 & & \downarrow & & \downarrow & & & & \\
 & & 0 & & 0 & & & &
\end{array}
$$

(which is completely analogous to the diagram used in the proof of Theorem 1.2). The first long row implies that $\operatorname{Ker}(z) = H^0(Y, F(-1))$ (the functor H^0 is left exact!). On the other hand, the first column yields the exact sequence

$$0 \to H^0(Y, \mathcal{O}_Y(-1)) \to H^0(Y, F(-1)) \to H^0(Y, T_Y(-1)).$$

Since $\mathcal{O}_Y(1)$ is ample and $\dim(Y) \geq 1$, $H^0(Y, \mathcal{O}_Y(-1)) = 0$. We claim that $H^0(Y, T_Y(-1)) = 0$. Indeed, assume first $\dim(Y) \geq 2$; if this space is $\neq 0$, by Theorem 1.3 Y would be a cone. But since Y is smooth, this is possible only if Y is a linear subspace, and by non-degeneratedness, $Y = \mathbb{P}^n$, contradicting $\operatorname{codim}_{\mathbb{P}^n}(Y) \geq 2$. If $\dim(Y) = 1$, $H^0(Y, T_Y(-1)) \neq 0$ immediately implies that Y is a line or a conic, which again contradicts $\operatorname{codim}_{\mathbb{P}^n}(Y) \geq 2$. Therefore $H^0(Y, T_Y(-1)) = 0$, which implies $\operatorname{Ker}(z) = H^0(Y, F(-1)) = 0$. □

Remark 3.2. Proposition 3.1 is in general false in positive characteristic. The two examples above also yield counterexamples to Proposition 3.1 in positive characteristic.

Corollary 3.3. *Under the hypotheses of Proposition* 3.1 *assume furthermore that the characteristic of k is arbitrary,* $\dim(Y) \geq 2$ *and*

$$H^1(Y, \mathcal{O}_Y(-1)) = H^1(Y, T_Y(-1)) = 0$$

(the first vanishing always holds in characteristic zero by the Kodaira vanishing theorem). Then the Zak map

$$z\colon H^0(Y, (n+1)\mathcal{O}_Y) \to H^0(Y, N_{Y|\mathbb{P}^n}(-1))$$

is surjective.

Proof. Indeed, the first column of the diagram from the proof of Proposition 3.1 yields the exact sequence

$$0 = H^1(Y, \mathcal{O}_Y(-1)) \to H^1(Y, F(-1)) \to H^1(Y, T_Y(-1)) = 0.$$

It follows that $H^1(Y, F(-1)) = 0$. On the other hand, from the cohomology sequence associated to the first row of the same diagram we get

$$H^0(Y, (n+1)\mathcal{O}_Y) \xrightarrow{z} H^0(Y, N_{Y|\mathbb{P}^n}(-1)) \longrightarrow H^1(Y, F(-1)) = 0,$$

i.e. z is surjective. □

Examples 3.1. i) Let $v_s : \mathbb{P}^r \hookrightarrow \mathbb{P}^{n(r,s)}$ be the s-fold Veronese embedding of $\mathbb{P}^r$, with $n(r,s) = \binom{r+s}{r} - 1$. Set $Y := v_s(\mathbb{P}^r)$, and assume $r, s \geq 2$. Then $\mathcal{O}_Y(1) := \mathcal{O}_{\mathbb{P}^{n(r,s)}}(1)|Y$ coincides to $\mathcal{O}_{\mathbb{P}^r}(s)$.

Claim. $H^1(Y, T_Y(-1)) = 0$ for every $r \geq 3$ or for $r = s = 2$. Moreover,

$$H^1(Y, \mathcal{O}_Y(-1)) = 0 \;\text{ for every }\; r \geq 2.$$

Indeed, the second statement comes from the explicit cohomology of the projective space. For the first, the Euler sequence for $\mathbb{P}^r$ yields the cohomology sequence

$$H^1(\mathbb{P}^r, (r+1)\mathcal{O}_{\mathbb{P}^r}(1-s)) \to H^1(\mathbb{P}^r, T_{\mathbb{P}^r}(-s)) \cong H^1(Y, T_Y(-1)) \to H^2(\mathbb{P}^r, \mathcal{O}_{\mathbb{P}^r}(-s)).$$

The first space is zero because $r \geq 2$, while the last one is zero either if $r \geq 3$, or if $r = s = 2$.

Therefore by Corollary 3.3, the Zak map of Y in $\mathbb{P}^{n(r,s)}$ is surjective. Therefore the conclusion of Theorem 1.2 holds true in this case. This result goes back to C. Segre and G. Scorza, see [131].

Notice that if in Example 3.1, i) we assume $r \geq 3$, we get $H^1(Y, \mathcal{O}_Y(i)) = H^1(Y, T_Y(i)) = 0$ for all $i \in \mathbb{Z}$.

ii) Let $i : \mathbb{P}^1 \times \mathbb{P}^r \hookrightarrow \mathbb{P}^{2r+1}$ be the Segre embedding of $\mathbb{P}^1 \times \mathbb{P}^r$, with $r \geq 2$. Set $Y = i(\mathbb{P}^1 \times \mathbb{P}^r)$. Then $\mathcal{O}_Y(1) = \mathcal{O}(1,1)$, where

$$\mathcal{O}(1,1) := p_1^*(\mathcal{O}_{\mathbb{P}^1}(1)) \otimes p_2^*(\mathcal{O}_{\mathbb{P}^r}(1)),$$

where p_1 and p_2 are the first and the second projections of $\mathbb{P}^1 \times \mathbb{P}^r$ respectively.

Claim. $H^1(Y, \mathcal{O}_Y(i)) = H^1(Y, T_Y(i)) = 0$ for every $i \in \mathbb{Z}$.

Again the first statement is trivial. For the second we have $T_Y \cong p_1^*(T_{\mathbb{P}^1}) \oplus p_2^*(T_{\mathbb{P}^r}) = p_1^*(\mathcal{O}_{\mathbb{P}^1}(2)) \oplus p_2^*(T_{\mathbb{P}^r})$. Then the conclusion follows easily from Künneth's formula and from the Euler sequence of $\mathbb{P}^r$.

In particular, by Corollary 3.3 the Zak map of $Y = i(\mathbb{P}^1 \times \mathbb{P}^r)$ in $\mathbb{P}^{2r+1}$ $(r \geq 2)$ is surjective. Therefore the conclusion of Theorem 1.2 holds true in this case. This result goes back to G. Scorza, see [132].

Another application of Theorem 1.3 is the following result.

Theorem 3.4 (Fujita [50]). *Let Y be a smooth projective variety over $\mathbb{C}$ which is embedded in the normal projective variety X as an ample Cartier divisor. Assume that* $\dim(Y) \geq 2$ *and* $H^1(Y, T_Y \otimes N_{Y|X}^{-i}) = 0$ *for every* $i \geq 1$, *where* $N_{Y|X}$ *is the normal bundle of Y in X. Then X is isomorphic to the cone over the polarized variety* $(Y, N_{Y|X})$.

Proof. Set $L := N_{Y|X}$. The normal sequence of Y in X

$$0 \to T_Y \to T_X|Y \to L \to 0$$

yields the cohomology sequence $(i \geq 1)$

$$H^1(Y, T_Y \otimes L^{-i}) \to H^1(Y, T_X|Y \otimes L^{-i}) \to H^1(Y, L^{1-i}). \tag{3.1}$$

For every $i \geq 2$ the first space is zero by hypotheses, and the third space is also zero by the Kodaira vanishing theorem ($\dim(Y) \geq 2$). It follows that $H^1(Y, T_X|Y \otimes L^{-i}) = 0$ for every $i \geq 2$. Therefore the exact sequence

$$0 \to T_X \otimes \mathcal{O}_X(-(i+1)Y) \to T_X \otimes \mathcal{O}_X(-iY) \to T_X|Y \otimes L^{-i} \to 0 \tag{3.2}$$

yields for every $i \geq 2$ a surjection

$$H^1(X, T_X \otimes \mathcal{O}_X(-(i+1)Y)) \to H^1(X, T_X \otimes \mathcal{O}_X(-iY)). \tag{3.3}$$

On the other hand, since X is normal of dimension ≥ 2, Y is an ample Cartier divisor on X, and T_X is a reflexive sheaf, a result of Enriques–Severi–Zariski–Serre [134], or also [73], Corollary 7.8, page 244, shows that

$$H^1(X, T_X \otimes \mathcal{O}_X(-iY)) = 0 \text{ for every } i \gg 0. \tag{3.4}$$

Then (3.3) and (3.4) and an induction on i yield $H^1(T_X \otimes \mathcal{O}_X(-2Y)) = 0$. Therefore the exact sequence (3.2) (with $i = 1$) yields the surjection

$$H^0(X, T_X \otimes \mathcal{O}_X(-Y)) \to H^0(T_X|Y \otimes L^{-1}) \to 0. \tag{3.5}$$

Finally, the exact sequence (3.1) (for $i = 1$) yields the cohomology sequence

$$H^0(Y, T_X|Y \otimes L^{-1}) \to H^0(Y, \mathcal{O}_Y) \to H^1(Y, T_Y \otimes L^{-1}), \tag{3.6}$$

in which the last space is zero by hypothesis. Therefore the first map of (3.6) is surjective. Recalling also the surjection (3.5), we get a surjection

$$H^0(X, T_X \otimes \mathcal{O}_X(-Y)) \to H^0(Y, \mathcal{O}_Y) \neq 0,$$

and in particular, $H^0(X, T_X \otimes \mathcal{O}_X(-Y)) \neq 0$. At this point we can apply Theorem 1.3 to the normal polarized variety $(X, \mathcal{O}_X(Y))$ to get the conclusion. □

Remark 3.5. We shall generalize Theorem 3.4 to higher codimension in Chapter 11, see Theorem 11.25.

Examples 3.2. In all these examples we shall assume that $\mathrm{char}(k) = 0$.

i) $(Y, L) = (\mathbb{P}^n, \mathcal{O}_{\mathbb{P}^n}(s))$, with $n \geq 3$ and $s \geq 1$. Then by Example 3.1, i),

$$H^1(Y, T_Y \otimes L^i) = 0, \quad \text{for every } \ i \in \mathbb{Z}.$$

ii) Let Y be an abelian variety of dimension $d \geq 2$. Then for every ample line bundle L on Y, we have $H^1(Y, T_Y \otimes L^{-i}) = 0$ for every $i \geq 1$. Indeed, in this case T_Y is isomorphic to the trivial bundle of rank d, and the assertion follows from the Kodaira vanishing theorem for example.

iii) Let $Y = Y_1 \times Y_2$ be a product of two smooth projective varieties Y_i with $\dim(Y_i) \geq 2$, $i = 1, 2$. Let L_i be an ample line bundle on Y_i, $i = 1, 2$ and set $L := p_1^*(L_1) \otimes p_2^*(L_2)$, where $p_i \colon Y \to Y_i$ is the canonical projection on Y_i, $i = 1, 2$. Then L is ample on Y. Moreover, $T_Y = p_1^*(T_{Y_1}) \oplus p_2^*(T_{Y_2})$. Then for every $i > 0$,

$$\begin{aligned}&H^1(Y, T_Y \otimes L^{-i})\\ &= H^1(Y, p_1^*(T_{Y_1} \otimes L_1^{-i}) \otimes p_2^*(L_2^{-i})) \oplus H^1(Y, p_1^*(L_1^{-i}) \otimes p_2^*(T_{Y_2} \otimes L_2^{-i})),\end{aligned}$$

and using Künneth's formulae we get $H^1(Y, T_Y \otimes L^{-i}) = 0$ for every $i \geq 1$.

iv) Let $Y = Y_1 \times Y_2 \times Y_3$ be a product of three smooth projective varieties Y_1, Y_2 and Y_3 each of dimension ≥ 1, and let $L := p_1^*(L_1) \otimes p_2^*(L_2) \otimes p_3^*(L_3)$, with L_i an ample line bundle on Y_i, $i = 1, 2, 3$. Then by arguments similar to those in the previous example we have $H^1(Y, T_Y \otimes L^{-i}) = 0$ for all $i \geq 1$.

v) Let Y be a hyperelliptic surface. This is a surface with invariants $b_2 = 2$, $p_g = 0$, $q = 1$ and $\chi(\mathcal{O}_Y) = 0$. Moreover, there are two elliptic curves B_0 and B_1, a finite subgroup $A \subset B_1$, an injective homomorphism $\alpha \colon A \to \mathrm{Aut}(B_0)$, and a free action of A on $B_1 \times B_0$ of the form $a(b_1, b_0) = (b_1 + a, \alpha(a)(b_0))$ (see e.g. [12], (10.25)). Then $Y = (B_1 \times B_0)/A$. Let $f \colon Z := B_1 \times B_0 \to Y$ be the canonical étale morphism. Then $f^*(T_Y) = T_Z = 2\mathcal{O}_Z$ (since Z is an abelian surface), and $f^*(L)$ is ample on Z. Since $\mathrm{char}(k) = 0$ the vanishing of $H^1(T_Y \otimes L^{-i})$ (for $i \geq 1$) follows from the vanishing $H^1(f^*(T_Y) \otimes f^*(L^{-i})) = H^1(Z, 2f^*(L^{-i})) = 0$ (Z is an abelian surface and $f^*(L)$ is ample on Z, see Example 3.2, ii).

To give another application of Theorem 3.4 we need the following result (see [74], page 110):

Theorem 3.6 (Hartshorne). *Let Y be an effective Cartier divisor on complete algebraic variety X such that the normal bundle $N_{Y|X}$ is ample. Then there exist a birational projective morphism $f\colon X \to Z$ and an open subset $U \subseteq X$ containing Y with the property that $f|U\colon U \to f(U)$ is an isomorphism and $Y' = f(Y)$ is an ample Cartier divisor on Z. In particular, X is a projective variety.*

Proof. We claim that the following three statements hold for every $n \gg 0$:

i) $N_{Y|X}^n = \mathcal{O}_Y(nY)$ is very ample,

ii) The restriction map $H^0(X, \mathcal{O}_X(Y)) \to H^0(Y, N_{Y|X}^n)$ is surjective, and

iii) The complete linear system $|nY|$ has no base points.

i) is obvious. For ii) and iii) we adjust part of the proof (due to Kleiman) of the Nakai–Moishezon criterion, see e.g. [74]. Since $N_{Y|X}$ is ample and Y is a complete scheme, $H^1(Y, N_{Y|X}^n) = 0$ for $n \gg 0$. Therefore from the cohomology exact sequence

$$H^1(X, \mathcal{O}_X((n-1)Y)) \xrightarrow{\alpha_n} H^1(X, \mathcal{O}_X(nY)) \longrightarrow H^1(Y, N_{Y|X}^n)$$

associated to the exact sequence

$$0 \to \mathcal{O}_X((n-1)Y) \to \mathcal{O}_X(nY) \to N_{Y|X}^n \to 0,$$

the map α_n is surjective for $n \gg 0$. Since X is complete, $H^1(X, \mathcal{O}_X((n-1)Y))$ is a finite dimensional vector space, it follows that the maps α_n actually become isomorphisms for $n \gg 0$. Therefore the cohomology exact sequence of the above short exact sequence yields ii). Then i) and ii) immediately imply iii).

Now let $\varphi = \varphi_{|nY|} : X \to P = \mathbb{P}(H^0(X, \mathcal{O}_X(nY))^*)$ be the morphism defined by the base-point-free complete linear system $|nY|$ for $n \gg 0$. Set $Z_1 = \varphi(X)$. Then $Y = \varphi^*(Y_1)$, with Y_1 a very ample divisor on Z_1. By the above claim, $\varphi|Y\colon Y \cong Y_1$ and $Y_{\mathrm{red}} = \varphi^{-1}((Y_1)_{\mathrm{red}})$. Let $\varphi = g \circ f$ be the Stein factorization of the projective morphism φ, with $f\colon X \to Z$ a proper morphism with $f_*(\mathcal{O}_X) = \mathcal{O}_Z$, and $g\colon Z \to Z_1$ a finite surjective morphism (see [66] III, (4.3.3)). Notice that $f|Y\colon Y \to Y' := g^*(Y_1)$ is an isomorphism and $Y_{\mathrm{red}} = f^{-1}(Y'_{\mathrm{red}})$, so by Zariski's Main Theorem (see [66] III, (4.4.1)), f is an isomorphism in a neighbourhood of Y onto an open subset of Z. Moreover since $\varphi = g \circ f$ is a projective morphism, f is also projective by a simple general property (see [66] II, (5.5.5)). Finally, $Y' = g^*(Y_1)$ is an ample Cartier divisor on Z because Y_1 is ample on Z_1 and g is finite. □

Definition 3.7. Let $i\colon Y \hookrightarrow X$ and $i' : Y' \hookrightarrow X'$ be two closed embeddings of the projective varieties Y and Y' into the projective varieties X and X' respectively.

We shall say that i and i' are *Zariski equivalent* if there exist two Zariski open subsets $U \subseteq X$ and $U' \subseteq X'$ containing Y and Y' respectively, and an isomorphism $\varphi\colon U \cong U'$ such that $\varphi(Y) = Y'$.

Now Theorems 3.4 and 3.6 together imply the following:

Corollary 3.8. *Let (Y, L) be a smooth polarized variety over $\mathbb{C}$ of dimension ≥ 2 such that $H^1(Y, T_Y \otimes L^{-i}) = 0$ for every $i \geq 1$. Assume that Y admits two embeddings $Y \hookrightarrow X$ and $Y \hookrightarrow X'$ in the normal projective varieties X and X' as effective Cartier divisors such that $N_{Y|X} \cong L \cong N_{Y|X'}$. Then the embeddings $Y \hookrightarrow X$ and $Y \hookrightarrow X'$ are Zariski equivalent.*

Chapter 4

The Zak Map of a Curve. Gaussian Maps

We first notice that the condition $H^1(T_Y \otimes L^{-1}) = 0$ of Corollary 3.3 is never fulfilled for any smooth projective curve Y embedded in $\mathbb{P}^n$ such that Y is non-degenerate and of codimension ≥ 2. The aim of this chapter is to provide an interpretation of the Zak map of a linearly normal smooth curve $Y \subset \mathbb{P}^n$ in terms of the so-called Gaussian maps. The advantage of Gaussian maps comes from the fact that in certain cases there are methods to check their surjectivity, see e. g. [147], [148], [38] and the references therein.

General Gaussian maps

Definition 4.1 (Wahl [148]). Let Y be a smooth projective variety, and let L, M be two line bundles on Y. Consider the canonical map

$$\mu_{L,M} : H^0(Y, L) \otimes H^0(Y, M) \to H^0(Y, L \otimes M). \tag{4.1}$$

Set $\mathcal{R}(L, M) := \mathrm{Ker}(\mu_{L,M})$. We are going to define *the Gaussian map*

$$\varphi_{L,M} : \mathcal{R}(L, M) \to H^0(Y, \Omega^1_{Y|k} \otimes L \otimes M) \tag{4.2}$$

associated to (L, M) on Y in the following way.

Let $\alpha = \sum\limits_i l_i \otimes m_i \in \mathcal{R}(L, M)$. Let U be an arbitrary affine open subset of Y such that $L|U \cong \mathcal{O}_U \cdot S$ and $M|U \cong \mathcal{O}_U \cdot T$, so that (over U), $l_i = a_i S$, $m_i = b_i T$, with $a_i, b_i \in \Gamma(U, \mathcal{O}_Y)$. Since $\alpha \in \mathcal{R}(L, M)$, we have $\sum\limits_i a_i b_i = 0$.

Then we set

$$\varphi_{L,M}(\alpha)|U := \sum_i (a_i db_i - b_i da_i) S \otimes T \in H^0(U, \Omega^1_{X|k} \otimes L \otimes M). \tag{4.3}$$

Replacing S by $\tilde{S} := u^{-1}S$ and T by $\tilde{T} := v^{-1}T$, with u and v units in $\Gamma(U, \mathcal{O}_Y)$, we have $l_i = \tilde{a}_i\tilde{S}$ and $m_i = \tilde{b}_i\tilde{T}$, where $\tilde{a}_i := ua_i$ and $\tilde{b}_i = vb_i$. Then

$$\begin{aligned}\sum_i(\tilde{a}_i d\tilde{b}_i - \tilde{b}_i d\tilde{a}_i)\tilde{S}\otimes\tilde{T} &= \sum_i(ua_ivdb_i + ua_ib_idv - vb_iuda_i - vb_ia_idu)u^{-1}v^{-1}S\otimes T\\ &= \sum_i(a_idb_i - b_ida_i)S\otimes T + (\sum_i a_ib_i)(udv - vdu)\\ &= \sum_i(a_idb_i - b_ida_i)S\otimes T,\end{aligned}$$

because $\sum_i a_ib_i = 0$. It follows that $\varphi_{L,M}(\alpha)$ is independent of the choices made, whence it is a well-defined element in $H^0(Y, \Omega^1_{Y|k}\otimes L\otimes M)$ as soon as we have checked that

$$\sum_i l_i\otimes m_i = 0 \implies \varphi_{L,M}(\alpha) = 0.$$

To this end, let $\{n_j\}$ be a basis of $H^0(Y, M)$. Then $m_i = \sum_j \beta_{ij}n_j$, with $\beta_{ij}\in k$. Then

$$\sum_i l_i\otimes m_i = \sum_i l_i\otimes(\sum_j\beta_{ij}n_j) = \sum_j(\sum_i\beta_{ij}l_i)\otimes n_j = 0,$$

whence (taking into account that $\{n_j\}$ is a basis of $H^0(Y, M)$ and of the properties of the tensor product), $\sum_i\beta_{ij}l_i = 0$. Writing $n_j = c_jT$, with $c_j\in\Gamma(U, \mathcal{O}_Y)$, we have $b_i = \sum_j\beta_{ij}c_j$ and $\sum_i\beta_{ij}a_i = 0$, so

$$\begin{aligned}\sum_i(a_idb_i - b_ida_i) &= \sum_{i,j}a_i\beta_{ij}dc_j - \sum_{i,j}\beta_{ij}c_jda_i = \sum_j(\sum_i\beta_{ij}a_i)dc_j - \sum_{i,j}\beta_{ij}c_jda_i\\ &= -\sum_{i,j}\beta_{ij}c_jda_i = -\sum_j c_jd(\sum_i\beta_{ij}a_i) = -\sum_j c_jd0 = 0,\end{aligned}$$

so that $\varphi_{L,M}(\alpha) = 0$, as required.

Properties

i) Let $f\colon X\to Y$ be a morphism of smooth projective varieties, and let L, M be two line bundles on Y. Then there is a natural commutative diagram

$$\begin{array}{ccc}\mathcal{R}(L,M) & \xrightarrow{\varphi_{L,M}} & H^0(\Omega^1_{Y|k}\otimes L\otimes M)\\ \downarrow & & \downarrow\\ \mathcal{R}(f^*(L), f^*(M)) & \xrightarrow[\varphi_{f^*(L),f^*(M)}]{} & H^0(\Omega^1_{X|k}\otimes f^*(L)\otimes f^*(M))\end{array}$$

ii) If N is a third line bundle on Y, there is a natural commutative diagram

$$\begin{array}{ccc} \mathcal{R}(L,M)\otimes H^0(Y,N) & \xrightarrow{\varphi_{L,M}\otimes \mathrm{id}} & H^0(\Omega^1_{Y|k}\otimes L\otimes M)\otimes H^0(Y,N) \\ \downarrow & & \downarrow \\ \mathcal{R}(L,M\otimes N) & \xrightarrow[\varphi_{L,M\otimes N}]{} & H^0(\Omega^1_{Y|k}\otimes L\otimes M\otimes N) \end{array}$$

iii) For every two line bundles L and M on a smooth projective variety Y we have the anti-commutative square

$$\begin{array}{ccc} \mathcal{R}(L,M) & \xrightarrow{\varphi_{L,M}} & H^0(Y,\Omega^1_{Y|k}\otimes L\otimes M) \\ \downarrow & & \downarrow \\ \mathcal{R}(M,L) & \xrightarrow[\varphi_{M,L}]{} & H^0(Y,\Omega^1_{Y|k}\otimes M\otimes L) \end{array}$$

i.e. $\varphi_{M,L}\circ u = -v\circ\varphi_{L,M}$, where u and v are the isomorphisms given by the commutativity of the tensor product. In particular, the Gaussian map $\varphi_{L,M}$ is surjective if and only if $\varphi_{M,L}$ is surjective.

iv) Assume $L = M$ and consider the natural map $\omega_L : \wedge^2 H^0(Y,L)\to\mathcal{R}(L,L)$ defined by

$$\omega_L(l_1\wedge l_2) := l_1\otimes l_2 - l_2\otimes l_1.$$

We get a composite

$$w_L := \varphi_{L,L}\circ\omega_L : \wedge^2 H^0(Y,L)\to H^0(Y,\Omega^1_{Y|k}\otimes L^2),$$

which is called *the Wahl map* associated to L on Y. Then we have the following:

Claim. If $\mathrm{char}(k)\neq 2$ then $\mathrm{Im}(w_L) = \mathrm{Im}(\varphi_{L,L})$, and in particular, $\mathrm{Coker}(w_L) = \mathrm{Coker}(\varphi_{L,L})$.

Indeed, the inclusion $\mathrm{Im}(w_L)\subseteq\mathrm{Im}(\varphi_{L,L})$ is obvious. Conversely, let $\alpha = \sum_i l_i\otimes m_i\in\mathcal{R}(L,L)$. Then define $\beta := \frac{1}{2}[\sum_i l_i\otimes m_i + \sum_i(-m_i)\otimes l_i]$. Clearly, $\beta\in\mathrm{Im}(\omega_L)$ and $\varphi_{L,L}(\alpha) = \varphi_{L,L}(\beta)$.

Of particular interest is the Wahl map

$$w_Y := w_{\omega_Y} : \wedge^2 H^0(Y,\omega_Y)\to H^0(Y,\Omega^1_{Y|k}\otimes\omega_Y^2) = H^0(Y,\omega_Y^3)$$

associated to a smooth projective curve Y of genus $g \geq 2$. Its main interest comes from the fact that w_Y is a map intrinsically associated to the curve Y. As we shall see below, one of the fundamental questions related to w_Y is whether w_Y is surjective. A necessary condition for the surjectivity of w_Y is obviously that

$$\dim_k \wedge^2 H^0(Y, \omega_Y) \geq \dim_k H^0(Y, \omega_Y^3).$$

Using Riemann–Roch, this condition amounts to

$$\frac{1}{2} g(g-1) \geq 5g - 5,$$

or else, $g \geq 10$.

Gaussian Maps of Polarized Curves

Let Y be a smooth projective curve and L a very ample line bundle on Y. Let $i = i_{|L|} : Y \hookrightarrow P := \mathbb{P}(H^0(Y, L)^*)$ be the linearly normal embedding into the projective space P given by the complete linear system $|L|$. In particular, $i^*(\mathcal{O}_P(1)) \cong L$. Consider the evaluation map $e \colon H^0(Y, L) \otimes \mathcal{O}_Y \to L$. Since L is very ample, the map e is surjective. Moreover, its kernel is identified to $\Omega^1_{P|k}(1)|Y = \Omega^1_{P|k} \otimes \mathcal{O}_P(1) \otimes \mathcal{O}_Y$, whence we get the exact sequence

$$0 \to \Omega^1_{P|k}(1)|Y \to H^0(Y, L) \otimes \mathcal{O}_Y \to L = \mathcal{O}_Y(1) \to 0, \tag{4.4}$$

which is nothing but the dual of the Euler sequence of P restricted to Y. Let M be another line bundle on Y. Tensoring (4.4) by M and taking cohomology we get the exact sequence

$$0 \longrightarrow H^0(Y, \Omega^1_{P|k}(1)|Y \otimes M) \to H^0(Y, L) \otimes H^0(Y, M) \xrightarrow{\mu_{L,M}} \tag{4.5}$$

$$\xrightarrow{\mu_{L,M}} H^0(Y, L \otimes M) \to H^1(Y, \Omega^1_{P|k}(1)|Y \otimes M) \xrightarrow{\alpha} H^0(Y, L) \otimes H^1(Y, M).$$

In particular, we get the identification

$$\mathcal{R}(L, M) = H^0(Y, \Omega^1_{P|k}(1)|Y \otimes M). \tag{4.6}$$

On the other hand, the dual normal sequence of Y in P yields the exact sequence

$$0 \longrightarrow N^*_{Y|P} \otimes L \otimes M \xrightarrow{b} \Omega^1_{P|k}(1)|Y \otimes M \xrightarrow{a} \Omega^1_{Y|k} \otimes L \otimes M \longrightarrow 0,$$

whence, taking (4.6) into account, the cohomology sequence

$$\mathcal{R}(L, M) \xrightarrow{\varphi_{L,M} = H^0(a)} H^0(Y, \Omega^1_{Y|k} \otimes L \otimes M) \longrightarrow \tag{4.7}$$

$$\longrightarrow H^1(Y, N^*_{Y|k} \otimes L \otimes M) \xrightarrow{H^1(b)} H^1(Y, \Omega^1_{P|k}(1)|Y \otimes M).$$

In particular from (4.7) we get

$$\operatorname{Coker}(\varphi_{L,M}) \cong \operatorname{Ker}(H^1(b)). \tag{4.8}$$

Therefore (4.8) implies the following:

Lemma 4.2. *Under the hypotheses of property* iii) *above the Gaussian map* $\varphi_{L,M}$ *is surjective if* $H^1(Y, N^*_{Y|P} \otimes L \otimes M) = 0$.

Corollary 4.3. *Assume furthermore that* $H^1(Y, L^{-1} \otimes M) = 0$ *and that the very ample line bundle* L *is normally presented, i.e. the graded* k*-algebra* $R(Y, L) := \bigoplus_{i=0}^{\infty} H^0(Y, L^i)$ *is generated by its homogeneous part of degree one and the ideal*

$$\operatorname{Ker}[S(H^0(X, L)^*) = \bigoplus_{i=0}^{\infty} H^0(P, \mathcal{O}_P(i)) \to R(Y, L)]$$

is generated by its homogeneous part of degree 2, *where* $S(V)$ *denotes the symmetric* k*-algebra associated to a* k*-vector space* V. *Then the Gaussian map* $\varphi_{L,M}$ *is surjective.*

Proof. Since L is normally presented, there exists a surjection of the form

$$m\mathcal{O}_P(-2) \to \mathcal{I}_Y \to 0,$$

where $\mathcal{I}_Y$ is the ideal sheaf of Y in $\mathcal{O}_P$. This yields the surjection

$$(m\mathcal{O}_P(-2)) \otimes L \otimes M \cong m(L^{-1} \otimes M) \to \mathcal{I}_Y \otimes L \otimes M \cong N^*_{Y|P} \otimes L \otimes M \to 0.$$

Since Y is a curve we get therefore a surjection

$$H^1(Y, m(L^{-1} \otimes M)) \to H^1(Y, N^*_{Y|P} \otimes L \otimes M) \to 0.$$

Using the hypothesis that $H^1(Y, L^{-1} \otimes M) = 0$ and Lemma 4.2 we get the conclusion of our corollary. □

If in Corollary 4.3 we take Y a non-hyperelliptic curve of genus $g \geq 3$, which is neither trigonal, nor a plane quintic, then by a theorem of Max Noether–Petri (see [3]), the canonical class $L = \omega_Y$ is very ample and normally presented. Take M of degree $\geq 4g - 3$. Then

$$\deg(L^{-1} \otimes M) = \deg(M) - \deg(L) \geq (4g - 3) - (2g - 2) = 2g - 1.$$

By Riemann–Roch, $H^1(Y, L^{-1} \otimes M) = 0$. Therefore Corollary 4.3 yields:

Theorem 4.4 (Lazarsfeld). *Let* Y *be a non-hyperelliptic curve of genus* $g \geq 3$ *which is neither trigonal, nor a plane quintic. Let* M *be a line bundle on* Y *of degree* $\geq 4g - 3$. *Then the Gaussian map* $\varphi_{\omega_Y, M}$ *is surjective.*

Remark 4.5. The conclusion of Theorem 4.4 still holds if we take M of degree $4g-4$, provided $M \not\cong \omega_Y^2$.

Corollary 4.6. *Under the hypotheses of Theorem 4.4 the map φ_{M,ω_Y} is surjective.*

Proof. The conclusion follows from Theorem 4.4 and from property iii) above. □

Theorem 4.7 (Wahl). *Let Y be a smooth projective curve of genus $g > 0$ and let L be a very ample line bundle on Y. Let $i\colon Y \hookrightarrow P = \mathbb{P}(H^0(Y,L)^*)$ be the linearly normal embedding given by $|L|$ into the projective space P. Then there is a canonical isomorphism*

$$\mathrm{Coker}(z) \cong \mathrm{Coker}(\varphi_{L,\omega_Y})^*,$$

where $z\colon H^0(Y,L)^* \to H^0(Y, N_{Y|P}(-1))$ is the Zak map of Y in P. In particular, the Zak map z is surjective if and only if the Gaussian map φ_{L,ω_Y} is surjective.

Proof. The proof that follows is taken from [39]. Using (4.8) we get

$$\mathrm{Coker}(\varphi_{L,\omega_Y}) \cong \mathrm{Ker}[H^1(Y, N^*_{Y|P} \otimes L \otimes \omega_Y) \to H^1(Y, \Omega^1_{P|k}(1)|Y \otimes \omega_Y)],$$

or by duality on Y,

$$\mathrm{Coker}(\varphi_{L,\omega_Y})^* \cong \mathrm{Coker}[H^0(Y, T_P(-1)|Y) \to H^0(Y, N_{Y|P}(-1))]. \tag{4.9}$$

On the other hand, by a general classical statement due to Petri (see [148], or [39]) the map μ_{L,ω_Y} is surjective (recall that L is very ample and $g > 0$ by hypotheses). If in (4.5) we take $M = \omega_Y$, we get that the map

$$\alpha\colon H^1(Y, \Omega^1_{P|k}(1) \otimes \omega_Y) \to H^0(Y,L) \otimes H^1(Y, \omega_Y)$$

is injective. Therefore by duality the map

$$\alpha^* : H^0(Y,L)^* \otimes H^0(Y, \mathcal{O}_Y) = H^0(Y,L)^* \to H^0(Y, T_P(-1)|Y) \tag{4.10}$$

is surjective. Then the definition of the Zak map, (4.9) and the surjectivity of the map (4.10) yield the conclusion of the theorem. □

Corollary 4.8. *Let Y be a smooth projective non-hyperelliptic curve of genus $g \geq 3$ which is neither trigonal, nor a plane quintic. Let L be a line bundle on Y of degree $\geq 4g-4$ such that $L \not\cong \omega_Y^2$. Then the Zak map associated to the linearly normal embedding $i = i_{|L|} : Y \hookrightarrow \mathbb{P}^n$, with $n = h^0(Y,L) - 1 = \deg(L) - g$, is surjective. In particular, every extension in $\mathbb{P}^{n+1}$ of the embedded curve $Y \subset \mathbb{P}^n$ is a cone.*

Proof. Since $4g-4 \geq 2g+1$ for $g \geq 3$, L is very ample on Y. Then the corollary is a consequence of Theorem 4.7 and of Corollary 4.6. For the last statement apply Theorem 1.2. □

Now we come back to the Wahl map $w_Y : \wedge^2 H^0(Y, \omega_Y) \to H^0(Y, \omega_Y^3)$ of a curve Y of genus $g \geq 2$. As we saw in property iv) above, the surjectivity of w_Y implies $g \geq 10$. Then we have the following fundamental result:

Theorem 4.9 (Ciliberto–Harris–Miranda [38]). *The Wahl map w_Y of a general curve Y of genus $g \geq 10$ and $g \neq 11$ is surjective.*

The conclusion of Theorem 4.9 is false for $g = 11$ (see [107], via Theorem 4.11 below). Note that the surjectivity of the Wahl map w_Y is an open condition in the moduli space $\mathcal{M}_g$ of isomorphism classes of curves of genus g. For some genera $g \geq 10$ Wahl produced for the first time explicit examples of some complete intersection curves Y with w_Y surjective (see [147]). Therefore for those genera Theorem 4.9 is due to Wahl. The method of Ciliberto–Harris–Miranda is entirely different, the main idea being to study the Wahl map for certain degenerations of curves of genus g.

Corollary 4.10. *Let $Y \hookrightarrow \mathbb{P}^{g-1}$ be the canonical embedding of a general projective curve Y of genus $g \geq 10$, $g \neq 11$. Then every extension of Y in $\mathbb{P}^g$ is a cone.*

Proof. By Theorems 4.9 and 4.7 and the property iv) above, the Zak map of $Y \hookrightarrow \mathbb{P}^{g-1}$ is surjective. Then the conclusion follows from Theorem 1.2. □

Theorem 4.11 (Wahl [147]). *Let Y be a smooth non-hyperelliptic projective curve of genus $g \geq 3$ which is contained in a $K3$ surface S. Then the Wahl map w_Y of Y is not surjective.*

Proof. The genus formula and the fact that the canonical class K_S is trivial yield $(Y^2) > 0$, where $(Y^2) = (Y^2)_S$ denotes the self-intersection of Y on S. It follows that $(Y \cdot C) \geq 0$ for every irreducible curve C on S. Assume that there exists an irreducible curve C on S such that $(Y \cdot C) = 0$. Using the Hodge index theorem (see e.g. [26], or [25], or [12]) we get $(C^2) < 0$. Then from the genus formula we get $C \cong \mathbb{P}^1$ and $(C^2) = -2$. The Hodge index theorem also implies that there are only finitely many such (-2)-curves. Then a projective contractibility criterion of M. Artin (see [4], or also [12], chapter 3) shows that there exists a birational morphism $f\colon S \to X$, with the following properties: X is a normal projective surface having finitely many singular points (which are rational double points) such that the canonical divisor K_X of X is Cartier (in particular, all the singularities of X are Gorenstein), $f^*(K_X)$ is a canonical divisor K_S on S, and the canonical map $H^1(X, \mathcal{O}_X) \to H^1(S, \mathcal{O}_S)$ is an isomorphism. In our case, since S is a $K3$-surface, $K_S = 0$ and $q = h^1(S, \mathcal{O}_S) = 0$. It follows that $K_X = 0$ and $h^1(X, \mathcal{O}_X) = 0$. In other words, X is a singular $K3$-surface. Moreover, by construction, f defines an isomorphism $f|U\colon U \cong f(U)$ from a Zariski open neighbourhood U of Y in S (we can take $U = S \setminus E$, where E is the union of all irreducible (-2)-curves C such that $(Y \cdot C) = 0$). In particular, Y can also be embedded (via $f|Y$) in X as a Cartier divisor. Again by construction, $(Y \cdot D)_X > 0$ for every irreducible curve D on X. Since $f^*(Y) = Y$, it also follows that $(Y^2)_X = (Y^2)_S > 0$. Then by the

Nakai–Moishezon criterion of ampleness (see e.g. [74], or also [12], chapter 1) we infer that Y is an ample Cartier divisor on X.

Now the adjunction formula together with the fact that $\omega_X = \mathcal{O}_X(K_X) \cong \mathcal{O}_X$ yield $\mathcal{O}_X(Y)|Y \cong \omega_Y$. Therefore for every $n \geq 0$ we get the exact sequence

$$0 \to \mathcal{O}_X((n-1)Y) \to \mathcal{O}_X(nY) \to \omega_Y^n \to 0,$$

which yields the cohomology sequence

$$0 \to H^0(X, \mathcal{O}_X((n-1)Y)) \to H^0(X, \mathcal{O}_X(nY)) \to H^0(Y, \omega_Y^n) \to$$
$$\to H^1(X, \mathcal{O}_X((n-1)Y)) \to H^1(X, \mathcal{O}_X(nY)) \to H^1(Y, \omega_Y^n) \to$$
$$\to H^2(X, \mathcal{O}_X((n-1)Y)) \to H^2(X, \mathcal{O}_X(nY)) \to 0.$$

This exact sequence together with the equalities $H^1(X, \mathcal{O}_X) = 0$, $h^2(X, \mathcal{O}_X) = h^1(Y, \omega_Y) = 1$ and $H^2(X, \mathcal{O}_X(nY)) \cong H^0(X, \mathcal{O}_X(-nY)) = 0$, for all $n > 0$, show that $H^1(\mathcal{O}_X(nY)) = 0$ for every $n \geq 0$. Therefore we get the exact sequences

$$0 \to H^0(X, \mathcal{O}_X((n-1)Y)) \to H^0(X, \mathcal{O}_X(nY)) \to H^0(Y, \omega_Y^n) \to 0, \quad \forall n \geq 0.$$

In particular, if we set $A := \bigoplus_{n=0}^{\infty} H^0(X, \mathcal{O}_X(nY))$ and $R := \bigoplus_{n=0}^{\infty} H^0(Y, \omega_Y^n)$, we get $A/tA \cong R$, where $t \in A_1 = H^0(X, \mathcal{O}_X(Y))$ is a global equation of Y in X. Now, by a classical result of Max Noether (see [3]), the canonical ring R of a non-hyperelliptic curve Y of genus $g \geq 3$ is generated by R_1. It follows that A is also generated by A_1 (because $A/tA \cong R$ and $\deg(t) = 1$) and $\dim H^0(X, \mathcal{O}_X(Y)) = g+1$. In particular, X is an extension of Y in $\mathbb{P}^g$ in the sense of Definition 1.1. This extension cannot be trivial because otherwise X (and hence also S) would be birationally equivalent to $Y \times \mathbb{P}^1$. At this point we can apply Theorem 1.2 to deduce that the Zak map of Y in $\mathbb{P}^{g-1}$ cannot be surjective. Therefore by Theorem 4.7 and property iv) above the Wahl map w_Y cannot be surjective. □

Remarks 4.12. i) In the proof of Theorem 4.11 we could have used Theorem 3.6 instead of the more delicate contractibility Theorem of M. Artin. Wahl's proof of Theorem 4.11 given in [147] is completely different from the above one. Another proof of this result can be found in [27].

ii) The major interest of Theorem 1.2 consists in the fact that it holds also for the case when Y is a curve, in which case the cokernel of the Zak map can be interpreted (via Theorem 4.7 above) in terms of the cokernel of a certain Gaussian map. Moreover, the Theorems 4.4 and 4.9 above produce very interesting examples of surjective Gaussian maps.

Chapter 5

Quasi-homogeneous Singularities and Projective Geometry

Let X be a smooth variety and Y a closed subvariety having only one singularity $y \in Y$. Then we may consider the normal sequence of Y in X

$$0 \longrightarrow T_Y \longrightarrow T_X|Y \xrightarrow{\varphi} N_{Y|X}.$$

Since Y is smooth outside y and X is smooth, φ is surjective outside the point y. Therefore

$$T^1_{Y,y} := \mathrm{Coker}(\varphi)$$

is a coherent sheaf concentrated at the point y. In particular, $T^1_{Y,y}$ (sometimes also denoted by T^1_Y) is a finite-dimensional vector space over k (because it is a coherent sheaf concentrated at one point). Then $T^1_{Y,y}$ is called *the space of first order infinitesimal deformations* of the isolated singularity (Y, y). This space turns out to be an extremely important intrinsic invariant of (Y, y), i.e. $T^1_{Y,y}$ depends only on the singularity (Y, y), and not on the choice of the embedding $Y \hookrightarrow X$ into the smooth variety X (see [127], or [5], or [120], cf. also Corollary 5.3 and remark 5.4 below).

The definition of $T^1_{Y,y} = T^1_Y$ being local, we may assume that $Y = \mathrm{Spec}(A)$, with A a finitely generated k-algebra. Then we may write $A = k[T_1, \ldots, T_n]/I$, where I is an ideal of the polynomial k-algebra $P := k[T_1, \ldots, T_n]$, and therefore we may take $X = \mathrm{Spec}(k[T_1, \ldots, T_n]) = \mathbb{A}^n$, and as $Y \hookrightarrow X$ the closed subvariety $V(I)$ of X defined by the ideal I. Then the canonical exact sequence (see [73], Proposition 8.12, page 176)

$$I/I^2 \to \Omega^1_{P|k} \otimes_P A \to \Omega^1_{A|k} \to 0$$

yields for every A-module M the exact sequence

$$0 \longrightarrow \mathrm{Hom}_A(\Omega^1_{A|k}, M) = \mathrm{Der}_k(A, M) \rightarrow \mathrm{Hom}_A(\Omega^1_{P|k} \otimes_P A, M) =$$

$$= \mathrm{Der}_k(P, M) \xrightarrow{\varphi_M} \mathrm{Hom}_A(I/I^2, M) \rightarrow T^1(A|k, M) \longrightarrow 0,$$

where by definition $T^1(A|k, M) := \operatorname{Coker}(\varphi_M)$. The A-module $T^1(A|k, M)$ is called the *module of deformations of* $A|k$ *by* M. Taking $M = A$ we get that $T^1_{Y,y}$ is the sheaf associated to the A-module $T^1(A|k, A)$.

Definition 5.1. Let A be a commutative k-algebra of finite type, and M an A-module. An *extension of* A *over* k *(or of* A/k*) by* M is an exact sequence

$$\mathbf{E}:\ 0 \longrightarrow M \longrightarrow E \xrightarrow{\ j\ } A \longrightarrow 0\,,$$

where E is a commutative k-algebra, j is a surjective homomorphism of k-algebras, $i(M) = \operatorname{Ker}(j)$ is a square-zero ideal of E (in particular, $i(M)$ becomes an A-module), and i defines an isomorphism of A-modules between M and $i(M)$.

Two extensions $\mathbf{E}$ and

$$\mathbf{E}':\ 0 \longrightarrow M \longrightarrow E' \xrightarrow{\ j'\ } A \longrightarrow 0$$

of A/k by M are said to be *equivalent* if there exists a k-algebra homomorphism $u\colon E \to E'$ inducing a commutative diagram

$$\begin{array}{ccccc} M & \xrightarrow{\ i\ } & E & \xrightarrow{\ j\ } & A \\ \Big\downarrow{\scriptstyle \mathrm{id}_M} & & \Big\downarrow{\scriptstyle u} & & \Big\downarrow{\scriptstyle \mathrm{id}_A} \\ M & \xrightarrow[\ i'\ ]{} & E' & \xrightarrow[\ j'\ ]{} & A \end{array}$$

In this case it follows easily that u must be an isomorphism of k-algebras. The set of equivalence classes of extensions of A/k by M will be denoted by $\operatorname{Ex}^1(A|k, M)$.

Theorem 5.2. *With the above definitions and notation, there is a natural bijection*

$$\alpha : \operatorname{Ex}^1(A|k, M) \to T^1(A|k, M).$$

Proof. We shall follow [127] and [92]. Consider an extension

$$\mathbf{E}:\ 0 \longrightarrow M \longrightarrow E \xrightarrow{\ j\ } A \longrightarrow 0\,.$$

Since P is a polynomial k-algebra the canonical surjection $g\colon P \to A = P/I$ factors through a (not necessarily unique) k-algebra homomorphism $f\colon P \to E$, i.e. $j \circ f = g$. Since $g(I) = 0$, $f(I) \subseteq \operatorname{Ker}(j) = M$. However, $i(M)$ is a square-zero ideal of E, hence f induces a homomorphism of A-modules $h\colon I/I^2 \to M$, and thus, an element of $T^1(A|k, M)$. If $f' : P \to E$ is another lifting of g, then $i^{-1}(f - f') : P \to M$ is a k-derivation, thus the map

$$\alpha : \operatorname{Ex}^1(A|k, M) \to T^1(A|k, M)$$

given by $(E) \to \alpha(E) := \hat{h}$ is well defined.

Conversely, let $\hat{h}$ be an arbitrary element of $T^1(A|k, M)$, with h an element of $\mathrm{Hom}_A(I/I^2, M)$, i.e. $h\colon I \to M$, an A-homomorphism which vanishes on I^2. On the product $P \times M$ consider the usual addition and the multiplication defined by

$$(x, m)(x', m') := (xx', x'm + xm'), \ \forall (x, m), \ (x', m') \in P \times M,$$

where M is regarded as a P-module by the restriction of scalars via the canonical map of k-algebras $g\colon P \to A$. With this multiplication, $P \times M$ becomes a k-algebra, and it is easy to see that $\{(x, -h(x)) \mid x \in I\}$ is an ideal of $P \times M$. Set

$$E_h := (P \times M)/\{(x, -h(x)) \mid x \in I\}.$$

Let $j_h\colon E_h \to A$ be the map defined by $j_h(\widehat{x, m}) = x \mod I \in A$, $\forall (\widehat{x, m}) \in E_h$. Clearly, j_h is a well defined surjective homomorphism of k-algebras. For every $(\widehat{x, m})$, $(\widehat{x', m'}) \in \mathrm{Ker}(j_h)$ we have

$$(\widehat{x, m})(\widehat{x', m'}) = (xx', \widehat{x'm + xm'}) = (\widehat{xx', 0}) = (xx', \widehat{-h(xx')}) = (\widehat{0, 0}),$$

because $x, x' \in I$ and h vanishes on I^2. In other words, $\mathrm{Ker}(j_h)$ is a square-zero ideal of E_h. Moreover, it is easily checked that the map $(\widehat{x, m}) \to m + h(x)$ yields a well-defined isomorphism of A-modules $\mathrm{Ker}(j_h) \cong M$. In other words, in this way we get an extension $\mathbf{E}_h$ of A/k by M.

We prove now that if $\hat{h} = \hat{h}'$ (with $h' \in \mathrm{Hom}_A(I/I^2, M)$), the extensions $\mathbf{E}_h$ and $\mathbf{E}'_h$ (of A/k by M) are equivalent. The equality $\hat{h} = \hat{h}'$ means that there is a derivation $d \in \mathrm{Der}_k(P, M)$ such that $h' = h + d|I\colon I \to M$ (h and h' are regarded as k-linear maps which vanish on I^2). Then it is easily checked that the function $\varphi\colon E_h \to E_{h'}$ defined by $\varphi((\widehat{x, m})) := (x, \widehat{m + d(x)})$ yields an isomorphism of extensions $(E_h) \cong (E_{h'})$.

Therefore we get a well-defined map $\beta\colon T^1(A|k, M) \to \mathrm{Ex}^1(A|k, M)$ given by $\beta(\hat{h}) :=$ class of (E_h), which is easily checked to be the inverse of the map α. □

Corollary 5.3. *$T^1(A/k, M)$ is independent of the choice of the presentation $A = P/I$. In particular, if (Y, y) is an isolated singularity, $T^1_{Y,y}$ is independent of the embedding $Y \hookrightarrow \mathbb{A}^n$.*

Proof. The corollary follows from Theorem 5.2 because the set $\mathrm{Ex}^1(A|k, M)$ is independent of the presentation $A = P/I$. □

Remark 5.4. One can prove that the A-module structure of $T^1(A|k, M)$ is also independent of the choice of the presentation. Roughly speaking, this is done in the following way: one can intrinsically define an addition on $\mathrm{Ex}^1(A|k, M)$ and a scalar multiplication $A \times \mathrm{Ex}^1(A|k, M) \to \mathrm{Ex}^1(A|k, M)$ such that $\mathrm{Ex}^1(A|k, M)$ becomes an A-module (i.e. depending only on A/k and on M). Then one checks that the bijective map α is in fact an isomorphism of A-modules (see [127] for details).

Quasi-homogeneous singularities

Assume now that $A = \bigoplus_{i=0}^{\infty} A_i$ is a finitely generated graded k-algebra such that $Y := \mathrm{Spec}(A)$ has an isolated singularity y at the maximal irrelevant ideal $m_A := \bigoplus_{i=1}^{\infty} A_i$. The singularities (Y, y) of the form $Y = \mathrm{Spec}(A)$ and $y = m_A$, where A is a finitely generated graded k-algebra are called *quasi-homogeneous*. Since A is finitely generated there are homogeneous elements $a_1, \ldots, a_n \in A$ of positive degrees $d_1, \ldots, d_n$ respectively such that $A = k[a_1, \ldots, a_n]$. Consider the polynomial k-algebra $P := k[T_1, \ldots, T_n]$ graded by the conditions that $\deg(T_i) = d_i$, $\forall i = 1, \ldots, n$. Then the map $\varphi \colon P \to A$ of graded k-algebras defined by $\varphi(T_i) = a_i$, $\forall i = 1, \ldots, n$, becomes a surjective homomorphism of graded k-algebras. If $I := \mathrm{Ker}(\varphi)$, then I is a homogeneous ideal of P such that $A = P/I$. Let M be a graded A-module. By the definition of $T^1(A|k, M)$ we have the exact sequence

$$0 \longrightarrow \mathrm{Der}_k(A, M) \xrightarrow{u} \mathrm{Der}_k(P, M) \xrightarrow{v = \varphi_M}$$

$$\xrightarrow{v} \mathrm{Hom}_A(I/I^2, M) \longrightarrow T^1(A|k, M) \longrightarrow 0,$$

in which the first three A-modules are graded and u and v are homomorphisms of graded A-modules. In particular, $T^1(A|k, M)$ becomes a graded A-module.

On the other hand, by Corollary 5.3 and Remark 5.4, the A-module $T^1(A|k, M)$ is independent of the presentation $A = P/I$, i.e. $T^1(A|k, M)$ is independent of the choice of the system of homogeneous generators $a_1, \ldots, a_n$ of the graded k-algebra A. One can prove that the structure of a *graded* A-module of $T^1(A|k, M)$ thus obtained is also independent of the choice of homogeneous presentation $A = P/I$ (see [120]). In particular, the space $T^1_{Y,y} = T^1(A|k, A)$ of first order infinitesimal deformations of the quasi-homogeneous isolated singularity (Y, y) (corresponding to the irrelevant ideal m_A) has an intrinsic decomposition

$$T^1_{Y,y} = \bigoplus_{i \in \mathbb{Z}} T^1_{Y,y}(i)$$

which corresponds to the structure of the graded A-module of $T^1(A|k, A)$. The subspace $T^1_{Y,y}(i)$ is called *the space of first order infinitesimal deformations of weight* i of the quasi-homogeneous isolated singularity (Y, y). Since the singularity (Y, y) is isolated, $T^1_{Y,y}$ is a finite-dimensional k-vector space, and in particular $T^1_{Y,y}(i) \neq 0$ only for finitely many values of $i \in \mathbb{Z}$.

Cones over projectively normal varieties

Let Y be a closed smooth subvariety of the n-dimensional projective space $\mathbb{P}^n$ of dimension ≥ 1. We recall that Y is called *projectively normal* in $\mathbb{P}^n$ if the Serre map (see [134], or also [73])

$$\alpha\colon k[Y] \to R(Y) := \bigoplus_{i\in\mathbb{Z}} H^0(Y, \mathcal{O}_Y(i))$$

is an isomorphism, where $k[Y] := k[T_0, \ldots, T_n]/I(Y)$ is the homogeneous coordinate k-algebra of Y in $\mathbb{P}^n$ (with $T_0, \ldots, T_n$ $n+1$ variables over k and $I(Y)$ the saturated ideal of Y in $\mathbb{P}^n$). Geometrically, this means that Y is irreducible and for every $i \geq 1$ the linear system cut out on Y by all hypersurfaces of degree i is complete. The ring $R(Y)$ is normal since Y is smooth. On the other hand, the Serre map α of any closed irreducible subvariety $Y \subset \mathbb{P}^n$ is TN-surjective, i.e. the maps $\alpha_i\colon k[Y]_i \to H^0(Y, \mathcal{O}_Y(i))$ are isomorphisms for all $i \gg 0$ (see [134]). It follows that (for a not necessarily projectively normal closed subvariety $Y \subseteq \mathbb{P}^n$) the ring $R(Y)$ is the normalization of $k[Y]$ in its field of quotients. Therefore Y is projectively normal if and only if the ring $R := k[Y]$ is normal.

Set $C := \operatorname{Spec}(k[T_0, \ldots, T_n]) = \mathbb{A}^{n+1}$. Inside C we have the affine cone $C_Y := \operatorname{Spec}(R) = \operatorname{Spec}(k[Y])$ with vertex $x = (0, 0, \ldots, 0)$ (which corresponds to the maximal irrelevant ideal m_R). Since Y is smooth and projectively normal in $\mathbb{P}^n$, C_Y is a closed normal subvariety of C having an isolated singularity at x. Moreover the gradings of R and of $k[T_0, \ldots, T_n]$ ($\deg(T_i) = 1$, $i = 0, \ldots, n$) yield obvious $\mathbb{G}_m$-actions on C_Y and on C such that the closed embedding $C_Y \hookrightarrow C$ becames $\mathbb{G}_m$-invariant. Set $U_Y := C_Y \setminus \{x\}$ and $U := C \setminus \{x\}$. Then U_Y and U are $\mathbb{G}_m$-invariant open subsets of C_Y and of C respectively, such that the closed embedding $U_Y \hookrightarrow U$ is also $\mathbb{G}_m$-invariant. We have natural projections $\pi_Y\colon U_Y \to Y$ and $\pi\colon U \to \mathbb{P}^n$ such that the following diagram (in which the horizontal arrows are the natural inclusions)

$$\begin{array}{ccc} \pi^{-1}(Y) = U_Y & \longrightarrow & U \\ \downarrow{\scriptstyle \pi_Y} & & \downarrow{\scriptstyle \pi} \\ Y & \longrightarrow & \mathbb{P}^n \end{array} \tag{5.1}$$

is cartesian.

Our aim is to compute in geometric terms the space $T^1_{C_Y} = T^1_{C_Y,x}$ of first order infinitesimal deformations of the isolated singularity (C_Y, x). By the definition of $T^1_{C_Y}$ we have the exact sequence

$$0 \to T_{C_Y} \to T_C|C_Y \to N_{C_Y|C} \to T^1_{C_Y} \to 0.$$

Since C_Y is an affine variety, by passing to global sections we get the exact sequence

$$0 \to H^0(T_{C_Y}) \to H^0(T_C|C_Y) \to H^0(N_{C_Y|C}) \to H^0(T^1_{C_Y}) = T^1(R|k, R) \to 0.$$

On the other hand, since T_{C_Y}, $T_C|C_Y$ and $N_{C_Y|C}$ are reflexive sheaves on the normal variety C_Y of dimension ≥ 2, and since x is a point of C_Y, by Proposition 1.4 we get

$$H^0(T_{C_Y}) \cong H^0(U_Y, T_{C_Y}), \;\; H^0(T_C|C_Y) \cong H^0(U_Y, T_C|C_Y),$$
$$H^0(N_{C_Y|C}) \cong H^0(U_Y, N_{C_Y|C}).$$

Therefore the above exact sequence becomes

$$0 \to H^0(U_Y, T_{C_Y}) \to H^0(U_Y, T_C|C_Y) \to H^0(U_Y, N_{C_Y|C}) \to$$
$$\to H^0(T^1_{C_Y}) = T^1(R|k, R) \to 0. \tag{5.2}$$

Since $U_Y \subset U$ is a closed immersion of smooth varieties we have the normal exact sequence

$$0 \to T_{U_Y} \to T_U|U_Y \to N_{U_Y|U} \to 0, \tag{5.3}$$

whose cohomology sequence together with the exact sequence (5.2) yield in particular the inclusion

$$T^1_{C_Y} = H^0(T^1_{C_Y}) \subseteq H^1(U_Y, T_{U_Y}). \tag{5.4}$$

Now look at the commutative diagram with exact rows and columns

$$\begin{array}{ccccccccc}
 & & & & 0 & & 0 & & \\
 & & & & \downarrow & & \downarrow & & \\
0 & \longrightarrow & \mathcal{O}_{U_Y} & \longrightarrow & T_{U_Y} & \longrightarrow & \pi_Y^*(T_Y) & \longrightarrow & 0 \\
 & & \downarrow{\scriptstyle \mathrm{id}} & & \downarrow & & \downarrow & & \\
0 & \longrightarrow & \mathcal{O}_{U_Y} & \longrightarrow & T_U|U_Y & \longrightarrow & \pi^*(T_{\mathbb{P}^n})|U_Y = \pi_Y^*(T_{\mathbb{P}^n}|Y) & \longrightarrow & 0 \\
 & & & & \downarrow & & \downarrow & & \\
 & & & & N_{U_Y|U} & \xrightarrow{\;\cong\;} & \pi_Y^*(N_{Y|\mathbb{P}^n}) & & \\
 & & & & \downarrow & & \downarrow & & \\
 & & & & 0 & & 0 & &
\end{array}$$

The fact that the first two rows are exact comes from Lemma 2.1 (applied to π_Y and to π), the equality in the second row from the commutativity of diagram (5.1), the last column is the pull back of the normal sequence of Y in $\mathbb{P}^n$, and the middle column is just (5.3). In particular, $N_{U_Y|U} \cong \pi_Y^*(N_{Y|\mathbb{P}^n})$. Then (5.2) yields

$$H^0(U_Y, T_U|U_Y) \xrightarrow{\alpha} H^0(U_Y, \pi_Y^*(N_{Y|\mathbb{P}^n})) \longrightarrow T^1_{C_Y} \longrightarrow 0. \tag{5.5}$$

Now we have to examine the map α more closely.

Claim. There is a canonical identification $T_U \cong \pi^*((n+1)\mathcal{O}_{\mathbb{P}^n}(1))$ such that the map

$$\alpha_i \colon H^0(U_Y, T_U|U_Y)_i \to H^0(U_Y, \pi_Y^*(N_{Y|\mathbb{P}^n}))_i$$

is identified (up to multiplication by a non-zero constant) to the Zak map

$$z(i+1) : H^0(Y, (n+1)\mathcal{O}_Y(i+1)) \to H^0(Y, N_{Y|\mathbb{P}^n}(i))$$

(twisted by $i+1$).

Then (5.5) and the claim yield the following result:

Theorem 5.5 (Schlessinger [128], [129], [120]). *Let $Y \subset \mathbb{P}^n$ be a smooth projectively normal closed subvariety of $\mathbb{P}^n$ of dimension ≥ 1, and let C_Y be the affine cone over Y in $\mathbb{P}^n$. Then there is a natural identification*

$$T^1_{C_Y}(i) \cong \operatorname{Coker}(z(i+1)),$$

where $z(i+1) : H^0((n+1)\mathcal{O}_Y(i+1)) \to H^0(N_{Y|\mathbb{P}^n}(i))$ is the Zak map of Y in $\mathbb{P}^n$ twisted by $i+1$.

It remains to prove the above claim. To do this we shall make use of the following special cases of the well-known Bott's formulae:

$$H^0(\mathbb{P}^n, \Omega^1_{\mathbb{P}^n}) = 0,\ H^1(\mathbb{P}^n, \Omega^1_{\mathbb{P}^n}(i)) = 0,\ \forall i \neq 0,\ \text{and}\ H^1(\mathbb{P}^n, \Omega^1_{\mathbb{P}^n}) \cong k. \tag{5.6}$$

First observe that the isomorphism classes of vector bundle extensions

$$0 \to \mathcal{O}_{\mathbb{P}^n} \to E \to T_{\mathbb{P}^n} \to 0$$

of $T_{\mathbb{P}^n}$ by $\mathcal{O}_{\mathbb{P}^n}$ are classified by $H^1(P^n, \Omega^1_{\mathbb{P}^n})$ which by (5.6) is a one-dimensional vector space. Therefore either $E \cong \mathcal{O}_{\mathbb{P}^n} \bigoplus T_{\mathbb{P}^n}$, or the above exact sequence is isomorphic (up to multiplication of the second map by a non-zero constant) to the Euler sequence of $\mathbb{P}^n$.

On the other hand, the isomorphism classes of vector bundle extensions of $\pi^*(T_{\mathbb{P}^n})$ by $\pi^*(\mathcal{O}_{\mathbb{P}^n}) = \mathcal{O}_U$ are classified by $H^1(U, \pi^*(\Omega^1_U))$, which by Lemma 2.2 and (5.6) is isomorphic to $\bigoplus_{i\in\mathbb{Z}} H^1(\mathbb{P}^n, \Omega^1_{\mathbb{P}^n}(i)) \cong H^1(\mathbb{P}^n, \Omega^1_{\mathbb{P}^n}) \cong k$. We infer that every vector bundle extension of $\pi^*(T_{\mathbb{P}^n})$ by $\pi^*(\mathcal{O}_{\mathbb{P}^n}) = \mathcal{O}_U$ is the pull back via π of a vector bundle extension of $T_{\mathbb{P}^n}$ by $\mathcal{O}_{\mathbb{P}^n}$. In particular, considering the exact sequence

$$0 \to \mathcal{O}_U \to T_U \to \pi^*(T_{\mathbb{P}^n}) \to 0,$$

(given by Lemma 2.1) it follows that either $T_U \cong \mathcal{O}_U \oplus \pi^*(T_{\mathbb{P}^n})$, or

$$T_U \cong (n+1)\pi^*(\mathcal{O}_{\mathbb{P}^n}(1)). \tag{5.7}$$

Notice that $T_U \cong (n+1)\mathcal{O}_U$, or else $\Omega^1_U \cong (n+1)\mathcal{O}_U$. Then the first possibility is ruled out because otherwise Lemma 2.2 would imply

$$H^0(\mathbb{P}^n, (n+1)\mathcal{O}_{\mathbb{P}^n}) \cong H^0(\mathbb{P}^n, \mathcal{O}_{\mathbb{P}^n}) \oplus H^0(\mathbb{P}^n, \Omega^1_{\mathbb{P}^n}),$$

which contradicts the first equality of (5.6).

Therefore there is an isomorphism of the form (5.7). This fact together with the above remarks prove the above claim, and thereby Theorem 5.5, completely. □

In view of Theorem 5.5, Theorem 1.2 implies:

Corollary 5.6. *Under the hypotheses of Theorem 1.2 assume furthermore that Y is smooth and projectively normal in $\mathbb{P}^n$ and $T^1_{C_Y}(-1) = 0$, where C_Y is the affine cone over Y in $\mathbb{P}^n$. Then every extension of Y in $\mathbb{P}^{n+1}$ is trivial.*

Corollary 5.6 is especially interesting because it makes a connection between two completely different things: the deformation theory of the vertex of the cone over Y in $\mathbb{P}^n$ (which is of local nature), and a problem of global projective geometry (i.e. the classification of extensions of Y in $\mathbb{P}^{n+1}$).

Definition 5.7. Let (Y, y) be an isolated singularity of an affine variety Y (with Y smooth outside y). Then (Y, y) is called a *rigid singularity* if $T^1_{Y,y} = 0$, i.e. if the space of first order infinitesimal deformations of (Y, y) is zero. Clearly, (Y, y) is rigid if y is a smooth point of Y.

The following result provides a criterion for the vertex x of the affine cone C_Y over a smooth projectively normal closed subvariety Y in $\mathbb{P}^n$ to be rigid.

Proposition 5.8. *Let Y be a smooth projectively normal closed subvariety of $\mathbb{P}^n$ of dimension ≥ 1 such that $H^1(Y, \mathcal{O}_Y(i)) = H^1(Y, T_Y(i)) = 0$ for all $i \in \mathbb{Z}$. Then $T^1_{C_Y} = 0$, i.e. the vertex x of C_Y is a rigid singularity.*

Proof. By (5.4) it will be sufficient to prove that $H^1(U_Y, T_{U_Y}) = 0$. To do this consider the following cohomology sequence associated to the first row of the diagram of the proof of Theorem 5.5:

$$H^1(U_Y, \mathcal{O}_{U_Y}) \to H^1(U_Y, T_{U_Y}) \to H^1(U_Y, \pi_Y^*(T_Y)).$$

Therefore the middle space is zero if the first and the last spaces are both zero. But by Lemma 2.2 and our assumptions we have

$$H^1(U_Y, \mathcal{O}_{U_Y}) = \bigoplus_{i \in \mathbb{Z}} H^1(Y, \mathcal{O}_Y(i)) = 0, \text{ and}$$

$$H^1(U_Y, \pi_Y^*(T_Y)) = \bigoplus_{i \in \mathbb{Z}} H^1(Y, T_Y(i)) = 0. \qquad \square$$

By Proposition 5.8 and Examples 3.1, *i)* and *ii)*, the vertex of the affine cone over the Veronese variety $Y = v_s(\mathbb{P}^r)$, with $r \geq 3$ and $s \geq 2$, or over the Segre variety $Y = i(\mathbb{P}^1 \times \mathbb{P}^r)$, with $r \geq 2$, is a rigid singularity.

Remark 5.9. (Pinkham [120].) Under the hypotheses and notation of Definition 1.1, assume that X is a non-trivial extension in $\mathbb{P}^{n+1}$ of a smooth subvariety $Y \hookrightarrow \mathbb{P}^n$. Thus $Y = X \cap H$, with $H = \mathbb{P}^n$ a hyperplane in $\mathbb{P}^{n+1}$. Consider a linear embedding $j : \mathbb{P}^{n+1} \hookrightarrow \mathbb{P}^{n+2}$, and let $C_X \subset \mathbb{P}^{n+2}$ be the projective cone over X of vertex a point $x \in \mathbb{P}^{n+2} \setminus j(\mathbb{P}^{n+1})$. Then H is a 2-codimensional linear subspace in $\mathbb{P}^{n+2}$ which generates a pencil of hyperplanes $\{H_t\}_{t \in \mathbb{P}^1}$ (all hyperplanes of $\mathbb{P}^{n+2}$ containing H). If we denote by 0 the point $[1,0] \in \mathbb{P}^1$, we may assume that H_0 is the hyperplane of the pencil containing x. Finally, for every $t \in \mathbb{P}^1$ set $X_t := C_X \cap H_t$. Then we get a family $\{X_t\}_{t \in \mathbb{P}^1}$ of projective varieties parametrized by $\mathbb{P}^1$ such that X_0 is isomorphic to the cone C_Y over Y of vertex x and $X_t \cong X$ for every $t \in \mathbb{P}^1 \setminus \{0\}$. In particular, we get a non-trivial deformation of the singularity (C_Y, x). For example, if Y is projectively normal in $\mathbb{P}^n$ and X is smooth, then this family is flat and yields a "smoothing" of the vertex x of the cone C_Y. Thus any non-trivial extension X of Y in $\mathbb{P}^{n+1}$ produces an interesting deformation of the vertex of the projective cone C_Y over Y. This construction is called "sweeping out the cone with hyperplane sections" (see [120], page 46).

We close this chapter by proving Proposition 1.11.

Proof of Proposition 1.11. Let I be the saturated homogeneous ideal of Y in the polynomial k-algebra $P := k[T_0, T_1, \ldots, T_n]$, and set $A = P/I$ (the homogeneous coordinate k-algebra of Y in $\mathbb{P}^n$). Since Y is projectively normal in $\mathbb{P}^n$ we can apply Theorem 5.5 to get

$$T^1_{C_Y}(-2) = \operatorname{Coker}(H^0(Y, (n+1)\mathcal{O}_Y(-1)) \longrightarrow H^0(Y, N_{Y|\mathbb{P}^n}(-2))),$$

whence $T^1_{C_Y}(-2) = H^0(Y, N_{Y|\mathbb{P}^n}(-2))$ because $H^0(Y, (n+1)\mathcal{O}_Y(-1)) = 0$. Therefore we have to check that $T^1_{C_Y}(-2) = T^1(A|k, A)(-2) = 0$. Every element $t \in T^1(A|k, A)(-2)$ is represented by a homomorphism $\varphi \in \operatorname{Hom}_A(I/I^2, A)(-2)$ of degree -2. To prove the proposition it will be enough to show that under its hypotheses we have $\varphi = 0$.

By our hypotheses, there is a system of generators $f_1, \ldots, f_s$ of I of degree 2. We may assume that $f_1, \ldots, f_s$ is minimal. A relation r among $f_1, \ldots, f_s$ is a system of homogeneous polynomials $(r_1, \ldots, r_s)$ of the same degree such that $r_1 f_1 + \cdots + r_s f_s = 0$. The hypotheses also say that there is a generating set $r^1, \ldots, r^u$ of independent relations $r^i = (r^i_1, \ldots, r^i_s)$, with $\deg(r^i_j) = 1$.

Then the classes $\overline{f}_i$, $\forall i = 1, \ldots, s$ of f_i in I/I^2 form a system of generators of I/I^2. Therefore φ is perfectly determined by the constants $\varphi(\overline{f}_i)$, $\forall i = 1, \ldots, s$. Now, assume that $\varphi \neq 0$. Then there is an i such that $\varphi(\overline{f}_i) \neq 0$ (in k). We may assume $\varphi(\overline{f}_1) \neq 0$. Then for every $i = 2, \ldots, s$ we have $\varphi(\overline{f}_i) = a_i \varphi(\overline{f}_1)$, with $a_i \in k$. Replacing $f_1, \ldots, f_s$ by f_1, $f_2 - a_1 f_1$, $f_s - a_s f_1$, we may therefore assume that $\varphi(\overline{f}_1) \neq 0$ and $\varphi(\overline{f}_i) = 0$, $\forall i = 2, \ldots, s$.

On the other hand, we claim that there is a linear relation $r = (r_1, \ldots, r_s)$ among $f_1, \ldots, f_s$ such that $r_1 \neq 0$. Indeed, considering the obvious relation $f = (f_2, -f_1, 0, \ldots, 0)$ among $f_1, \ldots, f_s$, by hypothesis we know that there are linear forms $c_1, \ldots, c_u \in P_1$ such that $f = c_1 r^1 + \cdots + c_u r^u$. In particular, $f_2 = c_1 r_1^1 + \cdots + c_u r_1^u$, which forces $r_1^i \neq 0$ for at least one $i \in \{1, \ldots, u\}$ (because $f_2 \neq 0$).

Now, fixing a relation $r = (r_1, \ldots, r_s)$ among $f_1, \ldots, f_s$ with $r_1 \neq 0$, we have $0 = \varphi(r_1 f_1 + \cdots + r_s f_s) = r_1 \varphi(f_1)$. Since $\varphi(f_1)$ is a non-zero constant, it follows that $r_1 = 0$, a contradiction. In this way we have proved that $\mathrm{Hom}_A(I/I^2, A)_{-2} = 0$, which implies $T^1(A|k, A)(-2) = 0$. □

Chapter 6

A Characterization of Linear Subspaces

In this chapter we prove the following characterization of linear subspaces:

Theorem 6.1 (Van de Ven [144]). *Let Y be a smooth closed irreducible subvariety of $\mathbb{P}^n$ of dimension $d \geq 1$ over the field $\mathbb{C}$ of complex numbers. Then the normal sequence*

$$0 \to T_Y \to T_{\mathbb{P}^n}|Y \to N_{Y|\mathbb{P}^n} \to 0$$

splits if and only if Y is a linear subspace of $\mathbb{P}^n$.

One implication is easy. Namely, assume that Y is a linear subspace of $\mathbb{P}^n$. Let L be a linear subspace of $\mathbb{P}^n$ of dimension $n-d-1$ such that $Y \cap L = \varnothing$. Projecting from L we get a morphism $\pi : \mathbb{P}^n \setminus L \to Y = \mathbb{P}^d$ such that $\pi \circ i = \mathrm{id}_Y$, where $i\colon Y \hookrightarrow \mathbb{P}^n$ is the natural inclusion. Therefore we get a canonical map $T_{\mathbb{P}^n \setminus L} \to \pi^*(T_Y)$, which, restricted to Y, yields a map $T_{\mathbb{P}^n}|Y \to i^*\pi^*(T_Y) = T_Y$. This is the desired splitting of the normal sequence.

The other implication is non-trivial. The proof that follows is due to Mustaţă–Popa [115], and its main idea is to make use of the first infinitesimal neighbourhood $Y(1) = (Y, \mathcal{O}_{\mathbb{P}^n}/\mathcal{I}^2)$ of Y in $\mathbb{P}^n$, where $\mathcal{I}$ is the ideal sheaf of Y in $\mathcal{O}_{\mathbb{P}^n}$. The first step is to interpret the splitting of the normal bundle in terms of $Y(1)$. Precisely, we have the following general

Lemma 6.2. *Let Y be a closed subvariety of an algebraic variety X, and let $\mathcal{I}$ be the ideal sheaf of Y in $\mathcal{O}_X$. Let*

$$\mathcal{I}/\mathcal{I}^2 \xrightarrow{\ \delta\ } \Omega^1_X|Y \longrightarrow \Omega^1_Y \longrightarrow 0$$

be the dual of the normal sequence of Y in X. Then δ admits a left inverse (i.e. a map $s : \Omega^1_X|Y \to \mathcal{I}/\mathcal{I}^2$ of coherent $\mathcal{O}_Y$-modules such that $s \circ \delta = \mathrm{id}$) if and only if the canonical closed immersion $i\colon Y \hookrightarrow Y(1)$ (with $Y(1)$ the first infinitesimal neighbourhood of Y in X) admits a retraction, i.e. there exists a morphism of schemes $r\colon Y(1) \to Y$ such that $r \circ i = \mathrm{id}_Y$.

Remark 6.3. Lemma 6.2 shows that the normal sequence

$$0 \to T_Y \to T_X|Y \to N_{Y|X} \to 0$$

of a smooth closed subvariety Y of a smooth algebraic variety X splits if and only if the closed embedding $Y \hookrightarrow Y(1)$ admits a retraction. In Chapter 8 we shall encounter a somewhat similar problem. Precisely, if $Z \subset Y \subset X$ are closed embeddings of smooth algebraic varieties (with $\mathrm{codim}_Y(Z) = 1$) then the splitting of the canonical exact sequence of normal bundles

$$0 \to N_{Z|Y} \to N_{Z|X} \to N_{Y|X}|Z \to 0$$

can also be expressed in terms of the first infinitesimal neighbourhood $Y(1)$ of Y in X (see Lemma 8.1 below).

Proof of Lemma 6.2. The existence of a retraction π is equivalent to the existence of a map of sheaves of rings $\pi' : \mathcal{O}_Y = \mathcal{O}_X/\mathcal{I} \to \mathcal{O}_X/\mathcal{I}^2$ such that the composite

$$\mathcal{O}_X/\mathcal{I} \xrightarrow{\pi'} \mathcal{O}_X/\mathcal{I}^2 \xrightarrow{\text{canonical}} \mathcal{O}_X/\mathcal{I}$$

is the identity. On the other hand, the existence of a map $s : \Omega^1_X|Y \to \mathcal{I}/\mathcal{I}^2$ such that $s \circ \delta = \mathrm{id}$ is equivalent to the existence of a derivation $D : \mathcal{O}_X \to \mathcal{I}/\mathcal{I}^2$ such that $D|\mathcal{I}$ coincides to the canonical map $\mathcal{I} \to \mathcal{I}/\mathcal{I}^2$.

Given $D \in \mathrm{Der}(\mathcal{O}_X, \mathcal{I}/\mathcal{I}^2)$ such that $D|\mathcal{I} : \mathcal{I} \to \mathcal{I}/\mathcal{I}^2$ is the canonical map, we define $\pi' : \mathcal{O}_X \to \mathcal{O}_X/\mathcal{I}^2$ by $\pi' = \varphi - D$, where $\varphi : \mathcal{O}_X \to \mathcal{O}_X/\mathcal{I}^2$ is the canonical map. One checks easily that π' is a map of rings and $\pi'|\mathcal{I} = 0$. Therefore one obtains a map $\pi' : \mathcal{O}_X/\mathcal{I} \to \mathcal{O}_X/\mathcal{I}^2$, which composed with the canonical map $\mathcal{O}_X/\mathcal{I}^2 \to \mathcal{O}_X/\mathcal{I}$, is the identity. Conversely, given π' one easily gets back D by $D = \varphi - \pi'$. □

Proof of Theorem 6.1.
Step 1. We may assume that Y is non-degenerate in $\mathbb{P}^n$.

Indeed, let L be the linear subspace of $\mathbb{P}^n$ generated by Y. Then we have the commutative diagram of dual normal sequences

$$\begin{array}{ccccccccc}
0 & \longrightarrow & N^*_{Y|\mathbb{P}^n} & \longrightarrow & \Omega^1_{\mathbb{P}^n}|Y & \xrightarrow{u} & \Omega^1_Y & \longrightarrow & 0 \\
 & & \downarrow & & \downarrow{\scriptstyle \alpha} & & \downarrow{\scriptstyle \mathrm{id}} & & \\
0 & \longrightarrow & N^*_{Y|L} & \longrightarrow & \Omega^1_L|Y & \xrightarrow{v} & \Omega^1_Y & \longrightarrow & 0
\end{array}$$

If $s : \Omega^1_Y \to \Omega^1_{\mathbb{P}^n}|Y$ is a map such that $u \circ s = \mathrm{id}$, then $v \circ (\alpha \circ s) = u \circ s = \mathrm{id}$. In other words, if the top sequence splits so does the bottom one.

Henceforth we will assume that Y is non-degenerate and then we have to prove that $Y = \mathbb{P}^n$. We shall first consider the case $d = \dim(Y) \geq 2$.

Step 2. $H^1(Y, N^*_{Y|\mathbb{P}^n}) = 0$.

Indeed, from the splitting of the dual of the normal sequence of Y in $\mathbb{P}^n$ we get

$$H^1(Y, \Omega^1_{\mathbb{P}^n}|Y) \cong H^1(Y, N^*_{Y|\mathbb{P}^n}) \oplus H^1(Y, \Omega^1_Y). \tag{6.1}$$

On the other hand, the restricted Euler sequence

$$0 \to \Omega^1_{\mathbb{P}^n}|Y \to (n+1)\mathcal{O}_Y(-1) \to \mathcal{O}_Y \to 0$$

yields the cohomology sequence

$$H^0(Y, (n+1)\mathcal{O}_Y(-1)) \to H^0(Y, \mathcal{O}_Y) \to H^1(Y, \Omega^1_{\mathbb{P}^n}|Y) \to H^1(Y, (n+1)\mathcal{O}_Y(-1)).$$

The first space is clearly zero. Moreover, since $k = \mathbb{C}$ and $\dim(Y) \geq 2$, by the Kodaira vanishing theorem the last cohomology space is zero (because $d \geq 2$). Therefore $\dim H^1(Y, \Omega^1_{\mathbb{P}^n}|Y) = 1$. On the other hand, $\dim H^1(Y, \Omega^1_Y) \geq 1$ because by a theorem of Néron–Severi $H^1(Y, \Omega^1_Y)$ contains $NS(Y) \otimes_{\mathbb{Z}} \mathbb{C}$, where $NS(Y)$ is the Néron–Severi group of Y. These two observations together with (6.1) prove step 2.

Step 3. $H^0(Y, N^*_{Y|\mathbb{P}^n}(1)) = 0$.

Indeed, from the restricted Euler sequence

$$0 \longrightarrow \Omega^1_{\mathbb{P}^n}|Y(1) \to (n+1)\mathcal{O}_Y \xrightarrow{\beta} \mathcal{O}_Y(1) \longrightarrow 0$$

we get

$$H^0(Y, \Omega^1_{\mathbb{P}^n}|Y(1)) = \mathrm{Ker}(\zeta) \tag{6.2}$$

where $\zeta := H^0(\beta) : H^0(Y, (n+1)\mathcal{O}_Y) \to H^0(Y, \mathcal{O}_Y(1))$. On the other hand, in the commutative diagram

$$\begin{array}{ccc} H^0(\mathbb{P}^n, (n+1)\mathcal{O}_{\mathbb{P}^n}) & \xrightarrow{\cong} & H^0(\mathbb{P}^n, \mathcal{O}_{\mathbb{P}^n}(1)) \\ \cong \big\downarrow & & \big\downarrow \mathrm{inj} \\ H^0(Y, (n+1)\mathcal{O}_Y) & \xrightarrow{\zeta} & H^0(Y, \mathcal{O}_Y(1)) \end{array}$$

the first vertical map is an isomorphism, the first horizontal map is also an isomorphism, and the second vertical map is injective (because Y is non-degenerate in $\mathbb{P}^n$). It follows that the map ζ is injective. Recalling (6.2), this implies that $H^0(Y, \Omega^1_{\mathbb{P}^n}|Y(1)) = 0$. Using the splitting of the dual normal sequence we get

$$0 = H^0(Y, \Omega^1_{\mathbb{P}^n}|Y(1)) \cong H^0(Y, N^*_{Y|\mathbb{P}^n}(1)) \oplus H^0(Y, \Omega^1_Y(1)).$$

This proves step 3.

Step 4. Let $i\colon Y \to Y(1)$ be the natural inclusion of Y in its first infinitesimal neighbourhood in $\mathbb{P}^n$. Then the map $i^* : \mathrm{Pic}(Y(1)) \to \mathrm{Pic}(Y)$ is injective, where $\mathrm{Pic}(Z)$ denotes the Picard group of a scheme Z.

Indeed, from the truncated exponential sequence

$$0 \to N^*_{Y|\mathbb{P}^n} = \mathcal{I}/\mathcal{I}^2 \to \mathcal{O}^*_{Y(1)} \to \mathcal{O}^*_Y \to 0,$$

(where for any scheme Z, $\mathcal{O}^*_Z$ denotes the sheaf of multiplicative groups of germs of all nowhere zero functions of $\mathcal{O}_Z$, and where the map $\mathcal{I}/\mathcal{I}^2 \to \mathcal{O}^*_{Y(1)}$ is given by $u \to 1+u$), we get the cohomology sequence

$$H^1(Y, N^*_{Y|\mathbb{P}^n}) \to \mathrm{Pic}(Y(1)) \to \mathrm{Pic}(Y).$$

Then step 4 follows from $H^1(Y, N^*_{Y|\mathbb{P}^n}) = 0$ (by step 2).

Step 5. (Conclusion in case $\dim(Y) \geq 2$).

By Lemma 6.2 our hypothesis says that there is a retraction $r\colon Y(1) \to Y$ of the inclusion $i\colon Y \hookrightarrow Y(1)$. Set $L := r^*\mathcal{O}_Y(1)$. Since $r \circ i = \mathrm{id}$, $i^*(L) \cong \mathcal{O}_Y(1) = \mathcal{O}_{Y(1)}(1)|Y$. Since the map $i^* : \mathrm{Pic}(Y(1)) \to \mathrm{Pic}(Y)$ is injective (step 4), $L \cong \mathcal{O}_{Y(1)}(1)$, i.e. $r^*(\mathcal{O}_Y(1)) \cong \mathcal{O}_{Y(1)}(1)$.

On the other hand, in the exact sequence

$$0 \longrightarrow H^0(Y, N^*_{Y|\mathbb{P}^n}(1)) \longrightarrow H^0(Y(1), \mathcal{O}_{Y(1)}(1)) \xrightarrow{i^*} H^0(Y, \mathcal{O}_Y(1))$$

the first H^0 is zero by step 3. Therefore the map $i^* : H^0(Y(1), \mathcal{O}_{Y(1)}(1)) \to H^0(Y, \mathcal{O}_Y(1))$ is injective. Now, let $t_0, \dots, t_n$ be a basis of $H^0(\mathbb{P}^n, \mathcal{O}_{\mathbb{P}^n}(1))$, and set $s_i := r^*(t_i|Y) \in H^0(Y(1), \mathcal{O}_{Y(1)}(1))$, $i = 0, \dots, n$. Since i^* is injective, $s_i = t_i|Y(1)$, $i = 0, \dots, n$. Since $\mathcal{O}_{Y(1)}(1)$ is very ample we get that $i \circ r = i(1)$, where $i(1) : Y(1) \hookrightarrow \mathbb{P}^n$ is the canonical inclusion. But this last fact is impossible, unless $Y = \mathbb{P}^n$. This proves Theorem 6.1 in case $\dim(Y) \geq 2$.

Step 6. Theorem 6.1 holds true if $\dim(Y) = 1$.

The argument of this step is taken from [95]. The splitting of the normal bundle of Y in $\mathbb{P}^n$ implies that $T_{\mathbb{P}^n}|Y \cong T_Y \oplus N_{Y|\mathbb{P}^n}$. Since $T_{\mathbb{P}^n}$ is an ample vector bundle, its restriction $T_{\mathbb{P}^n}|Y$ is also ample, whence T_Y is ample. Since Y is a curve, this is possible only when $Y \cong \mathbb{P}^1$. Set $e := \deg(Y)$. Then $\mathcal{O}_Y(1) \cong \mathcal{O}_{\mathbb{P}^1}(e)$. It remains to prove that $e = 1$. To this extent look at the commutative diagram with exact rows and columns

$$\begin{array}{ccccccccc}
 & & & & 0 & & 0 & & \\
 & & & & \downarrow & & \downarrow & & \\
0 & \longrightarrow & N^*_{Y|\mathbb{P}^n} & \longrightarrow & \Omega^1_{\mathbb{P}^n}|Y & \xrightarrow{a} & \Omega^1_Y = \mathcal{O}_{\mathbb{P}^1}(-2) & \longrightarrow & 0 \\
 & & \downarrow{\scriptstyle \mathrm{id}} & & \downarrow & & \downarrow & & \\
0 & \longrightarrow & N^*_{Y|\mathbb{P}^n} & \longrightarrow & (n+1)\mathcal{O}_{\mathbb{P}^1}(-e) & \longrightarrow & E & \longrightarrow & 0 \\
 & & & & \downarrow & & \downarrow & & \\
 & & & & \mathcal{O}_Y & \xrightarrow{\mathrm{id}} & \mathcal{O}_Y & & \\
 & & & & \downarrow & & \downarrow & & \\
 & & & & 0 & & 0 & &
\end{array}$$

in which the first row is the dual of the normal bundle of Y in $\mathbb{P}^n$, and the middle column is the dual of the Euler sequence of $\mathbb{P}^n$ restricted to Y. The splitting of the top row yields a map $s : \Omega^1_Y = \mathcal{O}_{\mathbb{P}^1}(-2) \to \Omega^1_{\mathbb{P}^n}|Y$ such that $a \circ s = \mathrm{id}$. In particular, s is injective. The injectivity of s and the above diagram yield an injective map $\mathcal{O}_{\mathbb{P}^1}(-2) \hookrightarrow (n+1)\mathcal{O}_Y(-1) = (n+1)\mathcal{O}_{\mathbb{P}^1}(-e)$. This forces $e \leq 2$.

If $e = 2$, Y is a plane conic. Then the dual of the Euler sequence

$$0 \to \Omega^1_{\mathbb{P}^2}|Y(1) \to 3\mathcal{O}_Y \to \mathcal{O}_Y(1) = \mathcal{O}_{\mathbb{P}^1}(2) \to 0$$

yields the exact sequence

$$0 \to H^0(Y, \Omega^1_{\mathbb{P}^2}|Y(1)) \to H^0(Y, 3\mathcal{O}_Y) \to H^0(Y, \mathcal{O}_{\mathbb{P}^1}(2)).$$

Since Y is a plane conic, the last map is an isomorphism. Therefore

$$H^0(Y, \Omega^1_{\mathbb{P}^2}|Y(1)) = 0 = H^0(\mathbb{P}^1, \Omega^1_{\mathbb{P}^2}|Y \otimes \mathcal{O}_{\mathbb{P}^1}(2)).$$

But this contradicts the existence of the injective map $s : \mathcal{O}_{\mathbb{P}^1}(-2) \to \Omega^1_{\mathbb{P}^2}|Y$. Therefore $e = 1$, so the proof of step 6 (and thereby the proof of Theorem 6.2) is complete. □

Remark 6.4. The main reason why we included the above proof of Theorem 6.2 here consists primarily in the use of the first infinitesimal neighbourhood of a closed subvariety in an ambient variety. This illustrates very well the philosophy that using schemes with nilpotents in the study of some rather classical questions of projective geometry could sometimes provide natural proofs. In chapter 8, following Ellingsrud, Gruson, Peskine and Strømme [46], we shall apply again this method (of using the first infinitesimal neighbourhoods) in connection with a problem of complete intersection.

Chapter 7

Cohomological Dimension and Connectedness Theorems

Cohomological Dimension

Definition 7.1. Let Z be an irreducible algebraic variety over k. We shall denote by $\operatorname{Coh}(Z)$ the category of all coherent sheaves $\mathcal{O}_Z$-modules. According to Hartshorne [75], or also [74], we define *the cohomological dimension*, $\operatorname{cd}(Z)$, of Z by

$$\operatorname{cd}(Z) = \max\{n \geq 0 \mid \exists F \in \operatorname{Coh}(Z) \text{ such that } H^n(Z, F) \neq 0\}.$$

A general result says that $0 \leq \operatorname{cd}(Z) \leq \dim(Z)$ (see [134], or [66] III, or [73]); moreover, a result of Serre asserts that $\operatorname{cd}(Z) = 0$ if and only if Z is affine (see [134] and [136], or also [66] III, or [73]).

The aim of this chapter is to prove the following special case of a basic result due to Hartshorne–Lichtenbaum (see [65], or also [75] or [90]). This special form of the theorem will be sufficient to prove a generalization of the connectedness theorem of Fulton–Hansen (see Theorem 7.14 below).

Theorem 7.2 (Hartshorne–Lichtenbaum). *Let X be an irreducible projective variety of dimension $n \geq 1$, let Y be a closed subset of X, and set $U := X \setminus Y$. Then $\operatorname{cd}(U) = n$ if and only if $Y = \varnothing$.*

Before proceeding to the proof (which follows Kleiman [90] closely) we need some preparation.

Lemma 7.3. *For every irreducible quasi-projective variety V of dimension n the following two conditions are equivalent:*

(i) $H^n(V, F) = 0$ *for all* $F \in \operatorname{Coh}(V)$.

(ii) $H^n(V, L) = 0$ *for all invertible sheaves* L *on* V.

Proof. Clearly (i)$\Longrightarrow$(ii). Conversely, assume that (ii) holds. Since V is a quasi-projective variety, every $F \in \mathrm{Coh}(V)$ is a quotient of direct sums of invertible sheaves. Indeed, pick an ample line bundle L on V. Since L is ample, $F \otimes L^m$ is generated by its global sections for $m \gg 0$. Since $F \otimes L^m$ is coherent, it follows that there is a surjection $p\mathcal{O}_V \to F \otimes L^m \to 0$ for some $p \geq 1$, and therefore a surjection $E := pL^{-m} \to F \to 0$. Thus we get an exact sequence of the form

$$0 \to G \to E \to F \to 0,$$

where E is a finite direct sum of line bundles and G a coherent sheaf on V. This yields the cohomology sequence

$$H^n(V, E) \to H^n(V, F) \to H^{n+1}(V, G).$$

By (ii) the first space is zero (the functor H^n commutes with the direct sums), and since $n + 1 > \dim(V) = n$ the third space is also zero, whence the middle space is zero as well. □

Lemma 7.4. *Let W be a closed subvariety of an algebraic variety V. For a fixed integer $n \geq 1$ the following two conditions are equivalent:*

(i) $H^n(W, F) = 0$ *for all coherent $\mathcal{O}_W$-modules F.*

(ii) $H^n(V, F) = 0$ *for all quasi-coherent $\mathcal{O}_V$-modules F with* $\mathrm{Supp}(F) \subseteq W$.

Proof. The implication (ii)$\Longrightarrow$(i) is obvious. Conversely, let F be a quasi-coherent $\mathcal{O}_V$-module with support in W. F is the direct limit of all its coherent $\mathcal{O}_V$-submodules G. Since the functor H^n commutes with direct limits, $H^n(V, F)$ is the direct limit of the $H^n(V, G)$'s. Hence we may assume F coherent. Now, let Z be the closed subscheme of V defined by the annihilator ideal of F. By hypothesis, Z_{red} is a closed subvariety of W, where Z_{red} denotes the (reduced) subscheme of Z defined by the nilpotent radical I of $\mathcal{O}_Z$. Since Z is a Noetherian quasi-compact scheme, there is an integer $m > 0$ such that $I^m = 0$. Then for any coherent $\mathcal{O}_Z$-module F consider the exact sequences

$$0 \to I^{i+1}F \to I^iF \to I^iF/I^{i+1}F \to 0, \quad i = 0, 1, \ldots, m-1,$$

where $I^0 := \mathcal{O}_Z$. Therefore we get the cohomology sequences

$$H^n(V, I^{i+1}F) \to H^n(V, I^iF) \to H^n(V, I^iF/I^{i+1}F), \quad i = 0, 1, \ldots, m-1. \tag{7.1}$$

Clearly the quotients $I^iF/I^{i+1}F$ are $\mathcal{O}_{Z_{\mathrm{red}}}$-modules, and hence also $\mathcal{O}_W$-modules. Therefore by (i),

$$H^n(V, I^iF/I^{i+1}F) = H^n(W, I^iF/I^{i+1}F) = 0 \quad \text{for all } i = 0, 1, \ldots, m-1.$$

In particular, $H^n(V, I^{m-1}F) = H^n(V, I^{m-1}F/I^mF) = 0$. Using this, (7.1) and an obvious descending induction on i we get $H^n(V, F) = 0$, as desired. □

Proof of Theorem 7.2. Assume first that $Y = \varnothing$, i.e. $U = X$ is projective. We have to find a coherent sheaf $F \in \mathrm{Coh}(U)$ such that $H^n(U, F) \neq 0$. Choose an ample line bundle L on U. Since U is projective, $H^n(U, L^{-m}) \neq 0$ for every $m \gg 0$ (for example use duality on U).

Conversely, assume $Y \neq \varnothing$. Then we have to prove that $H^n(U, F) = 0$ for every $F \in \mathrm{Coh}(U)$. We proceed by induction on $n = \dim(X)$. If $n = 1$ then U is an affine curve and the conclusion follows from a well-known vanishing result of Serre (see [134], or also [73]). Assume therefore $n \geq 2$. By Lemma 7.3 we may assume that F is invertible, and in particular, a torsion free sheaf.

Now we need the following:

Claim. Under the hypotheses of Theorem 7.2 assume $n \geq 2$. Then there is a closed irreducible subscheme Z of U of dimension $n-1$ such that $U \setminus Z$ is affine. Moreover, the scheme Z is quasi-projective, but not projective.

We first prove the above claim. Blow up X along Y to get the birational morphism $f\colon X' \to X$ with exceptional locus $Y' = f^{-1}(Y)$. Then Y' is an effective Cartier divisor on X'. Since X is projective and irreducible, X' is also projective and irreducible. Choose a projective embedding $X' \hookrightarrow \mathbb{P}^m$. Since $n \geq 2$, by Bertini (see e.g. [87]) we can find an irreducible hyperplane section T of X'. Because T is a very ample Cartier divisor on X' ($T \neq Y'$) and Y' is a Cartier divisor on X', $Y' + aT$ is very ample for $a \gg 0$. Therefore $X' \setminus (T \cup Y')$ is an affine open subset of X'. Set $Z' := f(T \cup Y')$ and $Z := U \cap Z'$. Then Z is irreducible because $Z = f(T) \cap U$ and T is irreducible. Moreover, $U \setminus Z = X \setminus Z' \cong X' \setminus (T \cup Y')$, and in particular $U \setminus Z$ is affine. Finally, Z is not projective because $Z = f(T) \setminus Y$ and $Y \cap f(T) \neq \varnothing$ ($T \cap Y' \neq \varnothing$ because T is a hyperplane section on X' and $\dim(Y') = n - 1 > 0$). Thus the claim is proved.

To conclude the proof of Theorem 7.2 in case $n \geq 2$, apply the above claim to find Z as above. Let $i\colon U \setminus Z \hookrightarrow U$ be the canonical inclusion. Then i is an affine morphism because for every affine open subset V of U, $i^{-1}(V) = V \cap (U \setminus Z)$ (on a separated scheme the intersection of any two affine open subsets is again affine!). Now, starting with a torsion free coherent sheaf F on U, consider the canonical map $\alpha\colon F \to i_*i^*(F)$. Since $\mathrm{Supp}(\mathrm{Ker}(\alpha)) \subseteq Z$, $\mathrm{Ker}(\alpha)$ is a torsion subsheaf of F, and since F is torsion free, $\mathrm{Ker}(\alpha) = 0$. Therefore we get the exact sequence

$$0 \to F \to i_*i^*(F) \to G \to 0$$

(where $G := \mathrm{Coker}(F \to i_*i^*(F))$) which yields the cohomology sequence

$$H^{n-1}(U, G) \to H^n(U, F) \to H^n(U, i_*i^*(F)). \tag{7.2}$$

Note that $i_*i^*(F)$ (and hence also G) is only a quasi-coherent $\mathcal{O}_U$-module.

Since i is an affine morphism $H^n(U, i_*i^*(F)) \cong H^n(U \setminus Z, i^*(F))$, and the latter space is zero because $n > 0$ and $U \setminus Z$ is affine, by Serre's Theorem. Therefore

$$H^n(U, i_*i^*(F)) = 0. \tag{7.3}$$

On the other hand, $\mathrm{Supp}(G) \subseteq Z$. Since Z is an irreducible quasi-projective (but not projective) variety of dimension $n-1$, by the inductive hypothesis, $H^{n-1}(Z, H) = 0$ for every coherent $\mathcal{O}_Z$-module H. Then by Lemma 7.4 we get

$$H^{n-1}(U, G) = 0. \tag{7.4}$$

Finally, (7.2), (7.3) and (7.4) yield $H^n(U, F) = 0$. □

Remark 7.5. Theorem 7.2 can be restated by saying that $\mathrm{cd}(U) \leq n-1$ if and only if $Y \neq \varnothing$. The above inductive proof of Theorem 7.2 required considering effectively the category of quasi-coherent (and not only of coherent) sheaves.

We illustrate now the use of the notion of cohomological dimension by proving the following useful result, which is going to play a crucial role in the proof of the connectedness Theorem of Fulton–Hansen:

Theorem 7.6. *Under the hypotheses of Theorem* 7.2 *assume* $n \geq 2$ *and* $\mathrm{cd}(U) \leq n-2$. *Then* Y *is connected.*

Proof. Assume that Y is not connected, i.e. Y can be written as $Y = Y_1 \cup Y_2$, with Y_1 and Y_2 non-empty closed subsets of Y such that $Y_1 \cap Y_2 = \varnothing$. Because X is projective of dimension n, by Theorem 7.2 and Lemma 7.3, there is an invertible sheaf L on X such that $H^n(X, L) \neq 0$.

In the exact sequence

$$H^n_{Y_i}(X, L) \longrightarrow H^n(X, L) \longrightarrow H^n(X \setminus Y_i, L)$$

the last space is zero by Theorem 7.2 because $Y_i \neq \varnothing$ for $i = 1, 2$. It follows that $h^n_{Y_i}(X, L) \geq h^n(X, L)$ for $i = 1, 2$, where $h^i_W(V, F) := \dim_k H^i_W(V, F)$ for every coherent sheaf F on an algebraic variety V and for every closed subvariety W of V. Moreover, since $Y = Y_1 \cup Y_2$ and $Y_1 \cap Y_2 = \varnothing$, by Mayer–Vietoris we have $H^n_Y(X, L) \cong H^n_{Y_1}(X, L) \oplus H^n_{Y_2}(X, L)$, and hence, $h^n_Y(X, L) \geq 2h^n(X, L)$.

On the other hand, in the exact sequence

$$H^{n-1}(X \setminus Y, L) \longrightarrow H^n_Y(X, L) \longrightarrow H^n(X, L) \longrightarrow H^n(X \setminus Y, L)$$

the first and the last spaces are zero by $\mathrm{cd}(X \setminus Y) \leq n-2$, whence $h^n_Y(X, L) = h^n(X, L)$, contradicting the inequality $h^n_Y(X, L) \geq 2h^n(X, L)$, because $h^n(X, L) \geq 1$. □

We shall use Theorem 7.6 to prove two results (Corollaries 7.7 and 7.11) due to Grothendieck (see [68], exposé XIII).

Corollary 7.7. *Let* X *be an irreducible algebraic variety of dimension* $n \geq 2$ *over* k, *and let* $f\colon X \to \mathrm{Proj}(S)$ *be a finite morphism, where* S *is a finitely generated graded* k*-algebra. Let* $t_1, \ldots, t_r \in S_+$ *be homogeneous elements of positive degrees. If* $r \leq n-1$ *(resp.* $r \leq n$*) then* $f^{-1}(V_+(t_1, \ldots, t_r))$ *is connected (resp. non-empty), where* $V_+(t_1, \ldots, t_r)$ *is the locus of* $\mathrm{Proj}(S)$ *of equations* $t_1 = \cdots = t_r = 0$.

Proof. In the standard notation of [66] II, set $Y := f^{-1}(V_+(t_1,\dots,t_r))$ and $P :=$ Proj(S). Since $P \setminus V_+(t_1,\dots,t_r) = D_+(t_1)\cup\dots\cup D_+(t_r)$, $D_+(t_i) := \mathrm{Spec}(S_{(t_i)})$ is affine $\forall i = 1,\dots,r$, and f is finite, it follows that $X \setminus Y$ is the union of the affine open subsets $f^{-1}(D_+(t_i))$, $i = 1,\dots,r$. Therefore by Čech, $\mathrm{cd}(X\setminus Y) \le r-1 \le n-2$ (resp. $\mathrm{cd}(X\setminus Y)\le n-1$ if $r\le n$). The conclusion follows from Theorem 7.6 (resp. from Theorem 7.2). □

The above corollary can be slightly refined. First we need the following:

Definition 7.8. Let V be an algebraic variety over k, and let $d \ge 0$ be a non-negative integer. V is said to be *d-connected* if every irreducible component of V is of dimension $\ge d+1$ and if $V\setminus W$ is connected for every closed subvariety W of V of dimension $< d$.

For example, every irreducible variety X of dimension $n \ge 1$ is $(n-1)$-connected. An algebraic variety X is 0-connected if and only if X is connected and every irreducible component of X is of dimension ≥ 1.

Example 7.1. Let X be the closed subvariety of the affine space $\mathbb{A}^4$ having two irreducible components X_1 and X_2, where X_1 is the plane of equations $x_1 = x_2 = 0$ and X_2 the plane of equations $x_3 = x_4 = 0$. Then $X_1\cap X_2 = \{x\}$, where $x = (0,0,0,0)$. Thus X is 0-connected. On the other hand, X is not 1-connected because $X\setminus\{x\}$ has two connected components $X_1\setminus\{x\}$ and $X_2\setminus\{x\}$.

Definition 7.9. A sequence $Z_0, Z_1,\dots,Z_n$ of (not necessarily pairwise distinct) irreducible components of an algebraic variety V is called a *d-join within* V if $\dim(Z_i) \ge d+1$ for every $i = 0,1,\dots,n$ and if $\dim(Z_{j-1}\cap Z_j) \ge d$ for every $j = 2,\dots,n$.

The following elementary result (whose proof is left as an exercise to the reader) will not be used in the sequel, but it better explains the concept of d-connectedness.

Proposition 7.10. *An algebraic variety X is d-connected if and only if $X = Z_0 \cup Z_1\cup\dots\cup Z_n$ for some d-join $Z_0, Z_1,\dots,Z_n$ within X.*

Corollary 7.11. *Under the hypotheses of Corollary* 7.7 *assume that $r\le n-1$. Then $f^{-1}(V_+(t_1,\dots,t_r))$ is $(n-r-1)$-connected.*

Proof. Since S is a finitely generated k-algebra, Proj(S) is a projective scheme over k, whence X is a projective variety because f is a finite morphism. Let $X\subset\mathbb{P}^N$ be an arbitrary projective embedding of X, and let A be a general linear subspace of $\mathbb{P}^N$ of dimension $N+r-n+1$. Since $\dim(X) = n$ and $Y := f^{-1}(V_+(t_1,\dots,t_r))$ is locally given by r equations in X, every irreducible component Z of Y is of dimension $\ge n-r$. It follows that $\dim(Z\cap A) = \dim(Z)+\dim(A)-N \ge (n-r)+(N+r-n+1)-N = 1$, and in particular, A meets every irreducible component of Y.

Set $Y' := Y \cap A$. If A is defined by linear equations $s_{r+1} = \cdots = s_{n-1} = 0$ in $\mathbb{P}^N$ then $X \setminus Y'$ is the union of the $n-1$ affine open subsets $f^{-1}(D_+(t_1)),\ldots,$ $f^{-1}(D_+(t_r))$, $U_{r+1},\ldots,U_{n-1}$, where $U_i := \{x \in X \mid s_i(x) \neq 0\}$, $\forall i = r+1,\ldots,n-1$. It follows that $\mathrm{cd}(X \setminus Y') \leq n-2$, whence by Theorem 7.6, Y' is connected.

We saw above that every irreducible component of Y is of dimension $\geq n-r$. Therefore to show that Y is $(n-r-1)$-connected it will be sufficient to check that $Y \setminus W$ is connected for every closed subset W of Y of dimension $< n-r-1$. Assume to the contrary, i.e. there is a closed subscheme W of Y of dimension $< n-r-1$ such that $Y \setminus W$ is not connected. Since A is general, $\dim(W \cap A) = \dim(W) + \dim(A) - N < (n-r-1) + (N+r-n+1) - N = 0$, whence A does not meet W. Moreover, since A meets every irreducible component of Y, the fact that $Y \setminus W$ is not connected implies that $Y' = Y \cap A = (Y \setminus W) \cap A$ is also not connected, a contradiction. □

In the sequel we shall need the following more general version of Corollary 7.11:

Theorem 7.12. *Let S be a finitely generated graded k-algebra, $t_1,\ldots,t_r \in S_+$ homogeneous elements of positive degrees, and U a Zariski open subset of* $\mathrm{Proj}(S)$ *containing $L := V_+(t_1,\ldots,t_r)$. Let $f\colon X \to U$ be a finite morphism, with X an irreducible algebraic variety of dimension n over k. If $r \leq n-1$ then $f^{-1}(L)$ is $(n-r-1)$-connected.*

Proof. By passing to the normalization of X we may assume that X is normal. Let Z' be the closure of $X' := f(X)$ in $P := \mathrm{Proj}(S)$, and let $g\colon Z \to Z'$ be the normalization of Z' in the field $K(X)$ of rational functions of X (which makes sense because the dominant morphism $X \to Z'$ yields the finite field extension $K(Z') = K(X') \to K(Z)$). Then we get a commutative diagram of the form

$$\begin{array}{ccc} X & \xrightarrow{\ i\ } & Z \\ {\scriptstyle f}\downarrow & & \downarrow{\scriptstyle g} \\ X' & \xrightarrow[\ i'\ ]{} & Z' \end{array}$$

in which i and i' are open immersions (i is an open immersion because X is normal), and g is a finite morphism. Since $L \subset U$ and $Z' \cap U = X'$ (X' is closed in U because f is finite), then $X' \cap L = Z' \cap L$, whence $f^{-1}(L) = g^{-1}(L)$. Then Theorem 7.12 follows from Corollary 7.11 applied to the composite of the closed immersion $Z' \hookrightarrow P$ with the finite morphism $g\colon Z \to Z'$. □

Weighted Projective Spaces

Definition 7.13. Let $k[T_0, T_1, \ldots, T_n]$ be the polynomial k-algebra in $n+1$ variables T_0, T_1,...,T_n (with $n \geq 1$). An $(n+1)$-uple $(e_0, e_1, \ldots, e_n) \in \mathbb{Z}^{n+1}$ of positive integers is called a *system of weights* if $e_i \geq 1$, $\forall i = 0, 1, \ldots, n$. Given a system of weights $e = (e_0, e_1, \ldots, e_n)$, grade $k[T_0, T_1, \ldots, T_n]$ by the conditions $\deg(T_i) = e_i$, $\forall i = 0, 1, \ldots, n$. In this way we get a finitely generated graded k-algebra (depending of $e = (e_o, e_1, \ldots, e_n)$), and set

$$\mathbb{P}^n(e) = \mathbb{P}^n(e_0, e_1, \ldots, e_n) := \text{Proj}(k[T_0, T_1, \ldots, T_n]).$$

Then $\mathbb{P}^n(e) = \mathbb{P}^n(e_0, e_1, \ldots, e_n)$ is a normal projective variety of dimension n which is called *the weighted projective space of weights* $e = (e_0, e_1, \ldots, e_n)$. Of course, $\mathbb{P}^n(1, 1, \ldots, 1)$ coincides with the usual projective space $\mathbb{P}^n$.

Example 7.2. As a non-trivial example of weighted projective space, take $e = (1, \ldots, 1, s)$, with $s \geq 2$ and $n \geq 2$. Then $\mathbb{P}^n(1, 1, \ldots, 1, s)$ is isomorphic to the projective cone over the Veronese embedding $v_s : \mathbb{P}^{n-1} \hookrightarrow \mathbb{P}^{N(n,s)}$, with $N(n, s) := \binom{n-1+s}{s} - 1$. Indeed, $\mathbb{P}(1, 1, \ldots, 1, s) = \text{Proj}(k[T_0, \ldots, T_n])$, with $\deg(T_i) = 1$, $\forall i = 0, 1, \ldots, n-1$, and $\deg(T_n) = s$. Then using the general elementary properties of Proj (see e.g. [66] II), we have canonical isomorphisms

$$\text{Proj}(k[T_0, \ldots, T_n]) \cong \text{Proj}(k[T_0, \ldots, T_n])^{(s)} \cong \text{Proj}(k[T_0, \ldots, T_{n-1}]^{(s)}[T]),$$

with T a variable of degree 1. We adopted the standard notation according to which $S^{(s)}$ denotes the graded k-algebra obtained from a graded k-algebra S by putting $S^{(s)}_m := S_{ms}$, $\forall m \geq 0$. The above isomorphisms and the definition of the projective cone over the Veronese embedding v_s yield the assertion.

An alternate description of the weighted projective space $\mathbb{P}^n(e)$ is the following. $\mathbb{P}^n(e)$ is the geometric quotient $(k^{n+1} \setminus \{(0, 0, \ldots, 0)\})/\mathbb{G}_m$, where the action of the multiplicative group $\mathbb{G}_m = k \setminus \{0\}$ on $k^{n+1} \setminus \{(0, 0, \ldots, 0)\}$ is given by $\lambda(t_0, \ldots, t_n) := (\lambda^{e_0} t_0, \ldots, \lambda^{e_n} t_n)$, for all $\lambda \in \mathbb{G}_m$ and $(t_0, \ldots, t_n) \in k^{n+1} \setminus \{(0, 0, \ldots, 0)\}$. Then the orbit of $(t_0, t_1, \ldots, t_n) \in k^{n+1} \setminus \{(0, 0, \ldots, 0)\}$ (regarded as a point of $\mathbb{P}^n(e)$) will be denoted by $[t_0, t_1, \ldots, t_n]$.

We refer the reader to [45] or to [28] for the basic properties of weighted projective spaces.

Connectedness Theorem

With these definitions we can prove the following generalization of a connectedness theorem of Fulton–Hansen (see [52], or also [53] or [87]) to weighted projective spaces:

Theorem 7.14. *Let $f\colon X \to \mathbb{P}^n(e) \times \mathbb{P}^n(e)$ be a finite morphism from the d-dimensional irreducible variety X to the product of the weighted projective space $\mathbb{P}^n(e)$ of weights $e = (e_0, e_1, \ldots, e_n)$ by itself. If $d > n$ then $f^{-1}(\Delta)$ is $(d-n-1)$-connected, where Δ is the diagonal of $\mathbb{P}^n(e) \times \mathbb{P}^n(e)$.*

Proof. We shall show that a construction used by Deligne (see [44], or also [53]) to simplify the proof of Fulton–Hansen connectedness Theorem [52] can be easily generalized to weighted projective spaces. Having the system $e = (e_0, e_1, \ldots, e_n)$ of weights fixed, consider the weighted projective space

$$\mathbb{P}^{2n+1}(e,e) = \mathrm{Proj}(k[T_0, \ldots, T_n; U_0, \ldots, U_n]),$$

of weights $(e,e) := (e_0, e_1, \ldots, e_n, e_0, e_1, \ldots, e_n)$, where $T_0, \ldots, T_n, U_0, \ldots, U_n$ are $2n+2$ independent indeterminates over k such that $\deg(T_i) = \deg(U_i) = e_i$ for every $i = 0, 1, \ldots, n$. Consider the closed subschemes

$$L_1 = V_+(T_0, \ldots, T_n) \text{ and } L_2 = V_+(U_0, \ldots, U_n)$$

of $P := \mathbb{P}^{2n+1}(e,e)$. Then $L_1 \cap L_2 = \varnothing$. Set $U := P \setminus (L_1 \cup L_2)$. Since $T_i - U_i$ is a homogeneous element of degree e_i, it makes sense to consider also the closed subscheme $H := V_+(T_0 - U_0, \ldots, T_n - U_n)$ of P. Clearly, $H \subset U$. The two natural inclusions $k[T_0, \ldots, T_n] \subset k[T_0, \ldots, T_n; U_0, \ldots, U_n]$ and $k[U_0, \ldots, U_n] \subset k[T_0, \ldots, T_n; U_0, \ldots, U_n]$ yield two rational maps $g_i : \mathbb{P}^{2n+1}(e,e) \dashrightarrow \mathbb{P}^n(e)$, $i = 1, 2$, which give rise to the rational map

$$g : \mathbb{P}^{2n+1}(e,e) \dashrightarrow \mathbb{P}^n(e) \times \mathbb{P}^n(e).$$

Then g is defined precisely in the open subset U of $\mathbb{P}^{2n+1}(e,e)$. Alternatively, if we interpret $\mathbb{P}^n(e)$ as the geometric quotient $(k^{n+1} \setminus \{(0, \ldots, 0)\})/\mathbb{G}_m$ mentioned above, then the map g is defined by

$$g([t_0, \ldots, t_n; u_0, \ldots, u_n]) = ([t_0, \ldots, t_n], [u_0, \ldots, u_n]).$$

It is clear that g/H defines an isomorphism $H \cong \Delta$. Consider the commutative diagram

$$\begin{array}{ccccc}
X' & \xrightarrow{\;f'\;} & U & \xleftarrow{\;\supset\;} & H \\
\Big\downarrow{\scriptstyle g'} & & \Big\downarrow{\scriptstyle g} & & \Big\downarrow{\scriptstyle \cong} \\
X & \xrightarrow{\;f\;} & \mathbb{P}^n(e) \times \mathbb{P}^n(e) & \xleftarrow{\;\supset\;} & \Delta
\end{array},$$

where the left square is cartesian, the horizontal arrows of the right square are the canonical closed immersions, and the right vertical arrow is an isomorphism. In our situation the variety $X' = X \times_{\mathbb{P}^n(e) \times \mathbb{P}^n(e)} U$ is irreducible of dimension $d+1$ because X is irreducible and all the fibers of g (and hence also of g') are isomorphic to $\mathbb{G}_m$.

Therefore we can apply Theorem 7.12 to the finite morphism $f' : X' \to U \subset \mathbb{P}^{2n+1}(e,e)$ and $L := H$, with $r = n+1 < d+1 = \dim(X')$, to deduce that ${f'}^{-1}(H)$ is $(d-n-1)$-connected. On the other hand, since $f^{-1}(\Delta) \cong {f'}^{-1}(H)$ we conclude the proof of our theorem. □

Corollary 7.15. (i) *Let $f\colon X \to \mathbb{P}^n(e) \times \mathbb{P}^n(e)$ be a proper morphism from an irreducible variety X such that* $\dim(f(X)) > n$. *Then $f^{-1}(\Delta)$ is connected.*

(ii) *Let $f\colon X \to \mathbb{P}^n(e)$ be a proper morphism from an irreducible variety X, and let Y be a closed irreducible subvariety of $\mathbb{P}^n(e)$ such that* $\dim(f(X)) + \dim(Y) > n$. *Then $f^{-1}(Y)$ is connected.*

Proof. (i) Let $f = g \circ h$ be the Stein decomposition of f, with $h\colon X \to X'$ a proper surjective morphism with connected fibers and $g\colon X' \to \mathbb{P}^n(e) \times \mathbb{P}^n(e)$ a finite morphism. Then X' is irreducible of dimension equal to the dimension of $f(X)$. By Theorem 7.14, $g^{-1}(\Delta)$ is $(\dim(X)-n-1)$-connected, whence connected. Since h is proper with connected fibers $f^{-1}(\Delta) = h^{-1}(g^{-1}(\Delta))$ is also connected.

To prove (ii) apply the first part to the morphism $f \times i\colon X \times Y \to \mathbb{P}^n(e) \times \mathbb{P}^n(e)$, where $i\colon Y \hookrightarrow \mathbb{P}^n(e)$ is the canonical inclusion, and use the fact that $f^{-1}(Y) \cong (f \times i)^{-1}(\Delta)$. □

Remarks 7.16. i) In the case $e = (1,1,\dots,1)$ (i.e. in the case of ordinary projective spaces) Corollary 7.15 is just the Fulton–Hansen connectedness theorem (see [52], or also [53]). However, the above proof is substantially different from the proofs of [52] or [53]. In fact, the present proof allowed us to give this generalized version of the result of Fulton–Hansen.

ii) The idea of using the $\mathbb{G}_m$-bundle $g : U \to \mathbb{P}^n \times \mathbb{P}^n$ in the proof of Theorem 7.14 is due to Deligne [44]; this simplifed the original proof given in [52].

iii) A special case of Fulton–Hansen's theorem was discovered in 1969 by Barth in [23], with arguments almost identical to those in [52].

At least in characteristic zero the weighted projective space $\mathbb{P}^n(e)$ appears as the quotient of $\mathbb{P}^n$ by the action of the finite group $G = \mu_{e_0} \times \mu_{e_1} \times \cdots \times \mu_{e_n}$ (where μ_m is the cyclic group of all roots of order m of 1) via the action given by

$$(\lambda_0, \lambda_1, \dots \lambda_n) \cdot [t_0, t_1, \dots, t_n] := [\lambda_0 t_0, \lambda_1 t_1, \dots, \lambda_n t_n],$$

$\forall(\lambda_0, \lambda_1, \dots \lambda_n) \in G$, and $\forall[t_0, t_1, \dots, t_n] \in \mathbb{P}^n$. One may ask whether the connectedness theorem (Theorem 7.14 or Corollary 7.15) is valid for an arbitrary quotient $\mathbb{P}^n/G$ of $\mathbb{P}^n$ by the action of a finite group G. The following example shows that in general this is not the case.

Example 7.3. Consider the action of the group $G = \mu_5$ of roots of order 5 of 1 on $\mathbb{P}^3$ ($\mathrm{char}(k) \neq 5$) given by

$$g \cdot [t_0, t_1, t_2, t_3] = [t_0, gt_1, g^2 t_2, g^3 t_3], \ \forall g \in G, \ \forall [t_0, t_1, t_2, t_3] \in \mathbb{P}^3.$$

Denote by P the quotient $\mathbb{P}^3/G$. Then G acts freely outside the four points $[1,0,0,0]$, $[0,1,0,0]$, $[0,0,1,0]$ and $[0,0,0,1]$. Consider the Fermat surface Y of equation $x_0^5+x_1^5+x_2^5+x_3^5=0$. Then Y is a G-invariant smooth surface on which G acts freely, and the quotient $S := Y/G$ (the Godeaux surface) is embedded in P. Let $\pi\colon Y \to S$ denote the canonical morphism, and let $f\colon X := Y \times Y \to P \times P$ be the composite of the inclusion $S \times S \hookrightarrow P \times P$ with the morphism $\pi \times \pi\colon Y \times Y \to S \times S$. Then f is a finite morphism, X is irreducible of dimension 4 and $\dim(P) = 3$. However, as is easily checked, $f^{-1}(\Delta)$ has five connected components, where Δ is the diagonal of $P \times P$.

We conclude this chapter by indicating a few applications of connectedness Theorem 7.14. We shall follow [53] closely.

Corollary 7.17. *Let Y and Z be two closed irreducible subvarieties of the weighted projective space $\mathbb{P}^n(e)$ such that* $\dim(Y)+\dim(Z) > n$. *Then $Y \cap Z$ is* $(\dim(Y)+\dim(Z)-n-1)$*-connected. More generally, let $u\colon Y \to \mathbb{P}^n(e)$ be a finite morphism from the irreducible variety Y, and let Z be a closed irreducible subvariety of $\mathbb{P}^n(e)$ such that* $\dim(Y) > \operatorname{codim}_{\mathbb{P}^n(e)}(Z)$. *Then $u^{-1}(Z)$ is* $(\dim(Y)+\dim(Z)-n-1)$*-connected.*

Proof. Set $X := Y \times Z$ and take as $f\colon X \to \mathbb{P}^n(e) \times \mathbb{P}^n(e)$ the product of the natural inclusions of Y and Z in $\mathbb{P}^n(e)$. Then apply Theorem 7.14 to get that $f^{-1}(\Delta) \cong Y \cap Z$ is $(\dim(Y)+\dim(Z)-n-1)$-connected.

For the second part one takes $X := Y \times Z$ and $f := u \times i\colon X = Y \times Z \to \mathbb{P}^n(e) \times \mathbb{P}^n(e)$, with $i\colon Z \hookrightarrow \mathbb{P}^n(e)$ the natural inclusion. Then by Theorem 7.14 $f^{-1}(\Delta) \cong u^{-1}(Z)$ is $(\dim(Y)+\dim(Z)-n-1)$-connected. □

Definition 7.18. Let $f\colon X \to Y$ be a morphism of algebraic schemes over k. The morphism f is said to be *unramified* (resp. *unramified at the point* $x \in X$) if $\Omega^1_{X|Y} = 0$ (resp. if $(\Omega^1_{X|Y})_x = 0$). Since by definition $\Omega^1_{X|Y}$ is $\mathcal{I}/\mathcal{I}^2$, where $\mathcal{I}$ is the ideal sheaf of the closed (diagonal) immersion $\Delta_{X|Y} : X \to X \times_Y X$, then one sees immediately that saying that f is unramified is the same as saying that the diagonal immersion $\Delta_{X|Y}$ is also an open immersion. In other words, if $f\colon X \to Y$ is unramified then $\Delta_{X|Y}(X)$ is a connected component of $X \times_Y X$. A morphism $f\colon X \to Y$ is said to be *étale* if f is unramified and flat. As a trivial observation, if $f\colon X \to Y$ is an étale morphism, with Y irreducible, and if Z is an irreducible component of X, then the restriction $f|Z\colon Z \to Y$ is unramified.

Corollary 7.19. *Let $f\colon X \to \mathbb{P}^n(e)$ be a finite unramified morphism from an irreducible projective variety X such that* $\dim(X) > \frac{n}{2}$. *Then f is a closed embedding.*

Proof. Apply Theorem 7.14 to $f \times f\colon X \times X \to \mathbb{P}^n(e) \times \mathbb{P}^n(e)$ to deduce that $X \times_{\mathbb{P}^n(e)} X$ is connected. On the other hand, since f is unramified, the diagonal Δ_X of $X \times X$ is a connected component of $X \times_{\mathbb{P}^n(e)} X$, whence $\Delta_X = X \times_{\mathbb{P}^n(e)} X$. Therefore f is injective. But an injective finite unramified morphism is a closed embedding (see [66] IV, (8.11.5) and (17.2.6)). □

Corollary 7.20. *Let Y be a closed irreducible subvariety of $\mathbb{P}^n(e)$ of dimension $> \frac{n}{2}$. If Y is not normal, then the normalization morphism $f\colon Y' \to Y$ must be ramified.*

Corollary 7.21. *Let Y be a closed irreducible subvariety of $\mathbb{P}^n(e)$ of dimension $> \frac{n}{2}$. Then Y is algebraically simply connected, i.e. every finite étale morphism $u\colon Y' \to Y$, with Y' connected, is an isomorphism.*

Proof. Let $u\colon Y' \to Y$ be a connected finite étale morphism, and let Z be an arbitrary irreducible component of Y'. If we set $v := u|Z\colon Z \to Y$, then the morphism v is finite and unramified. By Corollary 7.19, v is a closed embedding, i.e. v defines an isomorphism $Z \cong Y$. In particular, Z is a connected component of Y', whence $Z = Y'$ because Y' is connected. □

Another application of the Fulton–Hansen connectedness theorem we prove a beautiful theorem of Zak on tangencies. Let $Y \subset \mathbb{P}^n$ be a smooth closed irreducible subvariety of dimension $d \geq 1$ of $\mathbb{P}^n$. Henceforth (until the end of this chapter) we shall assume that Y is *non-degenerate* in $\mathbb{P}^n$. For every point $y \in Y$ let us denote by T_y the projective tangent space to Y at y. Let L be a linear subspace of $\mathbb{P}^n$. One says that L is *tangent* to Y at y if $T_y \subseteq L$. It follows that L is tangent to Y at y if and only if y is a singular point of the (scheme-theoretic) intersection $Y \cap L$. Then Zak's theorem on tangencies is the following:

Theorem 7.22 (Zak). *Under the above hypotheses, fix a linear subspace L of $\mathbb{P}^n$ of dimension e, with $d \leq e \leq n-1$. Then the closed subset $\{y \in Y \mid T_y \subseteq L\}$ has dimension $\leq e - d$.*

Proof. Assume to the contrary: there is an irreducible component $X \subseteq \{y \in Y \mid T_y \subseteq L\}$ of dimension $> e - d$. Then we claim that there exists a linear subspace $V \subseteq \mathbb{P}^n$ of codimension $e+1$ such that $V \cap (Y \cup L) = \varnothing$ and such that there exist two points $x \in X$ and $y \in Y$, $x \neq y$, with $\pi_V(x) = \pi_V(y)$ (i.e. the line xy intersects V). Indeed, since Y is non-degenerate in $\mathbb{P}^n$, $Y \not\subseteq L$, and in particular, there exists a point $y \in Y \setminus L$; if $x \in X$ is an arbitrary point of X then $x \in L$, whence $x \neq y$. Because T_x does not contain the line yx, yx cannot lie in Y. Then picking a point $z \in yx \setminus Y$, we may take as V a general linear subspace of $\mathbb{P}^n$ of dimension $n - e - 1$ through the point z.

Since $\dim(Y \times X) > e$ we may apply the connectedness Theorem 7.14 to the finite morphism $f := (\pi_V|Y) \times (\pi_V|X) : Y \times X \to \mathbb{P}^e \times \mathbb{P}^e$ to get that $f^{-1}(\Delta) = Y \times_{\mathbb{P}^e} X$ is connected. By the choice of V, the diagonal $\Delta_X \subseteq Y \times X$ is strictly contained in $Y \times_{\mathbb{P}^e} X$. The connectedness of $Y \times_{\mathbb{P}^e} X$ implies then that there exists a smooth curve T and a morphism $T \to Y \times_{\mathbb{P}^e} X$ whose image meets, but is not contained in Δ_X. In particular (restricting T if necessary), we get a family of pairs $\{(y_t, x_t)\}_{t \in T} \subseteq Y \times_{\mathbb{P}^e} X$ parametrized by T and a point $t_0 \in T$ such that $y_t \neq x_t$ for all $t \in T \setminus \{t_0\}$, and $y_{t_0} = x_{t_0} =: u$. As $t \to t_0$ the secant lines $y_t x_t$ degenerate to a tangent line $\omega \subseteq T_u$ to Y. On the other hand, for $t \neq t_0$ the secant line $y_t x_t$ meets the center of projection V, and hence ω also meets V. But $\omega \subseteq T_u \subseteq L$ (because $u \in X$), and L was disjoint from V, a contradiction. □

Here are two immediate consequences of Theorem 7.22.

Corollary 7.23. *Under the hypotheses of Theorem 7.22, the Gauss map*

$$Y \to \mathrm{Grass}(\mathbb{P}^n, \mathbb{P}^d)$$

defined by $y \to T_y$, is a finite morphism.

Proof. Take $e = d$ in Theorem 7.22. □

Corollary 7.24. *Under the hypotheses of Theorem 7.22, let X be an arbitrary hyperplane section of Y. Then X is smooth in codimension $2d-n-1$. In particular, if X is everywhere reduced if $d > \frac{n}{2}$. If moreover $d \geq \frac{n+2}{2}$ then X is irreducible and normal.*

Proof. For the first part take $e = n-1$ in Theorem 7.22. If $d > \frac{n}{2}$ then $2d-n-1 \geq 0$, whence X is smooth in codimension 0 by the first part, i.e. X is generically reduced. Since X is a hyperplane section of the smooth projective variety Y of dimension $d \geq 2$, then X satisfies Serre's condition S_1, whence X is everywhere reduced by [1], Proposition (2.2), page 131.

If $d \geq \frac{n+2}{2}$ then $2d - n - 1 \geq 1$, whence X is smooth in codimension 1; moreover, $d \geq 3$ in this case, whence X also satisfies Serre's condition S_2 (see [133] or [1]). Then the local normality of X follows from Serre's criterion of normality (see [133], Theorem 11, page IV-44, or also [1], Theorem (2.11), page 135). Then the irreducibility of X follows from the fact that X is connected (as a hyperplane section of Y), taking into account that every local ring of X is normal, and in particular, an integral domain. □

The last consequence of Theorem 7.22 is the following result, which gives a lower bound for the dimension of the dual variety of a projective subvariety of $\mathbb{P}^n$:

Corollary 7.25 (Zak). *Let Y be a smooth, irreducible and non-degenerate subvariety in $\mathbb{P}^n$ of dimension d. Let $Y^* \subseteq (\mathbb{P}^n)^*$ be the dual variety of Y. Then $\dim(Y^*) \geq d$.*

Proof. Consider the incidence correspondence

$$P := \{(y, L) \mid T_y \subseteq L\} \subseteq Y \times (\mathbb{P}^n)^*.$$

The first projection makes P a $\mathbb{P}^{n-d-1}$-bundle over Y, and in particular P is smooth, irreducible of dimension $n-1$. The dual variety Y^* is the image of P under the second projection. By Theorem 7.22 all fibers of $P \to Y^*$ have dimension $\leq n-d-1$. Then the conclusion follows from the theorem of dimension of fibers (see e.g. [138], Theorem 7, page 76). □

Remarks 7.26. i) For further applications of the Fulton–Hansen connectivity theorem as well as for some beautiful classical aspects of projective geometry we refer the reader to the recent Ph. D. lecture notes of Russo [126].

ii) In Chapter 11 we are going to reinterpret the connectedness results proved in this chapter from the point of view of formal-rational functions. For example, we shall include Zak's original proof of Corollary 7.23 (the finiteness of the Gauss map) using formal functions. This will illustrate in particular the deep connection between most of the material presented in Part 1 with the formal geometry presented in Part 2.

Chapter 8

A Problem of Complete Intersection

Let X be a smooth irreducible closed subvariety of dimension ≥ 2 of the smooth irreducible algebraic variety P. Let Y be an effective Cartier divisor of X. In this chapter, roughly speaking, we want to study the following:

Problem. Find conditions under which there exists a hypersurface H of P such that the scheme Y coincides with the scheme-theoretic intersection $X \cap H$.

The main result proved here (Theorem 8.4 below) is due to Ellingsrud, Gruson, Peskine and Strømme (see [46], cf. also [32], [33]). We shall follow these papers closely. We shall also apply the techniques of [46] to prove geometrically the following weaker form of a theorem of Barth–Larsen (see Theorem 8.7 below): $\operatorname{Pic}(X) \cong \mathbb{Z}$ for every closed smooth subvariety X of the complex projective space $\mathbb{P}^n$ with $\dim(X) \geq \frac{n+2}{2}$.

Coming back to the above problem, assume that such a hypersurface H of P does exist. First we want to find a (rather obvious) necessary condition in terms of the canonical exact sequence of normal bundles

$$0 \to N_{Y|X} \to N_{Y|P} \to N_{X|P}|Y \to 0. \tag{8.1}$$

From the equality $Y = X \cap H$ (in the scheme-theoretical sense) and from the above assumptions it follows that Y is a local complete intersection in H, hence we also have the canonical exact sequence of normal bundles

$$0 \to N_{Y|H} \to N_{Y|P} \to N_{H|P}|Y \to 0. \tag{8.2}$$

By general elementary statements, the fact that Y is a proper intersection of X with H implies that there are canonical isomorphisms

$$N_{Y|H} \cong N_{X|P}|Y \text{ and } N_{H|P}|Y \cong N_{Y|X}.$$

Therefore the exact sequence (8.2) yields a splitting of the exact sequence (8.1).

In other words, given a triple (Y, X, P) as at the beginning, a necessary condition for the existence of a hypersurface H of P such that $Y = X \cap H$ (scheme-theoretically), is the splitting of the exact sequence (8.1).

This is why we start this chapter by expressing the splitting of (8.1) in terms of the first infinitesimal neighbourhood $X(1)$ of X in P. Therefore we are going to study a question somewhat similar to the splitting condition of the normal sequence of a smooth closed subvariety of a smooth variety (see Lemma 6.2).

The dual of (8.1) is the exact sequence

$$0 \longrightarrow \mathcal{J}/\mathcal{I}\mathcal{J} \xrightarrow{\ \beta\ } \mathcal{I}/\mathcal{I}^2 \xrightarrow{\ \alpha\ } \mathcal{I}/(\mathcal{J}+\mathcal{I}^2) \longrightarrow 0 \tag{8.3}$$

where $\mathcal{I} = \mathcal{I}_Y$ (resp. $\mathcal{J} = \mathcal{I}_X$) is the ideal sheaf of Y (resp. of X) in $\mathcal{O}_P$. Since $Y \subset X$, $\mathcal{J} \subset \mathcal{I}$, and the maps α and β are defined as follows: α is the canonical map induced by the identity of $\mathcal{I}$ (taking into account that $\mathcal{I}^2 \subseteq \mathcal{J}+\mathcal{I}^2$), while β is the map induced by the inclusion $\mathcal{J} \subset \mathcal{I}$ (taking into account that $\mathcal{I}\mathcal{J} \subseteq \mathcal{I}^2$).

Lemma 8.1. *Under the above hypotheses, the exact sequence* (8.1) *splits if and only if there exists an effective Cartier divisor Y' on the first infinitesimal neighbourhood $X(1)$ of X in P such that $Y' \cap X = Y$ (scheme-theoretically).*

Remark 8.2. If an effective Cartier divisor Y' on $X(1)$ as in Lemma 8.1 does exist one gets the cartesian diagram of natural inclusions

$$\begin{array}{ccc} Y & \xrightarrow{\ \subset\ } & X \\ \cap\big\downarrow & & \big\downarrow\cap \\ Y' & \xrightarrow[\ \subset\ ]{} & X(1) \end{array}$$

Proof. (See [33]) Clearly (8.1) splits if and only if (8.3) does. So, from now on it will be more convenient to work with the exact sequence (8.3). Assume first that (8.3) splits, i.e. there exists a map $\sigma : \mathcal{I}/\mathcal{I}^2 \to \mathcal{J}/\mathcal{I}\mathcal{J}$ of $\mathcal{O}_Y$-modules such that $\sigma \circ \beta = \mathrm{id}$. If $\pi : \mathcal{I} \to \mathcal{I}/\mathcal{I}^2$ is the canonical surjection, set $\gamma := \sigma \circ \pi : \mathcal{I} \to \mathcal{J}/\mathcal{I}\mathcal{J}$ and $\mathcal{I}' := \mathrm{Ker}(\gamma)$. Clearly $\mathcal{J}^2 \subseteq \mathcal{I}^2 \subseteq \mathcal{I}' \subseteq \mathcal{I}$, whence $\mathcal{I}'/\mathcal{J}^2 \subseteq \mathcal{O}_P/\mathcal{J}^2 = \mathcal{O}_{X(1)}$ defines a subscheme $Y' \subseteq X(1)$. We shall prove that Y' is an effective Cartier divisor on $X(1)$ and that $Y' \cap X = Y$ (scheme-theoretically). The latter property is equivalent to

$$\mathcal{I}' + \mathcal{J} = \mathcal{I},$$

which is a consequence of the definition of $\mathcal{I}'$ plus the equality $\sigma \circ \beta = \mathrm{id}$. It remains therefore to check that Y' is a Cartier divisor on $X(1)$. This is a local calculation. Let $x \in Y$ be an arbitrary point and set $R := \mathcal{O}_{P,x}$, $I := \mathcal{I}_x$ and $J := \mathcal{J}_x$. Then R is a regular local ring, and from our hypotheses it follows that there exists an R-sequence $f_1, \dots, f_{n-d}, f_{n-d+1}$ such that $J = Rf_1 + \dots + Rf_{n-d}$ and $I = Rf_1 + \dots + Rf_{n-d} + Rf_{n-d+1}$, with $n := \dim(P) = \dim(R)$ and $d := \dim(X)$. Then

$$J^2 = \sum_{i,j=1}^{n-d} Rf_if_j, \quad I^2 = J^2 + Jf_{n-d+1} + Rf_{n-d+1}^2, \quad \text{and } IJ = J^2 + Jf_{n-d+1}.$$

Set $\overline{f}_i := f_i \mod I^2$, $i = 1, \ldots, n-d+1$ and $\tilde{f}_i := f_i \mod IJ$, $i = 1, \ldots, n-d$. Since $f_1, \ldots, f_{n-d+1}$ is an R-sequence, $\overline{f}_1, \ldots, \overline{f}_{n-d+1}$ is a basis of the R/I-module I/I^2, and from the hypotheses, $\tilde{f}_1, \ldots, \tilde{f}_{n-d}$ is a basis of the R/I-module J/IJ.

Since $\beta(\tilde{f}_i) = \overline{f}_i$, $i = 1, \ldots, n-d$ and $\sigma \circ \beta = \mathrm{id}$, $\sigma(\overline{f}_i) = \tilde{f}_i$, $i = 1, \ldots, n-d$. Moreover,

$$\sigma(\overline{f}_{n-d+1}) = \sum_{i=1}^{n-d} G_i \tilde{f}_i, \text{ with } G_i \in R/I.$$

Claim 1. $\mathrm{Ker}(\sigma) = (R/I)\overline{F}$, where $\overline{F} := \overline{f}_{n-d+1} - \sum_{i=1}^{n-d} G_i \overline{f}_i \in I/I^2$ (with $F \in I$).

Clearly, $\overline{F} \in \mathrm{Ker}(\sigma)$. Conversely, let $\overline{H} \in \mathrm{Ker}(\sigma)$, with $\overline{H} = \sum_{i=1}^{n-d} F_i \overline{f}_i$, with $F_i \in R/I$. Then

$$0 = \sigma(\overline{H}) = \sum_{i=1}^{n-d} F_i \sigma(\overline{f}_i) = \sum_{i=1}^{n-d} F_i \tilde{f}_i + F_{n-d+1} \sum_{i=1}^{n-d} G_i \tilde{f}_i = \sum_{i=1}^{n-d} (F_i + F_{n-d+1} G_i) \tilde{f}_i.$$

Since $\tilde{f}_1, \ldots, \tilde{f}_{n-d}$ is a basis of the R/I-module J/IJ we get $F_i + F_{n-d+1} G_i = 0$, $i = 1, \ldots, n-d$, and consequently

$$\overline{H} = \sum_{i=1}^{n-d+1} F_i \overline{f}_i = F_{n-d+1}(\overline{f}_{n-d+1} - \sum_{i=1}^{n-d} G_i \overline{f}_i) = F_{n-d+1} \overline{F},$$

and claim 1 is proved.

Claim 2. $I' = I^2 + RF$.

This is a direct consequence of claim 1 taking into account that $I' = \mathrm{Ker}(\gamma)$.

Claim 3. $I^2 + RF = J^2 + RF$.

Since $I^2 = J^2 + J f_{n-d+1} + R f_{n-d+1}^2$ the claim is equivalent to

$$f_i f_{n-d+1} \in J^2 + RF, \quad i = 1, \ldots, n-d+1. \tag{8.4}$$

Taking into account of the formula defining $\overline{F}$ we have $F = f_{n-d+1} - \sum_{i=1}^{n-d} g_i f_i$, where $g_i \in R$ such that $G_i = g_i \mod I$. Then

$$f_j F = f_j (f_{n-d+1} - \sum_{i=1}^{n-d} g_i f_i) = f_j f_{n-d+1} - \sum_{i=1}^{n-d} g_i f_i f_j, \; j = 1, \ldots, n-d+1. \tag{8.5}$$

In particular,

$$f_j f_{n-d+1} \in J^2 + RF, \; j = 1, \ldots, n-d. \tag{8.6}$$

Taking $j = n - d + 1$ in (8.5) we get

$$f_{n-d+1}F = f_{n-d+1}^2 - \sum_{i=1}^{n-d} g_i f_{n-d+1} f_i,$$

whence, using also (8.6) we get $f_{n-d+1}^2 \in J^2 + RF$. Claim 3 is proved.

By the above three claims, $I' = J^2 + RF = I^2 + RF$, where $I' = \mathcal{I}'_x$. In particular, the subscheme Y' of $X(1)$ is locally given by one equation. It remains to prove that this equation is a non-zero divisor. Again the verification is local, so that we have to prove that F is not a zero divisor in R/J^2. This can be done in the following way. Since $Y' \cap X = Y$ $F \mod J$ is a local equation of Y in X, and in particular, $F \mod J$ is not a zero divisor in R/J. If $F \mod J^2$ would be a zero divisor in R/J^2, then $F \mod J^2 \in J/J^2$ (R/J is a domain) which is not possible because we just remarked that $F \mod J$ was not a zero divisor in R/J.

Conversely, assume the existence of the effective Cartier divisor Y' on $X(1)$ such that $Y' \cap X = Y$ (scheme-theoretically). Then we have to find a splitting of (8.3). This means that we have an ideal $\mathcal{I}'$ containing $\mathcal{J}^2$ such that $\mathcal{I}' + \mathcal{J} = \mathcal{I}$, and in particular, $\mathcal{J}^2 \subseteq \mathcal{I}' \subseteq \mathcal{I}$.

Claim 4. $\mathcal{I}^2 \subseteq \mathcal{I}'$.

Again the verification is local. In the above notation, $I' = J^2 + RF$, with F a non-zero divisor modulo J^2. The equality $I' + J = I$ implies $J + RF = I$. Therefore we may assume that $F \equiv f_{n-d+1} \mod J$, i.e.

$$F = \sum_{i=1}^{n-d} g_i f_i + f_{n-d+1}. \tag{8.7}$$

Then proving claim 4 is equivalent to checking that

$$f_j f_{n-d+1} \in J^2 + RF,\ j = 1, \ldots, n - d + 1. \tag{8.8}$$

Multiplying (8.7) by f_j we get (8.8).

Claim 5. The exact sequence (8.3) splits.

Indeed, using claim 4 we may consider the map

$$\mu : \mathcal{I}'/\mathcal{I}^2 \oplus \mathcal{J}/\mathcal{I}\mathcal{J} \to \mathcal{I}/\mathcal{I}^2$$

defined by $\mu(f \mod \mathcal{I}^2, g \mod \mathcal{I}\mathcal{J}) = f + g \mod \mathcal{I}^2$. We claim that μ is an isomorphism. The surjectivity of μ comes from $\mathcal{I}' + \mathcal{J} = \mathcal{I}$. The verification of the injectivity of μ is local (along the same lines as above) and is left to the reader. The splitting $\sigma : \mathcal{I}/\mathcal{I}^2 \to \mathcal{J}/\mathcal{I}\mathcal{J}$ of (8.3) is then given by the second projection of the direct sum composed with μ^{-1}. □

If Y is a projective variety we shall denote by $\operatorname{Pic}^0(Y)$ (resp. by $\operatorname{Pic}^\tau(Y)$) the subgroup of $\operatorname{Pic}(Y)$ consisting of all isomorphism classes of line bundles which are algebraically (resp. numerically) equivalent to zero. Clearly, $\operatorname{Pic}^0(Y) \subseteq \operatorname{Pic}^\tau(Y)$, and a theorem of Matsusaka asserts that $\operatorname{Pic}^\tau(Y)/\operatorname{Pic}^0(Y)$ is a finite group (see e.g. [89]). The Néron–Severi group of Y is by definition $\operatorname{NS}(Y) := \operatorname{Pic}(Y)/\operatorname{Pic}^0(Y)$. By a result of Néron–Severi, $\operatorname{NS}(Y)$ is a finitely generated abelian group. Moreover, we also set $\operatorname{Num}(Y) := \operatorname{Pic}(Y)/\operatorname{Pic}^\tau(Y)$. By the definition $\operatorname{Num}(Y)$ is torsion free, whence $\operatorname{Num}(Y)$ is a free abelian group of finite rank since $\operatorname{NS}(Y)$ is a finitely generated abelian group by Néron–Severi's result.

Lemma 8.3. *Let X be a smooth irreducible closed subvariety of $\mathbb{P}^n$ over $\mathbb{C}$ of dimension ≥ 2. Let $X(1)$ be the first infinitesimal neighbourhood of X in $\mathbb{P}^n$. Then the image of the composite of natural maps*

$$\operatorname{Pic}(X(1)) \to \operatorname{Pic}(X) \to \operatorname{Num}(X)$$

is isomorphic to $\mathbb{Z}$.

Proof. Consider the (logarithmic derivative) map $\operatorname{dlog} : \operatorname{Pic}(X) \to H^1(X, \Omega^1_X)$ defined in the following way. If $[L] \in \operatorname{Pic}(X)$ is represented by the 1-cocycle $\{\xi_{ij}\}_{i,j}$ of $\mathcal{O}^*_X$ with respect to the affine covering $\{U_i\}$ of X (with $\xi_{ij} \in \Gamma(U_i \cap U_j, \mathcal{O}^*_X)$), then $\operatorname{dlog}(\{\xi_{ij}\})$ is by definition the cohomology class of the 1-cocycle $\{\frac{d\xi_{ij}}{\xi_{ij}}\}_{i,j}$ of Ω^1_X. Since $\operatorname{dlog}(\operatorname{Pic}^0(X)) = 0$ the map dlog yields the map $\operatorname{dlog} : \operatorname{NS}(X) = \operatorname{Pic}(X)/\operatorname{Pic}^0(X) \to H^1(X, \Omega^1_X)$. Moreover, since $\operatorname{Pic}^\tau(X)/\operatorname{Pic}^0(X)$ is a finite subgroup of $\operatorname{NS}(X)$ and the underlying abelian group of the $\mathbb{C}$-vector space $H^1(X, \Omega^1_X)$ is torsion free, we infer that $\operatorname{dlog}(\operatorname{Pic}^\tau(X)) = 0$. In other words, there is a unique map $\alpha : \operatorname{Num}(X) \to H^1(X, \Omega^1_X)$ such that $\operatorname{dlog} = \alpha \circ \beta$, where $\beta : \operatorname{Pic}(X) \to \operatorname{Num}(X)$ is the canonical surjection. Then it is a general fact that α induces an injective map $\alpha_{\mathbb{C}} : \operatorname{Num}(X) \otimes_{\mathbb{Z}} \mathbb{C} \to H^1(X, \Omega^1_X)$ (see [64], page 163). Using this, to prove the lemma it will be sufficient to show that the image of $\operatorname{Pic}(X(1))$ in $H^1(X, \Omega^1_X)$ (via the map α composed with $\operatorname{Pic}(X(1)) \to \operatorname{Num}(X)$) is contained in a one-dimensional complex vector subspace of $H^1(X, \Omega^1_X)$.

To do this we need the following three facts:

i) The canonical surjective map $\Omega^1_{\mathbb{P}^n}|X(1) \to \Omega^1_{X(1)}$ yields by restriction to X an isomorphism $\Omega^1_{\mathbb{P}^n}|X \cong \Omega^1_{X(1)}|X$.

ii) There is a natural map $H^1(X, \Omega^1_{\mathbb{P}^n}|X) \to H^1(X, \Omega^1_X)$.

iii) The $\mathbb{C}$-vector space $H^1(X, \Omega^1_{\mathbb{P}^n}|X)$ is one dimensional.

ii) is obvious because the map in question is induced by the canonical (surjective) map $\Omega^1_{\mathbb{P}^n}|X \to \Omega^1_X$, while iii) is just step 2 of the proof of Theorem 6.1. To prove i) consider the canonical exact sequence

$$\mathcal{I}^2_X/\mathcal{I}^4_X \to \Omega^1_{\mathbb{P}^n}|X(1) \to \Omega^1_{X(1)} \to 0,$$

and observe that after restricted to X the first map becomes zero.

Using i), ii) and iii), the fact that the image of $\mathrm{Pic}(X(1))$ in $H^1(X, \Omega^1_X)$ is contained in an one-dimensional complex vector subspace follows from the injectivity of $\alpha_{\mathbb{C}}$ and the commutative diagram

$$\begin{array}{ccccccc}
\mathrm{Pic}(X(1)) & \xrightarrow{\mathrm{dlog}} & H^1(X(1), \Omega^1_{X(1)}) & \longrightarrow & H^1(X, \Omega^1_{X(1)}|X) & \xleftarrow{\cong} & H^1(X, \Omega^1_{\mathbb{P}^n}|X) \\
\downarrow & & & & & & \downarrow \\
\mathrm{Pic}(X) & \longrightarrow & \mathrm{Num}(X) & \xrightarrow{\subset} & \mathrm{Num}(X) \otimes_{\mathbb{Z}} \mathbb{C} & \xrightarrow{\alpha_{\mathbb{C}}} & H^1(X, \Omega^1_X)
\end{array}$$

Lemma 8.3 is proved. □

Theorem 8.4 (Ellingsrud–Gruson–Peskine–Strømme [46], [32]). *Let X be a smooth projective complex surface embedded in $\mathbb{P}^n$ ($n \geq 3$) as a complete intersection. Let Y be a smooth connected curve in X such that the exact sequence of normal bundles*

$$0 \to N_{Y|X} \to N_{Y|\mathbb{P}^n} \to N_{X|\mathbb{P}^n}|Y \to 0$$

splits. Then there exists a hypersurface H of $\mathbb{P}^n$ such that $Y = X \cap H$ (scheme-theoretically).

Proof. Since X is a complete intersection in $\mathbb{P}^n$ the Lefschetz theorem on hyperplane sections (see e.g. [68], or [79], or also Appendix A below) implies that the restriction map $\mathrm{Pic}(\mathbb{P}^n) \to \mathrm{Pic}(X)$ is injective, $\mathrm{Pic}^{\tau}(X) = 0$, and the class of $\mathcal{O}_X(1)$ is not divisible in $\mathrm{Pic}(X)$. In particular, the canonical map $\mathrm{Pic}(X) \to \mathrm{Num}(X)$ is an isomorphism.

By Lemma 8.1 the splitting of the above sequence implies that there is an effective Cartier divisor Y' on $X(1)$ such that $Y' \cap X = Y$ (scheme-theoretically). In particular, the class of $\mathcal{O}_X(Y)$ is in the image of $\mathrm{Pic}(X(1)) \to \mathrm{Pic}(X) \cong \mathrm{Num}(X)$, which by Lemma 8.3 is isomorphic to $\mathbb{Z}$. It follows that there are two non-zero integers s and t such that $\mathcal{O}_X(sY) \cong \mathcal{O}_X(t)$. Since $\mathcal{O}_X(1)$ is not divisible in $\mathrm{Pic}(X)$ this implies that $\mathcal{O}_X(Y) \cong \mathcal{O}_X(d)$ for some $d > 0$.

On the other hand, since X is a complete intersection, X is projectively normal in $\mathbb{P}^n$. This implies that the restriction map

$$H^0(\mathbb{P}^n, \mathcal{O}_{\mathbb{P}^n}(d)) \to H^0(X, \mathcal{O}_X(d)) = H^0(X, \mathcal{O}_X(Y))$$

is surjective. In particular, there exists a section $a \in H^0(\mathbb{P}^n, \mathcal{O}_{\mathbb{P}^n}(d))$ such that $a|Y \in H^0(X, \mathcal{O}_X(Y))$ is a global equation of Y. In other words, there is a hypersurface H of degree d such that $X \cap H = Y$ (scheme-theoretically). □

Remark 8.5. In some sense Theorem 8.4 may be considered as a modern satisfactory answer to the so-called "problem of contact of surfaces along curves" studied by D. Gallarati and others especially in the 1940s and 1950s (see [56], and the bibliography therein; cf. also [57]).

Another application of Lemma 8.3 is a weak form of a theorem of Barth concerning the Picard group of small-codimensional smooth closed subvarieties of the complex projective space. In this application we shall use the following generalization of Kodaira's vanishing theorem due to Le Potier ([101], or also [130]):

Theorem 8.6 (Le Potier vanishing theorem). *Let X be a smooth complex projective variety of dimension $d \geq 2$, and let E be an ample vector bundle of rank r on X. Then $H^i(X, E^*) = 0$ for every $i \leq d - r$.*

Theorem 8.7. *Let X be a smooth closed subvariety of the complex projective space $\mathbb{P}^n$ of dimension $\geq \frac{n+2}{2}$ $(n \geq 4)$. Then* $\mathrm{Pic}(X) \cong \mathbb{Z}$.

Proof. Let $\mathcal{I}$ be the ideal sheaf of X in $\mathcal{O}_{\mathbb{P}^n}$. Then the truncated exponential sequence

$$0 \to \mathcal{I}/\mathcal{I}^2 = N^*_{X|\mathbb{P}^n} \to \mathcal{O}^*_{X(1)} \to \mathcal{O}^*_X \to 0$$

yields the cohomology sequence

$$H^1(X, N^*_{X|\mathbb{P}^n}) \to \mathrm{Pic}(X(1)) \to \mathrm{Pic}(X) \to H^2(X, N^*_{X|\mathbb{P}^n}).$$

Since the normal bundle $N_{X|\mathbb{P}^n}$ of X in $\mathbb{P}^n$ is a quotient of $T_{\mathbb{P}^n}|X$, $N_{X|\mathbb{P}^n}$ is ample of rank $= \mathrm{codim}_{\mathbb{P}^n}(X)$. The hypothesis $\dim(X) \geq \frac{n+2}{2}$ is equivalent to $\dim(X) - \mathrm{codim}_{\mathbb{P}^n}(X) \geq 2$, whence by Theorem 8.6 the first and the last cohomology group are both zero. Thus the restriction map $\mathrm{Pic}(X(1)) \to \mathrm{Pic}(X)$ is an isomorphism.

The hypothesis $\dim(X) \geq \frac{n+2}{2}$ also implies $\dim(X) > \frac{n}{2}$, whence by Corollary 7.21, X is algebraically simply connected. This implies $q = \dim_k H^1(X, \mathcal{O}_X) = 0$. Indeed, otherwise $\mathrm{Pic}^0(X) \neq 0$, and since $\mathrm{Pic}^0(X)$ is the underlying group of a complex torus, there exists a non-trivial line bundle L of finite order $m \geq 2$. Then L produces a connected non-trivial étale cover $X' \to X$ of order m, with $X' = \mathrm{Spec}(\bigoplus_{i=0}^{m-1} L^i)$ (see Appendix B at the end of this chapter), contradicting the fact that X is algebraically simply connected.

Since $\mathrm{Pic}^0(X) = 0$, by Matsusaka's theorem $\mathrm{Pic}^\tau(X)$ ([89]) is a finite subgroup of $\mathrm{Pic}(X)$. Again if there is a non-trivial $L \in \mathrm{Pic}^\tau(X)$, one gets a connected non-trivial étale cover X as above. Therefore $\mathrm{Pic}^\tau(X) = 0$, i.e. $\mathrm{Pic}(X) \cong \mathrm{Num}(X)$.

At this point, using the bijectivity of the restriction maps $\mathrm{Pic}(X(1)) \to \mathrm{Pic}(X)$ and $\mathrm{Pic}(X) \to \mathrm{Num}(X)$, we can conclude by applying Lemma 8.3. □

Remarks 8.8. i) Let X be a smooth closed subvariety of the complex projective space $\mathbb{P}^n$ $(n \geq 3)$ of dimension $> \frac{n}{2}$. Then the last part of the proof of Theorem 8.7 shows that $H^1(X, \mathcal{O}_X) = 0$ and $\mathrm{Pic}(X)$ is a free abelian group of finite rank. Indeed, $\mathrm{Pic}(X) = \mathrm{Num}(X)$ and $\mathrm{Num}(X)$ is a group of finite rank (by a result of Néron–Severi) with no torsion.

ii) Theorem 8.7 is a weak form of a famous result of Barth–Larsen [24], or also [22] which asserts (more precisely) that, under the hypotheses of Theorem 8.7,

$\mathrm{Pic}(X) \cong \mathbb{Z}$, generated by the class of $\mathcal{O}_X(1)$. We shall show later that an algebraic result of Faltings [49] (Theorem 10.3 below) implies — via a general statement (Proposition 10.10 below) — that the class of $\mathcal{O}_X(1)$ is not divisible in $\mathrm{Pic}(X)$, for every closed irreducible subvariety X of $\mathbb{P}^n$ over a field of characteristic zero, such that $\dim(X) > \frac{n}{2}$ (see Corollary 10.12 below). This fact together with Theorem 8.7 will imply (in an algebro-geometric way) the result of Barth–Larsen. See also [117] for an algebraic proof of a generalized version of the theorem of Barth–Larsen.

Appendix A

In this appendix we shall show how the following Lefschetz theorem for the Picard group (which was used in the proof of Theorem 8.4):

Theorem 8.9. *Let X be a smooth projective complex surface embedded in $\mathbb{P}^n$ ($n \geq 3$) as a complete intersection. Then the natural restriction map $a : \mathrm{Pic}(\mathbb{P}^n) \to \mathrm{Pic}(X)$ is injective and* $\mathrm{Coker}(a)$ *is torsion free. In other words, a is injective and the class of $\mathcal{O}_X(1)$ is not divisible in* $\mathrm{Pic}(X)$.

can be deduced from the following special case of the topological Lefschetz theorem for hyperplane sections (see e.g. Milnor [105]):

Theorem 8.10. *Let X be a smooth projective complex surface embedded in $\mathbb{P}^n$ ($n \geq 3$) as a complete intersection. Then the natural maps of singular integral cohomology $H^i(\mathbb{P}^n, \mathbb{Z}) \to H^i(X, \mathbb{Z})$ are isomorphisms for $i < 2$ and injective with torsion free cokernel for $i = 2$.*

Proof of Theorem 8.9. For a complex algebraic variety V denote by $\mathcal{O}_V^{\mathrm{an}}$ (resp. by $(\mathcal{O}_V^{\mathrm{an}})^*$, resp. by $\mathbb{Z}_V$) the sheaf of holomorphic functions on V (resp. the sheaf of nowhere vanishing holomorphic functions on V, resp. the constant sheaf $\mathbb{Z}$ on V). Then the commutative diagram of exponential sequences

$$\begin{array}{ccccccccc}
0 & \longrightarrow & \mathbb{Z}_{\mathbb{P}^n} & \longrightarrow & \mathcal{O}_{\mathbb{P}^n}^{\mathrm{an}} & \xrightarrow{\exp} & (\mathcal{O}_{\mathbb{P}^n}^{\mathrm{an}})^* & \longrightarrow & 0 \\
 & & \downarrow & & \downarrow & & \downarrow & & \\
0 & \longrightarrow & \mathbb{Z}_X & \longrightarrow & \mathcal{O}_X^{\mathrm{an}} & \xrightarrow{\exp} & (\mathcal{O}_X^{\mathrm{an}})^* & \longrightarrow & 0
\end{array}$$

yields the following commutative diagram with exact cohomology sequences

$$\begin{array}{ccccccc}
H^1(\mathbb{P}^n, \mathcal{O}_{\mathbb{P}^n}^{\mathrm{an}}) & \longrightarrow & H^1(\mathbb{P}^n, (\mathcal{O}_{P^n}^{\mathrm{an}})^*) & \longrightarrow & H^2(\mathbb{P}^n, \mathbb{Z}) & \longrightarrow & H^2(\mathbb{P}^n, \mathcal{O}_{\mathbb{P}^n}^{\mathrm{an}}) \\
\downarrow & & \downarrow & & \downarrow & & \\
H^1(X, \mathcal{O}_X^{\mathrm{an}}) & \longrightarrow & H^1(X, (\mathcal{O}_X^{\mathrm{an}})^*) & \longrightarrow & H^2(X, \mathbb{Z}) & &
\end{array}$$

By the GAGA results of Serre [135], $H^i(V, \mathcal{O}_V) \cong H^i(V, \mathcal{O}_V^{\mathrm{an}})$ for all $i \geq 0$ and $\mathrm{Pic}(V) = H^1(V, \mathcal{O}_V^*) \cong H^1(V, (\mathcal{O}_V^{\mathrm{an}})^*)$ for every complex projective variety V. Moreover, $H^i(\mathbb{P}^n, \mathcal{O}_{\mathbb{P}^n}) = 0$ for $i = 1, 2$, and $H^1(X, \mathcal{O}_X) = 0$ because X is a complete intersection surface in $\mathbb{P}^n$. Therefore the last commutative diagram becomes

$$\begin{array}{ccccccc}
0 & \longrightarrow & \mathrm{Pic}(\mathbb{P}^n) & \longrightarrow & H^2(\mathbb{P}^n, \mathbb{Z}) & \longrightarrow & 0 \\
 & & \Big\downarrow \alpha & & \Big\downarrow \beta & & \\
0 & \longrightarrow & \mathrm{Pic}(X) & \longrightarrow & H^2(X, \mathbb{Z}) & &
\end{array}$$

By Theorem 8.10, the map β is injective and $\mathrm{Coker}(\beta)$ has no torsion. This fact together with the last commutative diagram with exact rows imply that α is injective and $\mathrm{Coker}(\alpha)$ has no torsion. □

Appendix B

In this short appendix we recall briefly some basic facts about cyclic covers. Let X be an irreducible projective variety of dimension ≥ 1, and let $L \in \mathrm{Pic}(X)$ be a line bundle of finite order $n \geq 2$. In particular, there is an isomorphism

$$L^n \cong \mathcal{O}_X. \tag{8.9}$$

We shall assume that $\mathrm{char}(k)$ does not divide n. Using this isomorphism, we can endow the $\mathcal{O}_X$-module $\mathcal{A} := \bigoplus_{i=0}^{n-1} L^i$ with a structure of commutative $\mathcal{O}_X$-algebra in the following way. For any two local sections s of L^i and t of L^j we define the product st as follows:

- $st := s \otimes t$, which is a local section of L^{i+j} if $i + j \leq n - 1$, and
- st is the image of $s \otimes t$ (which is a section of L^{i+j}) under the isomorphism $L^{i+j} \to L^{i+j-n}$ deduced from (8.9), if $i + j \geq n$.

Then taking $X' := \underline{\mathrm{Spec}}(\mathcal{A})$ and $f \colon X' \to X$ the structural morphism of $\underline{\mathrm{Spec}}(\mathcal{A})$, we get an irreducible projective variety X' together with a canonical finite étale morphism $f \colon X' \to X$ of degree n (here the fact that the characteristic of k does not divide the order n of L is essential). By construction, $f^*(L) \cong \mathcal{O}_{X'}$.

Then the morphism f is called *the cyclic étale cover* of X associated to the line bundle $L \in \mathrm{Pic}(X)$ of order n.

More explicitly, X' is obtained as follows. Let $(\xi_{ij})_{i,j} \in Z^1(\mathcal{U}, \mathcal{O}_X^*)$ be a 1-cocycle of $\mathcal{O}_X^*$ with respect to a finite affine covering $\mathcal{U} = (U_i)_i$ of X which represents $L \in \mathrm{Pic}(X) = H^1(X, \mathcal{O}_X^*)$. Since $L^n \cong \mathcal{O}_X$ the 1-cocycle $(\xi_{ij}^n)_{i,j}$ is a 1-coboundary, i.e. we can write

$$\xi_{ij}^n = g_i/g_j, \text{ on } U_i \cap U_j, \text{ with } g_i \in \Gamma(U_i, \mathcal{O}_X^*).$$

Then define $f\colon X' \to X$ locally on U_i by taking f_i the restriction to the open subset

$$X_i := \{(x, z_i) \in U_i \times \mathbb{A}^1 \mid z_i(x)^n = g_i(x)\}$$

of the second projection of $U_i \times \mathbb{A}^1$. Then the morphisms f_i patch together to yield the étale morphism f with the above properties.

Part II

Formal Functions in Algebraic Geometry

Chapter 9

Basic Definitions and Results

One of the fundamental constructions in Commutative and Local Algebra is the completion of a commutative Noetherian (unitary) ring A with respect to an ideal I of A ($I \neq A$), see e.g. [116], or [133], or [121], or [7]. This allows one to associate to A another commutative Noetherian (unitary) ring $\hat{A}$ (which is called the I-adic completion of A) together with a flat homomorphism of unitary rings $\varphi\colon A \to \hat{A}$ such that:

i) $\hat{A}$ is a commutative Noetherian ring of Krull dimension $= \dim(A)$, which is complete with respect to the $\hat{I} := I\hat{A}$-adic topology of $\hat{A}$, and

ii) The canonical homomorphisms $A/I^n \to \hat{A}/\hat{I}^n$ induced by φ are isomorphisms for every positive integer $n \geq 1$.

The I-adic completion $\hat{A}$ is constructed by the formula

$$\hat{A} = \operatorname{inv}\lim_{n\geq 1} A/I^n.$$

The typical example is when A is the polynomial k-algebra $k[X_1, \ldots, X_n]$ in n variables $X_1,\ldots,X_n$ ($n \geq 1$) with coefficients in a field k and $I = AX_1 + \cdots + AX_n$. Then $\hat{A}$ is nothing but the formal power series ring $k[[X_1, \ldots, X_n]]$.

More generally, let X be a Noetherian scheme (or an algebraic variety over a field k), and let Y be a closed subscheme of X, corresponding to an ideal sheaf $\mathcal{I}$ of the structural sheaf $\mathcal{O}_X$ of X. Then the formal completion of $X_{/Y}$ of X along Y is a formal scheme (which is in particular a ringed space) whose underlying topological space coincides with the underlying topological space of Y, and whose structural sheaf $\mathcal{O}_{X_{/Y}}$ has the following property: if $U := \operatorname{Spec}(A)$ is an affine open subset of X such that $U \cap Y \neq \varnothing$, and if the closed subscheme $U \cap Y$ of U is defined by the ideal I, then

$$U_{/U\cap Y} = \operatorname{Spf}(\hat{A}),$$

with $\hat{A}$ the I-adic completion of A. In particular, $\Gamma(U_{/U\cap Y}, \mathcal{O}_{X_{/Y}}) = \hat{A}$. Moreover, there exists a canonical flat morphism of formal schemes $f\colon X_{/Y} \to X$, which set-theoretically coincides with the inclusion $Y \hookrightarrow X$. Note also that $f|U_{/U\cap Y}$: $U_{/U\cap Y} \to U = \operatorname{Spec}(A)$ corresponds to the canonical flat homomorphism of unitary rings $\varphi\colon A \to \hat{A}$.

The formal completion $X_{/Y}$ of X along Y is a fundamental object in algebraic geometry. It is an analogue of a tubular neighbourhood of Y in X, when X is a complex projective manifold and Y is a complex submanifold of X. Another way of defining $X_{/Y}$ is the following:

$$X_{/Y} = \operatorname{dir}\lim_{n\geq 0} Y(n) = \bigcup_{n=0}^{\infty} Y(n),$$

where $Y(n) = (Y, \mathcal{O}_X/\mathcal{I}^{n+1})$ is the infinitesimal neighbourhood of order n of Y in X ($Y = Y(0) \subset Y(1) \subset Y(2) \subset \cdots \subset X$). We refer the reader less familiar with the language of formal schemes to [66] I, chapter 10.

Now, if $X_{/Y}$ is the formal completion of X along Y, according to [81] it makes sense to define the sheaf $\mathcal{M}_{X_{/Y}}$ of formal-rational functions of X along Y. Actually this basic concept makes sense for an arbitrary locally Noetherian formal scheme $\mathcal{Z}$. For every affine open subset U of $\mathcal{Z}$ put

$$\mathcal{M}^0_{\mathcal{Z}}(U) := [\mathcal{O}_{\mathcal{Z}}(U)]_0,$$

where $[A]_0$ denotes the total ring of fractions of arbitrary commutative unitary ring A. We claim that in this way we obtain a presheaf on $\mathcal{Z}$. For this it is enough to show that for every two affine open subsets U and V of $\mathcal{Z}$ such that $V \subseteq U$ there is a natural restriction map $\mathcal{M}^0_{\mathcal{Z}}(U) \to \mathcal{M}^0_{\mathcal{Z}}(V)$. To show this, let f be a non-zero divisor of the ring $A = \mathcal{O}_{\mathcal{Z}}(U)$. Since $\mathcal{O}_{\mathcal{Z}}$ is a coherent sheaf (see [66] I, (10.10.3)) and since there exists an equivalence between the category of finitely generated A-modules and the category of coherent $\mathcal{O}_{\mathcal{Z}}|U$-modules (see [66] I, (10.10.2)), the multiplication by f defines an injective map $f : \mathcal{O}_{\mathcal{Z}}|U \to \mathcal{O}_{\mathcal{Z}}|U$, and hence $f|V$ is a non-zero divisor of $\mathcal{O}_{\mathcal{Z}}(V)$. Therefore the restriction map $\mathcal{O}_{\mathcal{Z}}(U) \to \mathcal{O}_{\mathcal{Z}}(V)$ yields the desired natural restriction map $\mathcal{M}^0_{\mathcal{Z}}(U) \to \mathcal{M}^0_{\mathcal{Z}}(V)$. Thus we proved that $\mathcal{M}^0_{\mathcal{Z}}$ is a presheaf on $\mathcal{Z}$.

Then denote by $\mathcal{M}_{\mathcal{Z}}$ the sheaf associated to the presheaf $\mathcal{M}^0_{\mathcal{Z}}$. Then $\mathcal{M}_{\mathcal{Z}}$ is a quasi-coherent sheaf of $\mathcal{O}_{\mathcal{Z}}$-modules containing $\mathcal{O}_{\mathcal{Z}}$ as a subsheaf. Define also

$$K(\mathcal{Z}) := H^0(\mathcal{Z}, \mathcal{M}_{\mathcal{Z}}).$$

Then $K(\mathcal{Z})$ is a commutative unitary ring which is called *the ring of formal-rational functions of* $\mathcal{Z}$. In particular, if Y is a closed subvariety of an algebraic variety X, we get:

$$K(X_{/Y}) := H^0(X_{/Y}, \mathcal{M}_{X_{/Y}}),$$

which is called *the ring of formal-rational functions of X along Y*. It is a k-algebra if X is an algebraic variety over k. Unlike the usual sheaf of rational functions of an irreducible algebraic variety X, the sheaf $\mathcal{M}_{X_{/Y}}$ is very far from being constant when $Y \neq X$. In particular, if $U = \operatorname{Spec}(A)$ is an affine open subset of X such that $U \cap Y \neq \varnothing$ and $U \cap Y \neq Y$, the canonical restriction ring homomorphism

$$\Gamma(X_{/Y}, \mathcal{M}_{X_{/Y}}) = K(X_{/Y}) \to \Gamma(U_{/U\cap Y}, \mathcal{M}_{X_{/Y}})$$

(which quite often is injective) can be very far from being surjective.

Coming back to a general locally Noetherian formal scheme $\mathcal{Z}$, let $\xi \in K(\mathcal{Z})$ be a non-zero formal-rational function on $\mathcal{Z}$, and consider the homomorphism of $\mathcal{O}_{\mathcal{Z}}$-modules $\lambda_\xi : \mathcal{O}_{\mathcal{Z}} \to \mathcal{M}_{\mathcal{Z}}$ defined by $\lambda_\xi(f) = f\xi$. Since $\mathcal{O}_{\mathcal{Z}}$ is a subsheaf of $\mathcal{M}_{\mathcal{Z}}$, we define *the pole sheaf* P_ξ *of* ξ by $P_\xi := \lambda_\xi^{-1}(\mathcal{O}_{\mathcal{Z}})$. By construction P_ξ is a sheaf of ideals of $\mathcal{O}_{\mathcal{Z}}$. We claim that P_ξ is also coherent. Indeed, for every $z \in \mathcal{Z}$ there exists a small affine open subset U of $\mathcal{Z}$ containing z such that $\xi \in \mathcal{M}^0_{\mathcal{Z}}(U)$. Thus $\xi = \frac{a}{b}$, with $a, b \in H^0(U, \mathcal{O}_{\mathcal{Z}})$ and b a non-zero divisor of $H^0(U, \mathcal{O}_{\mathcal{Z}})$. If $\mu_a : \mathcal{O}_U \to \mathcal{O}_U$ is the map $\mu_a(f) := af$, then clearly $P_\xi | U = \mu_a^{-1}(b\mathcal{O}_U)$. Therefore $P_\xi|U$ is coherent as the preimage of the coherent sheaf of ideals $b\mathcal{O}_U$.

One may also consider the sheaf $Q_\xi := \xi P_\xi$, which by definition of P_ξ is also a coherent sheaf of ideals of $\mathcal{O}_{\mathcal{Z}}$. Then ξ is naturally identified to a homomorphism $\xi \in \mathrm{Hom}_{\mathcal{O}_{\mathcal{Z}}}(P_\xi, Q_\xi)$ (multiplication by ξ).

Note also that if the formal scheme $\mathcal{Z}$ is regular, i.e. for every point $x \in \mathcal{Z}$ the local ring $\mathcal{O}_{\mathcal{Z},x}$ is regular, then the coherent sheaves of ideals P_ξ and Q_ξ are both invertible. Indeed, the problem is local, and one uses the fact that the local rings $\mathcal{O}_{\mathcal{Z},x}$ are factorial (by a result of Auslander–Buchsbaum, see e.g. [133], IV-39) and the definition of P_ξ to deduce that P_ξ is invertible. Since $Q_\xi = \xi P_\xi$, Q_ξ is also invertible.

Lemma 9.1. *If $\mathcal{Z}$ is an affine formal scheme, then the canonical map*

$$\mathcal{M}^0_{\mathcal{Z}}(\mathcal{Z}) = [H^0(\mathcal{Z}, \mathcal{O}_{\mathcal{Z}})]_0 \to \mathcal{M}_{\mathcal{Z}}(\mathcal{Z}) = K(\mathcal{Z})$$

is an isomorphism, i.e. $K(\mathcal{Z})$ coincides with the total ring of fractions of $H^0(\mathcal{Z}, \mathcal{O}_{\mathcal{Z}})$.

Proof. Let $\xi \in K(\mathcal{Z})$ be an arbitary formal-rational function of pole sheaf P_ξ. Since P_ξ is a coherent sheaf of ideals of $\mathcal{O}_{\mathcal{Z}}$ and $\mathcal{Z}$ is affine, P_ξ is generated by the ideal $P := H^0(\mathcal{Z}, P_\xi)$ of $A := H^0(\mathcal{O}_{\mathcal{Z}})$. If P contains a non-zero divisor of A, using the definitions we are done. If not, let $\mathrm{Ass}(A) = \{p_1, \ldots, p_n\}$. Then $P \subseteq p_1 \cup \cdots \cup p_n$. Since p_i is a prime ideal for every $i = 1, \ldots, n$, there exists an i such that $P \subseteq p_i =: q$. Since $q \in \mathrm{Ass}(A)$, there exists a non-zero $a \in A$ such that $q = (0 : a)$. If follows that P_ξ lies in the kernel of the map $a : \mathcal{O}_{\mathcal{Z}} \to \mathcal{O}_{\mathcal{Z}}$, and this fact implies that P_ξ cannot contain non-zero divisors even locally, which is impossible by the definition of P_ξ. □

A first question that arises is to find conditions under which $K(X_{/Y})$ is a field. A partial answer is given by the following (see [81]):

Proposition 9.2. *Let X be an irreducible algebraic variety, and let Y be a closed subvariety of X. Assume Y connected and X normal at every point of Y. Then $K(X_{/Y})$ is a field.*

Proof. For every $x \in Y$ the local rings $\mathcal{O}_{X_{/Y},x}$ and $\mathcal{O}_{X,x}$ have the same completion, namely the m_x-adic completion $\hat{\mathcal{O}}_{X,x}$ of $\mathcal{O}_{X,x}$ (where m_x is the maximal ideal of $\mathcal{O}_{X,x}$). Since X is normal at every point of Y, for every $x \in Y$, $\mathcal{O}_{X,x}$ is normal, whence by the analytic normality of normal rings (see e.g. [154], Vol II, pp. 313-320, or [116], 37.5, or also [112], page 413), $\hat{\mathcal{O}}_{X,x}$ is a domain, and in particular the subring $\mathcal{O}_{X_{/Y},x}$ of $\hat{\mathcal{O}}_{X,x}$ is also a domain (for every $x \in Y$). Therefore for every affine open subset U of X such that $U \cap Y$ is connected, the ring $H^0(U_{/U\cap Y}, \mathcal{O}_{X_{/Y}})$ is a domain. By Lemma 9.1, $K(U_{/U\cap Y}) = [H^0(U_{/U\cap Y}, \mathcal{O}_{X_{/Y}})]_0$, whence $K(U_{/U\cap Y})$ is a field for every affine open subset $U \subseteq X$ such that $Y \cap U$ is connected. Now, since Y is connected we can find affine open subsets $U_1,\dots,U_m$ of X with the following properties: $U_i \cap Y$ is connected for every $i = 1, \dots, m$, $U_i \cap U_{i+1} \cap Y \neq \varnothing$ for every $i = 1, \dots, m-1$, and $Y \subset U_1 \cup \dots \cup U_m$. Since $\mathcal{M}_{X_{/Y}}$ is a sheaf the natural sequence of maps

$$K(X_{/Y}) \longrightarrow \prod_{i=1}^{m} \mathcal{M}(U_{i/U_i\cap Y}) \rightrightarrows \prod_{i,j=1, U_i\cap U_j\cap Y\neq\varnothing}^{m} \mathcal{M}(U_i \cap U_{j/U_i\cap U_j\cap Y})$$

is exact, i.e.

$$K(X_{/Y}) = \operatorname{Ker}\left[\prod_{i=1}^{m} \mathcal{M}(U_{i/U_i\cap Y}) \rightrightarrows \prod_{i,j=1, U_i\cap U_j\cap Y\neq\varnothing}^{m} \mathcal{M}(U_i \cap U_{j/U_i\cap U_j\cap Y})\right].$$

Since $\mathcal{M}(U_{i/U_i\cap Y})$ is a field for every $i = 1, \dots, m$ and $U_i \cap U_{i+1} \cap Y \neq \varnothing$ for every $i = 1, \dots, m-1$, it follows easily that $K(X_{/Y})$ is also a field. □

In the setup of the beginning, there is a natural ring homomorphism

$$\alpha_{X,Y} : K(X) \to K(X_{/Y}). \tag{9.1}$$

The main reason why this map exists is that the canonical morphism $f\colon X_{/Y} \to X$ is flat. For example, when $X = \operatorname{Spec}(A)$ and Y is given by the ideal I, then $K(X) = [A]_0$ and $K(X_{/Y}) = [\hat{A}]_0$. Since φ is flat, φ maps every non-zero divisor of A into a non-zero divisor of $\hat{A}$, whence, by passing to total fractions, the homomorphism of rings $\varphi\colon A \to \hat{A}$ yields a unique homomorphism of rings $\alpha : [A]_0 \to [\hat{A}]_0$. Then using Lemma 9.1 we get the map $\alpha_{X,Y}$ in the case when X is affine.

If X is a non-affine algebraic variety, pick affine open subsets $U_1, \dots, U_m$ of X such that $U_i \cap Y \neq \varnothing$ and $Y \subset U_1 \cup \dots \cup U_m$. Clearly we may even assume that $X = U_1 \cup \dots \cup U_m$. Since U_i is affine, by what we have said above there exist the canonical maps

$$\alpha_{U_i, U_i\cap Y} : K(U_i) \to K(U_{i/U_i\cap Y}), \quad i = 1, \dots, m.$$

Moreover, since $U_{ij} := U_i \cap U_j$ is also affine for every i and j, we similarly have the canonical maps

$$\alpha_{U_{ij}, U_{ij}\cap Y} : K(U_{ij}) \to K(U_{ij/U_{ij}\cap Y}), \quad i,j = 1, \dots, m.$$

Now since $\mathcal{M}_X$ and $\mathcal{M}_{X_{/Y}}$ are sheaves we have

$$K(X) = \operatorname{Ker}\left[\prod_{i=1}^{m} K(U_i) \rightrightarrows \prod_{i,j=1}^{m} K(U_{ij})\right]$$

and

$$K(X_{/Y}) = \operatorname{Ker}\left[\prod_{i=1}^{m} K(U_{i/U_i \cap Y}) \rightrightarrows \prod_{i,j=1}^{m} K(U_{ij/U_{ij}\cap Y})\right].$$

Putting everything together we get the canonical map (9.1) we were looking for.

The basic tool in the study of formal-rational functions (and especially in the extension problem of formal-rational functions to rational functions) is the following fundamental existence theorem of Grothendieck (see [66] III, (5.1.4)).

Theorem 9.3 (Grothendieck). *Let I be an ideal of a commutative Noetherian ring A such that A is complete with respect to the I-adic topology. Let $Y := \operatorname{Spec}(A)$, $Y' := \operatorname{Spec}(A/I) = V(I)$, $f\colon X \to Y$ a proper morphism of schemes, and $X' := f^{-1}(Y')$. Denote by $g\colon X_{/X'} \to X$ the canonical morphism. Then the functor $F \to \hat{F} = g^*(F)$ is an equivalence of the category of coherent sheaves on X with the category of coherent sheaves on $X_{/X'}$.*

Here is an important consequence of Theorem 9.3 which will be used in the sequel (see [81], Proposition (2.2)):

Proposition 9.4. *Under the assumptions and notation of Theorem 9.3, the canonical map $\alpha_{X,X'} : K(X) \to K(X_{/X'})$ (see (9.1)) is an isomorphism.*

Proof. First we prove the injectivity of $\alpha := \alpha_{X,X'}$. Let $\xi \in K(X)$ be a rational function on X such that $\alpha(\xi) = 0$. Let P_ξ be the pole sheaf of ξ and set $Q_\xi := \xi P_\xi$. Then we saw above that P_ξ and Q_ξ are coherent sheaves of ideals on X and $\xi \in \operatorname{Hom}_{\mathcal{O}_X}(P_\xi, Q_\xi)$. The hypothesis says that $\hat{\xi} = \alpha(\xi)$ is zero as an element of $\operatorname{Hom}_{\mathcal{O}_{X_{/X'}}}(\hat{P}_\xi, \hat{O}_\xi)$. By [66] I, (10.8.11), there exists an open neighbourhood U of X' in X such that $\xi|U = 0$. To prove the injectivity of α it will be enough to show that $U = X$. Indeed, if not, set $F := X \setminus U$. Since f is proper, $f(F)$ is a closed subset of Y, and since $F \cap X' = \varnothing$ and $X' = f^{-1}(Y')$, it follows that $Y' \cap f(F) = \varnothing$. This last fact is impossible if $f(F)$ is closed and non-empty because A is I-adically complete.

Now we prove the surjectivity of α. Let therefore $\xi \in K(X_{/X'})$ be any non-zero formal-rational function, and let $\mathcal{P}_\xi$ be the pole sheaf of ξ. Set $\mathcal{Q}_\xi := \xi\mathcal{P}_\xi$. Since $\mathcal{P}_\xi$ and $\mathcal{Q}_\xi$ are coherent sheaves of ideals on $X_{/X'}$, by Theorem 9.3 there exist coherent sheaves of ideals P and Q on the (ordinary) scheme X such that

$\hat{P} = \mathcal{P}_\xi$ and $\hat{Q} = \mathcal{Q}_\xi$. Consider the commutative diagram

$$\begin{array}{ccc} \mathcal{O}_X & \xrightarrow{\beta} & \underline{\mathrm{Hom}}(P, \mathcal{O}_X) \\ \downarrow & & \downarrow \\ \mathcal{O}_{X_{/X'}} & \xrightarrow[\hat{\beta}]{} & \underline{\mathrm{Hom}}(\hat{P}, \mathcal{O}_{X_{/X'}}) \end{array}$$

in which the vertical arrows are completion maps, and β maps $\varphi \in \mathcal{O}_X(U)$ to the homomorphism $P(U) \to \mathcal{O}_X(U)$ induced by the multiplication by φ (and $\hat{\alpha}$ is defined similarly). Thus $\mathrm{Ker}(\hat{\beta})$ is the completion of $\mathrm{Ker}(\beta)$ by the equivalence of coherent categories given by Theorem 9.3. By the definition of $\mathcal{P}_\xi$, $\mathrm{Ker}(\hat{\beta}) = 0$, and hence $\mathrm{Ker}(\beta) = 0$. This means that P is locally generated by non-zero divisors.

Using Theorem 9.3 again we have $\mathrm{Hom}(P, Q) \cong \mathrm{Hom}(\mathcal{P}_\xi, \mathcal{Q}_\xi)$. Since $\xi \in \mathrm{Hom}(\mathcal{P}_\xi, \mathcal{Q}_\xi)$, we infer that there exists a unique $\mu \in \mathrm{Hom}(P, Q)$ such that $\hat{\mu} = \xi$. Let x be an arbitrary point of X and let U be an affine neighbourhood of x; then $P(U)$ contains non-zero divisors of $\mathcal{O}_X(U)$. Let $b \in P(U)$ be such a non-zero divisor and put $a := \mu(b) \in Q(U) \subseteq \mathcal{O}_X(U)$. Then the quotient $\frac{a}{b} \in \mathcal{M}_X(U)$ is independent of the choice of b, as one can immediately check. Therefore these local functions patch together to yield a rational function $\xi' \in K(X)$ such that $\xi = \alpha(\xi')$. □

Following [81] we now need to introduce some definitions.

Definition 9.5. Let $\varphi\colon A \to B$ be a homomorphism of commutative unitary rings. We say that φ is *admissible* if φ maps every non-zero divisor of A to a non-zero divisor of B. Let $f\colon X' \to X$ be a morphism of finite type of schemes. We say that f is *quasi-admissible* if for every point $x' \in X'$ there are affine open neighbourhoods $U' = \mathrm{Spec}(A')$ of x' and $U = \mathrm{Spec}(A)$ of $x = f(x')$ such that $f(U') \subseteq U$ and the following condition holds:

i) The ring homomorphism $\varphi\colon A \to A'$ (associated to this inclusion) is admissible.

It is easily checked that if f is quasi-admissible then the condition i) holds for every affine open subset $U' = \mathrm{Spec}(A')$ and $U = \mathrm{Spec}(A)$ such that $f(U') \subseteq U$. Therefore if f is quasi-admissible, then there exists a canonical homomorphism of sheaves of rings $\mathcal{M}_X \to f_*(\mathcal{M}_{X'})$, and in particular, a canonical homomorphism $K(X) \to K(X')$ between the rings of rational functions.

We say that f is *admissible* (resp. *locally birational*) if in the above definition the condition is the following stronger one:

ii) There is an element $a \in A$ such that a is a non-zero divisor in A and $\varphi(a)$ is a non-zero divisor in A', and such that the induced map $A_a \to A'_a$ is flat, where A_a and A'_a are the localizations of A and A' by powers of a (resp. by the still stronger condition:)

iii) There exists a non-zero divisor a in A and an injective homomorphism $\psi\colon A' \to A_a$ such that $\psi \circ \varphi$ coincides to the canonical homomorphism $A \to A_a$.

Since f is of finite type, the condition iii) above is also equivalent to saying that φ is admissible and the induced homomorphism $[A]_0 \to [B]_0$ is an isomorphism.

For example, by the theorem of generic flatness (see e.g. [67], IV (6.1)), every dominant morphism of irreducible varieties $f\colon X' \to X$ is admissible.

Finally we say that f is *birational* if it is locally birational and the canonical map $\mathcal{M}_X \to f_*(\mathcal{M}_{X'})$ is an isomorphism.

Lemma 9.6. *Consider a cartesian diagram of quasi-compact schemes*

$$\begin{array}{ccc} Z' = X' \times_X Z & \longrightarrow & X' \\ {\scriptstyle h}\downarrow & & \downarrow{\scriptstyle f} \\ Z & \xrightarrow{\quad g \quad} & X \end{array}$$

where f is of finite type and g is flat. Then h is admissible (resp. locally birational, resp. birational) if f is so.

Proof. Since the properties in question are local on X and on Z, we may assume that $X = \operatorname{Spec}(A)$ and $Z = \operatorname{Spec}(B)$ are affine. Moreover, we pick a finite affine covering $U_i = \operatorname{Spec}(A'_i)$ of X' for which ii) (resp. iii)) of Definition 9.5 holds. Assume f admissible. Then for every i there exists a non-zero divisor $a_i \in A$ such that the induced map $A_{a_i} \to (A'_i)_{a_i}$ is flat, whence

$$A_{a_i} \otimes_A B = B_{a_i} \to (A'_i)_{a_i} \otimes_A B = (A'_i \otimes_A B)_{a_i}$$

is also flat. Moreover, since $A \to B$ is flat, a_i remains a non-zero divisor in B. So h is admissible if f is so. Assume f is locally birational. Then there exist a_i as above and an inclusion $A'_i \hookrightarrow A_{a_i}$ whose restriction to A is just the canonical homomorphism $A \to A_{a_i}$. Since $A \to B$ is flat we get an inclusion $A'_i \otimes_A B \hookrightarrow A_{a_i} \otimes_A B = B_{a_i}$ whose composite with $B \to A'_i \otimes_A B$ is the canonical homomorphism $B \to B_{a_i}$. Thus, h is locally birational if f is so.

Now assume f birational; then we have a canonical exact diagram of rings

$$[A]_0 \longrightarrow \prod_i [A'_i]_0 \rightrightarrows \prod_{i,j} [A'_{ij}]_0,$$

where $A'_{ij} := \Gamma(U_i \cap U_j, \mathcal{O}_{X'})$. (Here we implicitly made use of Lemma 9.1 for ordinary schemes.) Put $K := [A]_0$. Since B is flat over A, $[B]_0$ is flat over K.

Therefore tensoring the above exact diagram with $[B]_0$ over K and noting that $[A'_i]_0 \cong K$ and $[A'_i]_0 \otimes_K [B]_0 \cong [B]_0 \cong [A'_i \otimes_A B]_0$, we get the exact diagram

$$[B]_0 \longrightarrow \prod_i [A'_i \otimes_A B]_0 \rightrightarrows \prod_{i,j} [A'_{ij}]_0 \otimes_K [B]_0.$$

Now, since $A \to A'_i$ is admissible, and $A'_i \to A'_{ij}$ is flat (because $\operatorname{Spec}(A'_{ij}) \hookrightarrow \operatorname{Spec}(A'_i)$ is an open immersion, and in particular, is a flat morphism!), hence admissible, $A \to A'_{ij}$ is also admissible (but it does not necessarily follow that the map $[A]_0 \to [A_{ij}]_0$ is an isomorphism!). Since $A \to B$ is flat, by the first part already proved, $B \to A'_{ij} \otimes_A B$ is also admissible. Moreover, $A'_{ij} \to A'_{ij} \otimes_A B$ is admissible since it is flat. It follows that $[A'_{ij}]_0 \otimes_K [B]_0 \subseteq [A_{ij} \otimes_A B]_0$. Thus

$$[B]_0 \longrightarrow \prod_i [A'_i \otimes_A B]_0 \rightrightarrows \prod_{i,j} [A'_{ij} \otimes_A B]_0$$

is exact, which proves that $K(Z) \to K(Z') = \Gamma(Z, h_*(\mathcal{M}_{Z'}))$ is an isomorphism. This implies (via Lemma 9.1 again) that the canonical map $\mathcal{M}_Z \to h_*(\mathcal{M}_{Z'})$ is an isomorphism. □

Lemma 9.7. *Let $f\colon X' \to X$ be a dominant morphism between two irreducible quasi-projective varieties X' and X. Let Y be a closed irreducible subvariety of X such that $f^{-1}(Y) \neq \varnothing$. Then the morphism f yields a canonical homomorphism $\hat{f}^* : K(X_{/Y}) \to K(X'_{/f^{-1}(Y)})$ rendering commutative the diagram*

$$\begin{array}{ccc} K(X) & \xrightarrow{f^*} & K(X') \\ \Big\downarrow{\scriptstyle \alpha_{X,Y}} & & \Big\downarrow{\scriptstyle \alpha_{X',f^{-1}(Y)}} \\ K(X_{/Y}) & \xrightarrow{\hat{f}^*} & K(X'_{/f^{-1}(Y)}) \end{array}$$

Proof. By what we remarked above, the dominant morphism f is generically flat, and hence admissible. Set $\hat{X} := X_{/Y}$ and $\hat{X}' := X'_{/f^{-1}(Y)}$.

Assume first that $X = \operatorname{Spec}(A)$ is affine. Let I be the ideal of Y in A, and let $\hat{A}$ be the I-adic completion of A. Since $f^{-1}(Y) \neq \varnothing$ we may consider the cartesian diagram

$$\begin{array}{ccccc} \hat{X}' = \widehat{X' \otimes_A \hat{A}} & \xrightarrow{a} & X' \otimes_A \hat{A} & \longrightarrow & X' \\ \Big\downarrow{\scriptstyle \hat{f}} & & \Big\downarrow{\scriptstyle f'} & & \Big\downarrow{\scriptstyle f} \\ \hat{X} = \operatorname{Spf}(\hat{A}) & \longrightarrow & \operatorname{Spec}(\hat{A}) & \xrightarrow{b} & X = \operatorname{Spec}(A) \end{array}$$

where $\widehat{X' \otimes_A \hat{A}}$ denotes the formal completion of $X' \otimes_A \hat{A}$ along ${f'}^{-1}(b^{-1}(Y))$ (which obviously coincides with $\hat{X}'$). Since the morphism b is flat (because A is Noetherian) then by Lemma 9.6 f' is admissible because f is so. In particular, f' yields the map

$$f'^* : K(\mathrm{Spec}(\hat{A})) \to K(X' \otimes_A \hat{A})$$

between the corresponding rings of rational functions. But by Lemma 9.1,

$$K(\mathrm{Spec}(\hat{A})) = [\hat{A}]_0 = K(\hat{X}).$$

On the other hand, we claim that the morphism a is flat. We shall first prove that the scheme $X' \otimes_A \hat{A}$ is locally Noetherian. This being a local problem, we may assume that $X' = \mathrm{Spec}(B)$ is also affine, where B is a finitely generated A-algebra (in fact, B is a finitely generated k-algebra). Then $X' \otimes_A \hat{A} = \mathrm{Spec}(B \otimes_A \hat{A})$, and the ring $B \otimes_A \hat{A}$ is Noetherian because first, $\hat{A}$ is Noetherian (as the completion of a Noetherian ring), and second, $B \otimes_A \hat{A}$ is a finitely generated $\hat{A}$-algebra (B is a finitely generated A-algebra). The hypothesis that $f^{-1}(Y) \neq \varnothing$ translates into $IB \neq B$. Then $\hat{X}' = \mathrm{Spf}(\hat{B})$, where $\hat{B}$ denotes the IB-adic completion of B. But $\hat{B}$ also coincides with the completion of $B \otimes_A \hat{A}$ with respect to the $B \otimes_A \hat{I}$-adic topology (or else, the completed tensor product of B with A, where B is endowed with the trivial 0-adic topology, and A with the I-adic topology). Thus the map a corresponds to the canonical map $B \otimes_A \hat{A} \to \hat{B}$ (passing to the completion), and since we saw that the ring $B \otimes_A \hat{A}$ is Noetherian, this latter map is flat, i.e. the morphism a is flat.

Now, since a is flat, it yields a canonical map

$$a^* : K(X' \otimes_A \hat{A}) \to K(\widehat{X' \otimes_A \hat{A}}) = K(\hat{X}').$$

Then the desired map (in the case when X is affine) is $\hat{f}^* := a^* \circ f'^*$.

Assume now that X is only quasi-projective. Pick a point $x \in f(X') \cap Y$ (recall that $f^{-1}(Y) \neq \varnothing$). Since X is quasi-projective we can find a finite covering $(U_i)_i$ of Y, with U_i affine open subsets of X all containing the (fixed) point x. This implies that for every i the restrictions $f_i := f|f^{-1}(U_i) : f^{-1}(U_i) \to U_i$ still have the property that $f_i^{-1}(Y \cap U_i) \neq \varnothing$. Set $U := \bigcup_i U_i$ and $U' := f^{-1}(U)$. Clearly, $U_{/Y} = \hat{X}$ and $U'_{/f^{-1}(Y)} = \hat{X}'$. Let us denote by $\mathcal{M}_{\hat{X}}$ and $\mathcal{M}_{\hat{X}'}$ the sheaves of formal-rational functions on $\hat{X}$ and $\hat{X}'$ respectively. Then $K(\hat{X}) = H^0(\hat{X}, \mathcal{M}_{\hat{X}}) = H^0(\hat{U}, \mathcal{M}_{\hat{X}})$ and $K(\hat{X}') = H^0(\hat{X}') = H^0(\hat{X}', \mathcal{M}_{\hat{X}'}) = H^0(\hat{U}', \mathcal{M}_{\hat{X}'})$. Since U_i is affine, by the previous case we have the well-defined canonical maps

$$\hat{f}_i^* : K(\hat{U}_i) = \mathcal{M}_{\hat{X}}(\hat{U}_i) \to K(\hat{U}_i') = \mathcal{M}_{\hat{X}}(\hat{U}_i'),$$

$$\hat{f}_{ij}^* : K(\widehat{U_i \cap U_j}) = \mathcal{M}_{\hat{X}}(\hat{U}_i \cap \hat{U}_j) \to K(\widehat{U_i' \cap U_j'}) = \mathcal{M}_{\hat{X}'}(\hat{U}_i' \cap \hat{U}_j'),$$

where $\hat{V}$ and $\hat{V}'$ denote the formal completion of any open subset V of X along $V \cap Y$ and the formal completion of $V' := f^{-1}(V)$ along $f^{-1}(V \cap Y)$ respectively. Since $\mathcal{M}_{\hat{X}}$ and $\mathcal{M}_{\hat{X}'}$ are sheaves, we get the exact sequences

$$K(\hat{X}) \longrightarrow \prod_i \mathcal{M}_{\hat{X}}(\hat{U}_i) \rightrightarrows \prod_{i,j} \mathcal{M}_{\hat{X}}(\hat{U}_i \cap \hat{U}_j),$$

$$K(\hat{X}') \longrightarrow \prod_i \mathcal{M}_{\hat{X}'}(\hat{U}_i') \rightrightarrows \prod_{i,j} \mathcal{M}_{\hat{X}'}(\hat{U}_i' \cap \hat{U}_j').$$

Moreover, by the previous case (i.e. when X was affine) we have the maps

$$\prod_i \hat{f}_i^* : \prod_i \mathcal{M}_{\hat{X}}(\hat{U}_i) \to \prod_i \mathcal{M}_{\hat{X}'}(\hat{U}_i'),$$

and

$$\prod_{i,j} \hat{f}_{ij}^* : \prod_{i,j} \mathcal{M}_{\hat{X}}(\hat{U}_i \cap \hat{U}_j) \to \prod_{i,j} \mathcal{M}_{\hat{X}'}(\hat{U}_i' \cap \hat{U}_j'),$$

making the obvious square commutative. This square yields the unique map $\hat{f}^* : K(\hat{X}) \to K(\hat{X}')$ we were looking for. The commutativity of the diagram from the statement of the lemma is obvious by the construction of $\hat{f}^*$. □

Corollary 9.8. *Let $f\colon X' \to X$ be a dominant morphism of irreducible quasi-projective algebraic varieties, and let $Y' \subset X'$ and $Y \subset X$ be non-empty closed subvarieties such that $f(Y') \subseteq Y$. Then there exists a canonical ring homomorphism $\tilde{f}^* : K(X_{/Y}) \to K(X'_{/Y'})$ rendering commutative the diagram*

$$\begin{array}{ccc} K(X) & \xrightarrow{f^*} & K(X') \\ \Big\downarrow{\scriptstyle \alpha_{X,Y}} & & \Big\downarrow{\scriptstyle \alpha_{X',Y'}} \\ K(X_{/Y}) & \xrightarrow{\tilde{f}^*} & K(X'_{/Y'}) \end{array}$$

Proof. By Lemma 9.7 we have the commutative diagram

$$\begin{array}{ccc} K(X) & \xrightarrow{f^*} & K(X') \\ \Big\downarrow{\scriptstyle \alpha_{X,Y}} & & \Big\downarrow{\scriptstyle \alpha_{X',f^{-1}(Y)}} \\ K(X_{/Y}) & \xrightarrow{\hat{f}^*} & K(X'_{/f^{-1}(Y)}) \end{array}$$

Since $Y' \subseteq f^{-1}(Y)$, there is a canonical flat morphism of formal schemes ψ : $X'_{/Y'} \to X'_{/f^{-1}(Y)}$, which induces a ring homomorphism φ: $K(X'_{/f^{-1}(Y)}) \to K(X'_{/Y'})$ (φ is just the restriction map of formal-rational functions). Then the desired map $\tilde{f}^*$ is $\varphi \circ \hat{f}^*$, and the commutativity of the above diagram yields the commutativity of the diagram of the corollary. □

Theorem 9.9 (Hironaka–Matsumura [81]). *Let $f\colon X' \to X$ be a proper birational morphism of irreducible algebraic varieties, let Y be a closed subvariety of X and let $Y' := f^{-1}(Y)$. Then the canonical morphism $\hat{f} : X'_{/Y'} \to X_{/Y}$ induces an isomorphism $\mathcal{M}_{X_{/Y}} \cong \hat{f}_*(\mathcal{M}_{X'_{/Y'}})$. In particular, $K(X_{/Y}) \cong K(X'_{/Y'})$.*

Proof. Since $\mathcal{M}_{X_{/Y}}$ is a sheaf, it is sufficient to prove the theorem when $X = \operatorname{Spec}(A)$ is affine. Let $\hat{A}$ be the I-adic completion of A, where I is the ideal of A defining Y, and set $Z := \operatorname{Spec}(\hat{A})$. Consider the cartesian diagram

$$\begin{array}{ccc} Z' = X' \times_X Z & \longrightarrow & X' \\ \Big\downarrow h & & \Big\downarrow f \\ Z & \xrightarrow{\;g\;} & X \end{array}$$

where $g\colon Z \to X$ is the canonical flat morphism induced by $A \to \hat{A}$. Clearly $X_{/Y} \cong Z_{/g^{-1}(Y)} =: \hat{Z}$ and $X'_{/Y'} \cong Z'_{/h^{-1}(g^{-1}(Y))} =: \hat{Z}'$. Since g is flat and f is proper and birational, by Lemma 9.6 the morphism h is also proper and birational. Therefore in the commutative diagram

$$\begin{array}{ccc} K(Z) & \longrightarrow & K(\hat{Z}) \\ \Big\downarrow & & \Big\downarrow \\ K(Z') & \longrightarrow & K(\hat{Z}') \end{array}$$

the first vertical map is an isomorphism by birationality, and the horizontal arrows are isomorphisms by Proposition 9.4. Therefore $K(\hat{Z}) \cong K(\hat{Z}')$, or else, $K(X_{/Y}) \cong K(X'_{/Y'})$. □

Using Theorem 9.9 we can improve Proposition 9.2 in the following way:

Corollary 9.10. *Let X be an irreducible algebraic variety, and let Y be a closed subvariety of X. Let $u\colon X' \to X$ be the (birational) normalization of X. Then $K(X_{/Y})$ is a field if and only if $u^{-1}(Y)$ is connected. If Y is a closed subvariety of an irreducible variety X, then $K(X_{/Y})$ is a finite product of fields.*

Proof. By Theorem 9.9, $K(X_{/Y}) \cong K(X'_{/u^{-1}(Y)})$ (since $K(X') = K(X)$). Hence by Proposition 9.2, if $u^{-1}(Y)$ is connected, $K(X'_{/u^{-1}(Y)})$ is a field. Conversely, if $u^{-1}(Y)$ is the disjoint union of two non-empty closed subsets Y_1 and Y_2, then clearly

$$K(X'_{/u^{-1}(Y)}) \cong K(X'_{/Y_1}) \times K(X'_{/Y_2}),$$

and in particular, $K(X'_{/u^{-1}(Y)})$ cannot be a field.

For the last statement, by what we said above, we may replace X by its normalization X' and Y by $u^{-1}(Y)$. Therefore we may assume X normal. Let $Y_1,\ldots,Y_m$ be the connected components of Y. Then

$$K(X_{/Y}) \cong K(X_{/Y_1}) \times \cdots \times K(X_{/Y_m}),$$

and $K(X_{/Y_i})$ is a field for every $i = 1,\ldots,m$. $\square$

Now, following [81], we can generalize Theorem 9.9 in the following way:

Theorem 9.11 (Hironaka–Matsumura [81]). *Let $f\colon X' \to X$ be a proper surjective morphism of irreducible varieties. Then for every closed subvariety Y of X there is a canonical isomorphism*

$$K(X'_{/f^{-1}(Y)}) \cong [K(X') \otimes_{K(X)} K(X_{/Y})]_0,$$

where $[A]_0$ denotes the total ring of fractions of a commutative unitary ring A.

Proof. We first prove the theorem when $X = \operatorname{Spec}(A)$ is affine. If I is the ideal of A defining Y, let $\hat{A}$ be the I-adic completion of A. Set $Z := \operatorname{Spec}(\hat{A})$ and $Z' := X' \times_X Z$. As in the proof of Theorem 9.9 we get the cartesian diagram

$$\begin{array}{ccccc} \hat{X}' = \hat{Z}' & \longrightarrow & Z' & \longrightarrow & X' \\ {\scriptstyle \hat{f} = \hat{h}}\downarrow & & \downarrow{\scriptstyle h} & & \downarrow{\scriptstyle f} \\ \hat{X} = \hat{Z} & \longrightarrow & Z & \xrightarrow{\ g\ } & X \end{array}$$

where $\hat{X} := X_{/Y}$ and $\hat{X}' := X'_{/f^{-1}(Y)}$. Since f is a surjective morphism between irreducible varieties, f is admissible by generic flatness. Therefore by Lemma 9.6 h is admissible because g is flat. Therefore we get a canonical homomorphism $K(Z) \to K(Z')$ making the following diagram commutative:

$$\begin{array}{ccc} K(X') & \longrightarrow & K(Z') \\ \downarrow & & \downarrow \\ K(X) & \longrightarrow & K(Z) \end{array}$$

By Proposition 9.4, $K(Z) \cong K(\hat{X})$ and $K(Z') \cong K(\hat{X}')$. Therefore, if $Z = \operatorname{Spec}(A)$ is affine we have only to prove that

$$[K(X') \otimes_{K(X)} K(Z)]_0 \cong K(Z').$$

Let $V = \operatorname{Spec}(B)$ be an affine open subset of X'. Then the map $A \to B$ is admissible (by generic flatness, because the restriction morphism $f|V: V \to X$ is dominant), so by Lemma 9.6 the induced map $\hat{A} \to B \otimes_A \hat{A}$ is also admissible. Moreover, since the map $B \to B \otimes_A \hat{A}$ is flat, it is admissible. Thus we get $[B \otimes_A \hat{A}]_0 = [[B]_0 \otimes_{[A]_0} [\hat{A}]_0]_0$. Since $K(X') = [B]_0$ (because X' is an irreducible variety), and since $V' = \operatorname{Spec}(B') \subseteq V = \operatorname{Spec}(B)$ is another non-empty open subset of X', the restriction map $[B \otimes_A \hat{A}]_0 \to [B' \otimes_A \hat{A}]_0$ is an isomorphism, and then we get exactly the desired isomorphism. So the theorem is proved when X is affine.

Now assume X arbitrary, and let $U_i = \operatorname{Spec}(A_i)$, $i = 1, \dots, n$, be a finite affine covering of X. Then we have the exact diagram

$$K(\hat{X}) \longrightarrow \prod_i \mathcal{M}_{\hat{X}}(\hat{U}_i) \rightrightarrows \prod_{i,j} \mathcal{M}_{\hat{X}}(\hat{U}_{ij}),$$

where $U_{ij} := U_i \cap U_j$. Set $L := K(X')$ and $K := K(X)$. Then L and K are fields, and therefore we get the exact diagram

$$L \otimes_K K(\hat{X}) \to \prod_i L \otimes_K \mathcal{M}_{\hat{X}}(\hat{U}_i) \rightrightarrows \prod_{i,j} L \otimes_K \mathcal{M}_{\hat{X}}(\hat{U}_{ij}).$$

If the field extension $K \subseteq L$ is is finite (i.e. if the morphism f is generically finite), using the last statement of Corollary 9.10 we see that all the rings in this last diagram are finite products of local Artin rings. In particular, if A is such a finite product then $A = [A]_0$. Therefore this exact diagram becomes

$$[L \otimes_K K(\hat{X})]_0 \to \prod_i [L \otimes_K \mathcal{M}_{\hat{X}}(\hat{U}_i)]_0 \rightrightarrows \prod_{i,j} [L \otimes_K \mathcal{M}_{\hat{X}}(\hat{U}_{ij})]_0.$$

By the case when X was affine (already proved) we get the exact diagram

$$[L \otimes_K K(\hat{X})]_0 \to \prod_i \mathcal{M}_{\hat{X}'}(f^{-1}(U_i)) \rightrightarrows \prod_{i,j} \mathcal{M}_{\hat{X}'}(f^{-1}(U_{ij})).$$

This means that $K(\hat{X}') \cong [L \otimes_K K(\hat{X})]_0$, as we wanted.

Therefore the theorem is proved in the following two cases: either if X is affine, or if the morphism $f: X' \to X$ is generically finite.

Let us pass to the general case, i.e. when the field extension $K \subset L$ is transcendental and X is not affine. Let $X = U_1 \cup \dots \cup U_n$ be the union of n open affine subsets ($n \geq 2$), and set $U := U_1 \cup \dots \cup U_{n-1}$ and $U' = U_n$. Then $X = U \cup U'$. Set $W := U \cap U'$, $\hat{U} := U_{/U \cap Y}$, $\hat{U}' := U_{/U' \cap Y}$, $\hat{W} := W_{/W \cap Y}$,

$\widehat{f^{-1}(U)} := f^{-1}(U)_{/Y'\cap f^{-1}(U)}$, $\widehat{f^{-1}(U')} := f^{-1}(U')_{/Y'\cap f^{-1}(U')}$, and $\widehat{f^{-1}(W)} := f^{-1}(W)_{/Y'\cap f^{-1}(W)}$. Using induction on n (the case $n = 1$ having been treated above), we may assume that the theorem holds for U, for U' and for W (note that W is the union of the affine open subsets $U_1 \cap U_n, \ldots, U_{n-1} \cap U_n$), i.e.

$$[L \otimes_K K(\hat{U})]_0 \cong K(\widehat{f^{-1}(U)}), \quad [L \otimes_K K(\hat{U}')]_0 \cong K(\widehat{f^{-1}(U')}), \quad \text{and}$$

$$[L \otimes_K K(\hat{W})]_0 \cong K(\widehat{f^{-1}(W)}).$$

Then we have the exact diagram

$$K(\hat{X}) \longrightarrow K(\hat{U}) \times K(\hat{U}') \rightrightarrows K(\hat{W}).$$

Taking into account the three isomorphisms above, all we need for our proof is to derive from this the exactness of the diagram

$$[L \otimes_K K(\hat{X})]_0 \to [L \otimes_K K(\hat{U})]_0 \times [L \otimes_K K(\hat{U}')]_0 \rightrightarrows [L \otimes_K K(\hat{W})]_0.$$

Since by the last part of Corollary 9.10 $K(\hat{U})$, $K(\hat{U}')$ and $K(\hat{W})$ are finite products of fields, the following claim will complete the proof of Theorem 9.11:

Claim. In the above notation, let K_1 and K_2 be subrings of the ring K'. Assume that K_1, K_2 and K' are finite products of fields containing K, and set $\hat{K} := K_1 \cap K_2$. Thus we have the exact diagram

$$\hat{K} \longrightarrow K_1 \times K_2 \rightrightarrows K',$$

where the double arrow consists of the two projections followed by the inclusions in K'. Then the induced diagram

$$[L \otimes_K \hat{K}]_0 \to [L \otimes_K K_1]_0 \times [L \otimes_K K_2]_0 \rightrightarrows [L \otimes_K K']_0$$

is also exact.

Proof of the claim. Let $t_1, \ldots, t_r \in L$ be a transcendence basis of L over K. Then $K(t_1, \ldots, t_r) \subseteq L$ is a finite algebraic extension. Thus we get the pure transcendental extensions $K \subset K(t_1) \subset \cdots \subset K(t_1, \ldots, t_r)$ and the finite algebraic extension $K(t_1, \ldots, t_r) \subseteq L$. Therefore it is sufficient to consider the following two cases:

i) L is a simple transcendental extension of K, i.e. $L = K(t)$, with t transcendental over K, and

ii) L is a finite algebraic extension of K.

In the second case, $[L \otimes_K A]_0 = L \otimes_K A$, for every ring A which is a finite product of fields, and the assertion follows from the fact that L is flat over A (note that this case is similar to the proof of Theorem 9.11 when the morphism f was generically finite).

Therefore from now on we shall assume that we are in the case i). Then we have to prove that

$$[K_1[t]]_0 \cap [K_2[t]]_0 = [\hat{K}[t]]_0, \quad \text{where } t \text{ is an indeterminate over } K'. \tag{9.2}$$

By hypotheses we can write $K' = F_1 \times \cdots \times F_s$, with $s \geq 1$ and F_i a field for every $i = 1, \ldots, s$. Let $1 = e_1 + \cdots + e_s$ be the corresponding decomposition of 1 into idempotents. In general, if F is a field, t is an indeterminate over F and $u(t) = \frac{f(t)}{g(t)} \in F(t)$, with $f(t), g(t) \in F[t]$, we shall say that $\frac{f(t)}{g(t)}$ is in the standard form if $f(t)$ and $g(t)$ are relatively prime in $F[t]$, and $g(t)$ is monic. Clearly the standard form is unique and invariant under the extension of the field of coefficients.

Let now $u \in [K_1[t]]_0 \cap [K_2[t]]_0$, with $u = (u_1, \ldots, u_s)$ and $u_i = \frac{f_i(t)}{g_i(t)} \in F_i(t)$ in the standard form $\forall i = 1, \ldots, s$, so that $u = \frac{f}{g}$, where $f(t) := (f_1(t), \ldots, f_s(t))$ and $g(t) := (g_1(t), \ldots, g_s(t))$. We assert that $f(t)$ and $g(t)$ belong to $K_i[t]$, $i = 1, 2$, which of course will prove the claim.

To prove this assertion write $K_1 = E_1 \times \cdots \times E_p$ as a product of fields, and let $1 = e'_1 + \cdots + e'_p$ be the corresponding decomposition of 1 into idempotents. Then, after reordering $\{e_1, \ldots, e_s\}$ suitably, we have $e'_1 = e_1 + \cdots + e_a$, $e'_2 = e_{a+1} + \cdots + e_b$, and so on. In particular, $E_1 = K_1 e'_1$ is a subfield of $F_1 \times \cdots \times F_a$. Let $\varphi_i \colon E_1 \to F_i$ be the natural projection, $\forall i = 1, \ldots, a$. Since $u \in [K_1[t]]_0 = E_1(t) \times \cdots \times E_p(t)$, ue'_1 has a standard form $ue'_1 = \frac{A(t)}{B(t)}$, with $A(t), B(t) \in E_1[t]$. Then $\frac{\varphi_i(A(t))}{\varphi_i(B(t))}$ is the standard form of u_i, and hence $\varphi_i(A(t)) = f_i(t)$ and $\varphi_i(B(t)) = g_i(t)$, $\forall i = 1, \ldots, a$. It follows that $(f_1(t), \ldots, f_a(t)) = A(t) \in E_1[t]$ and $(g_1(t), \ldots, g_a(t)) = B(t) \in E_1[t]$. The same is true for $E_2, \ldots, E_p$, whence $f(t), g(t) \in K_1[t]$.

In a similar way we have $f(t), g(t) \in K_2[t]$. This proves (9.2), and thereby Theorem 9.11. □

Now we need the following:

Definition 9.12 (Hironaka–Matsumura [81]). In the notation of the beginning, assume that X is a projective irreducible variety over an algebraically closed field k. We say that Y is $G3$ in X if the canonical map $\alpha_{X,Y} : K(X) \to K(X_{/Y})$ is an isomorphism of rings. (In particular, if Y is $G3$ in X, $K(X_{/Y})$ is a field.) We also say that Y is $G2$ in X if $K(X_{/Y})$ is a field and if the map $\alpha_{X,Y}$ makes $K(X_{/Y})$ a finite field extension of $K(X)$.

In other words, the fact that Y is $G3$ in X means that every formal-rational function of X along Y "can be extended" to a usual rational function on X. Note that obviously "Y $G3$ in X" implies "Y $G2$ in X".

An immediate consequence of Theorem 9.11 is the following:

Corollary 9.13. (i) *Under the hypotheses and notation of Theorem* 9.9, *assume that* Y *is G3 in* X. *Then* $f^{-1}(Y)$ *is G3 in* X'.

(ii) *Let* Y *be a closed subvariety of the irreducible variety* X *such that* $K(X_{/Y})$ *is a field. Then every finitely generated field subextension* $E|K(X)$ *of the field extension* $K(X_{/Y})|K(X)$ *is separable, i.e. there exists a transcendence basis* $\{t_1, \dots, t_r\}$ *of* E *over* $K(X)$ *such that* E *is a finite separable field extension of* $K(X)(t_1, \dots, t_r)$.

Proof. Part (i) is a direct consequence of Theorem 9.11. On the other hand, by [87], théorème 3.3, the statement of (ii) is equivalent to the fact that for every purely inseparable finite extension L of $K(X)$, the ring $K(X_{/Y}) \otimes_{K(X)} L$ is reduced. To check this latter condition, let $f\colon X' \to X$ be the normalization of X in the field L (note that since L is a finite field extension of $K(X)$, L is finitely generated over k). Then f is a finite morphism and $K(X') = L$, so by Theorem 9.11 there is a canonical isomorphism

$$K(X'_{/f^{-1}(Y)}) \cong [K(X_{/Y}) \otimes_{K(X)} L]_0.$$

By Corollary 9.10, $K(X'_{/f^{-1}(Y)})$ is a finite product of fields, and in particular, is a reduced ring. Therefore $[K(X_{/Y}) \otimes_{K(X)} L]_0$ is reduced, whence the subring $K(X_{/Y}) \otimes_{K(X)} L$ is also reduced, as required. □

As a beautiful application of Theorem 9.11 we prove the following fundamental result due to Hironaka and Matsumura [81] (see also [74]):

Theorem 9.14 (Hironaka–Matsumura [81]). *Every positive-dimensional connected closed subvariety* Y *of the* n*-dimensional projective space* $\mathbb{P}^n$ $(n \geq 2)$ *over an algebraically closed field* k *of arbitrary characteristic is G3 in* $\mathbb{P}^n$.

Proof. We will follow the proof given in [81].

Step 1. The theorem holds true if Y is a line in $\mathbb{P}^n$.

Proof of step 1. Since for every two lines L_1 and L_2 of $\mathbb{P}^n$ there is an automorphism σ of $\mathbb{P}^n$ such that $\sigma(L_1) = L_2$, there is no loss of generality in assuming that the line Y is defined by the equations $x_2 = \cdots = x_n = 0$, where $[x_0, \dots, x_n]$ is a homogeneous coordinate system of $\mathbb{P}^n$. Set $v_i := \frac{x_i}{x_0}$, $i = 1, \dots, n$, and $t_j := \frac{v_j}{v_1}$, $j = 2, \dots, n$. Consider the blow up morphism $f\colon X \to \mathbb{P}^n$ with center the point $Q = [1, 0, \dots, 0]$, and denote by $E = f^{-1}(Q)$ the exceptional locus of f and by Y' the strict transform of Y in X. Then $f^{-1}(Q) = E \cup Y'$. Denote $\hat{X} := X_{/E\cup Y'}$, $\hat{X}_1 := X_{/E}$ and $\hat{X}_2 := X_{/Y'}$. By Theorem 9.9 it will be sufficient to prove that $K(X) \cong K(\hat{X})$.

Then Y' is covered by two affine open subsets, $\mathrm{Spec}(A_0)$ and $\mathrm{Spec}(A_1)$, where $A_0 = k[v_1, t_2, \dots, t_n]$ and $A_1 = k[\frac{1}{v_1}, t_2, \dots, t_n]$, and in each such ring Y' is defined by the ideal generated by $t_2, \dots, t_n$. It follows that $\mathrm{Spec}(A_0) \cup \mathrm{Spec}(A_1) \cong \mathbb{P}^1 \times V$, where V is the affine $(n-1)$-dimensional space $\mathrm{Spec}(k[t_2, \dots, t_n])$, and under this isomorphism Y' is mapped to $\mathbb{P}^1 \times O$, where O is the origin of V (i.e. the point defined by $t_2 = \cdots = t_n = 0$). Therefore we can apply Theorem 9.11 to the projection $\mathbb{P}^1 \times V \to V$ to deduce that

$$K(\hat{X}_2) = k[[t_2, \ldots, t_n]]_0(v_1). \tag{9.3}$$

On the other hand, by Theorem 9.9 we get

$$K(\hat{X}_1) = k[[v_1, \ldots, v_n]]_0. \tag{9.4}$$

Let

$$f = \sum_{i_1,\ldots,i_n=0}^{\infty} a_{i_1\ldots i_n} v_1^{i_1} \ldots v_n^{i_n} \in k[[v_1, \ldots, v_n]]$$

be arbitrary formal power series. Since $v_j = t_j v_1$, $\forall j = 2, \ldots, n$ we can rewrite f as

$$\sum_{i_1,\ldots,i_n=0}^{\infty} a_{i_1\ldots i_n} v_1^{i_1+\cdots+i_n} t_2^{i_2} \cdots t_n^{i_n} = \sum_{h=0}^{\infty}\Big(\sum_{i_2+\cdots+i_n \le h} a_{h-i_2-\cdots-i_n, i_2, \ldots, i_n} t_2^{i_2} \cdots t_n^{i_n}\Big) v_1^h.$$

This proves the inclusion $k[[v_1, \ldots, v_n]] \subseteq k[t_2, \ldots, t_n][[v_1]]$, whence (9.4) implies $K(\hat{X}_1) \subseteq k(t_2, \ldots, t_n)((v_1))$, where for every field H and for every indeterminate v over H we set $H((v)) = H[[v]]_0$. Recalling also (9.3) we get

$$K(X) \subseteq K(\hat{X}) \subseteq K(\hat{X}_1) \cap K(\hat{X}_2) \subseteq k(t_2, \ldots, t_n)((v_1)) \cap L(v_1),$$

where $L := k[[t_2, \ldots, t_n]]_0$ and, of course, $K(X) = k(v_1, t_2, \ldots, t_n)$. Here the fields under consideration are all included in $L((v_1))$. Indeed, if x is the unique point of intersection of E and Y', and if $\hat{X}_3 := X_{/\{x\}}$, then $K(\hat{X})$, $K(\hat{X}_1)$ and $K(\hat{X}_2)$ are canonically embedded in $K(\hat{X}_3) = k[[v_1, t_2, \ldots, t_n]]_0 \subseteq L((v_1))$. Thus the above inclusions are taken in this latter field. Therefore, if we set $F := k(t_2, \ldots, t_n) \subset L := k[[t_2, \ldots, t_n]]_0$, step 1 will follow from the following general algebraic statement:

Step 2. For all fields F and L such that $F \subset L$, let v be an indeterminate over L. Then $F((v)) \cap L(v) = F(v)$.

Proof of step 2. Let $0 \neq \xi = \sum_{h=p}^{\infty} c_h v^h \in F((v)) \cap L(v)$, with $c_h \in F$ and $c_p \neq 0$ (p may be negative). Since $\xi \in L(v)$, $\xi = (\sum_{i=0}^{m} a_i v^i)/(\sum_{j=0}^{n} b_j v^j)$, with $a_i, b_j \in L$, we get

$$\Big(\sum_{j=0}^{n} b_j v^j\Big)\Big(\sum_{h=p}^{\infty} c_h v^h\Big) = \sum_{i=0}^{m} a_i v^i,$$

therefore $\sum_{j+h=i} b_j c_h = 0$ for all $i > m$. Let $\{e_\alpha\}_\alpha$ be a vector basis of L over F and write $b_j = \sum_\alpha b_{j\alpha} e_\alpha$, with $b_{j\alpha} \in F$. Thus for each α we have

$$\sum_{j+h=i} b_{j\alpha} c_h = 0, \quad \forall i > m.$$

This means that $(\sum_{j=0}^{n} b_{j\alpha} v^j)\xi \in F(v)$, $\forall \alpha$. Since not all $b_{j\alpha}$ are zero, this implies $\xi \in F(v)$.

This proves step 2, and thereby step 1.

Step 3. (Conclusion.) Let Y be an arbitrary connected closed subvariety of $\mathbb{P}^n$ of dimension ≥ 1. By Proposition 9.2 we know that $K(\mathbb{P}^n_{/Y})$ is a field. Let Y' be an irreducible curve contained in Y. Then we have natural maps

$$K(\mathbb{P}^n) \to K(\mathbb{P}^n_{/Y}) \to K(\mathbb{P}^n_{/Y'}).$$

The first map is $\alpha_{\mathbb{P}^n,Y}$, which is obviously injective. The second map is also injective because $K(\mathbb{P}^n_{/Y})$ is a field. Therefore it will be enough to prove that Y' is $G3$ in $\mathbb{P}^n$. In other words, without loss of generality we may assume that Y is an irreducible curve.

Let $Z = \mathbb{P}^1$ be a line in $\mathbb{P}^n$ and choose a linear subspace of dimension $n-2$ such that $L \cap Y = L \cap Z = \varnothing$. Set $V := \mathbb{P}^n \setminus L$ and denote by $\pi\colon V \to \mathbb{P}^1$ the linear projection of center L. Let Y_1 be the normalization of Y and let $g\colon Y_1 \to \mathbb{P}^1$ be the composite of $\pi|Y\colon Y \to \mathbb{P}^1$ with the normalization morphism. Clearly, g is a finite surjective morphism. Consider the fiber product

$$\begin{array}{ccc} W := V \times_{\mathbb{P}^1} Y_1 & \xrightarrow{\ f\ } & V \\ {\scriptstyle \pi'}\downarrow & & \downarrow{\scriptstyle \pi} \\ Y_1 & \xrightarrow[\ g\ ]{} & \mathbb{P}^1 \end{array}$$

Since π is the projection of the vector bundle $\mathbb{V}((n-1)\mathcal{O}_{\mathbb{P}^1}(-1))$ and the embedding $Z \hookrightarrow V$ is its zero section (see [66] II), it follows that $\pi' : W \to Y_1$ is the projection of the vector bundle $\mathbb{V}((n-1)g^*(\mathcal{O}_{\mathbb{P}^1}(-1)))$ and its zero section is $Z' := f^{-1}(Z)$. Note that f is a finite surjective morphism since g is so. Consider the "diagonal embedding" $Y_1 \hookrightarrow W$ induced by the composite of the inclusion $Y \subset V$ with $Y_1 \to Y$ and by id_{Y_1}. Then its image, Y' is another section of π'. Therefore there exists an automorphism $\sigma\colon W \to W$ of W (considered as an affine bundle over Y_1) such that $\sigma(Y') = Z'$. Then σ extends to a formal isomorphism $\hat{\sigma} : W_{/Y'} \to W_{/Z'}$, inducing an isomorphism of formal-rational functions

$\hat{\sigma} : K(W_{/Z'}) \to K(W_{/Y'})$ (and the subfield $K(W)$ is mapped onto itself by this isomorphism).

Now, by step 1, Z is $G3$ in V, and therefore, by Corollary 9.10, Z' is also $G3$ in W. By what we just said before, this implies that Y' is also $G3$ in W, i.e. $K(W) \cong K(W_{/Y'})$. Then there is a natural morphism of formal schemes $W_{/Y'} \to V_{/Y}$ (induced by f), hence an inclusion of fields

$$K(V_{/Y}) = K(\mathbb{P}^n{}_{/Y}) \subseteq K(W_{/Y'}) = K(W).$$

Since the field extension $f^* : K(V) \to K(W)$ is finite, this implies in particular that the field extension $\alpha_{\mathbb{P}^n,Y} : K(\mathbb{P}^n) \to K(\mathbb{P}^n{}_{/Y})$ is also finite, i.e. Y is $G2$ in $\mathbb{P}^n$.

To deduce that Y is $G3$ in X we proceed as follows. From what we just said the branch locus B of the field extension $K(\mathbb{P}^n{}_{/Y})|K(\mathbb{P}^n)$ is contained in the branch locus of $K(W)|K(V)$, which is just $\pi^{-1}(B')$, with B' the branch locus of $g\colon Y_1 \to \mathbb{P}^1$. Now, this argument is independent of the choice of the center of projection L and the line Z in $\mathbb{P}^n$. As these vary, the branch locus B of $K(\mathbb{P}^n{}_{/Y})|K(\mathbb{P}^n)$ is always contained in the subset $\pi^{-1}(B')$. Thus B must have codimension ≥ 2, so that by the purity of the branch locus (see e.g. [1], VI, Theorem (6.8), or also [68], exposé X, Theorem 3.4), $B = \varnothing$. But $\mathbb{P}^n$ is simply connected, so $K(\mathbb{P}^n{}_{/Y}) = K(\mathbb{P}^n)$. □

Corollary 9.15 (Hartshorne). *Let Y be an arbitrary closed connected subvariety of $\mathbb{P}^n$ $(n \geq 2)$ of dimension ≥ 1, and set $U := \mathbb{P}^n \setminus Y$. Then $H^{n-1}(U, F) = 0$ for every coherent sheaf F on U, i.e. $\mathrm{cd}(U) \leq n-2$, where $\mathrm{cd}(U)$ is the cohomological dimension of U (see Definition 7.1 above).*

The proof of Corollary 9.15 makes essential use of Theorem 9.14 together with an extra (non-trivial) argument, see [75], or also [74]. In fact, this corollary was an important discovery of Hartshorne in 1968, see [75]. Later we shall prove a more general result, see Theorem 11.21 below.

Now we give a (more general form of an) example of Hartshorne [75] of a closed subvariety Y of a projective variety X which is $G2$ but not $G3$.

Example 9.1. (Hartshorne if $n = 3$, [19] in general.) Start with the n-dimensional complex projective space $X' := \mathbb{P}^n$ $(n \geq 3)$ and consider the group G (of order $n+1$) generated by the automorphism $\sigma : \mathbb{P}^n \to \mathbb{P}^n$ given by

$$\sigma([x_0, x_1, \ldots, x_n]) = [x_0, \zeta x_1, \zeta^2 x_2, \ldots, \zeta^n x_n],$$

where $\zeta \in \mathbb{C}$ is a (fixed) primitive root of unity of order $n+1$. Let U' be the open subset of $X' := \mathbb{P}^n$ in which G acts freely. Consider the line L of X' given by the equations:

$$x_0 = x_1, \;\; x_2 = x_3, \;\; x_4 = \cdots = x_n = 0.$$

It is easily checked that the $n+1$ lines ξL (for each $\xi \in G$) are mutually disjoint and $\xi L \subset U'$, for all $\xi \in G$. In particular, the ramification locus $X' \setminus U'$ of the action of G on X' is of codimension ≥ 2 in X'. Let $f\colon X' \to Z := X'/G$ be the canonical morphism onto the quotient, and set $U := f(U')$. Thus $U' = f^{-1}(U)$ and the restriction $f' := f|U' : U' \to U$ is an étale morphism of degree $n+1$. In particular, U is smooth, and in fact U coincides precisely with the smooth locus of the normal projective variety Z. If we set $Y := f(L)$, it follows that $Y \subset U$ and the restriction $f|L\colon L = \mathbb{P}^1 \to Y$ is an isomorphism. Since f' is étale, by a general result (see e.g. [58], Theorem 4.2, or also Lemma 9.19 below) the morphism f yields an isomorphism $X'_{/L} \cong Z_{/Y}$ between formal completions. In particular, the map $\tilde{f}^* : K(Z_{/Y}) \to K(X'_{/L})$ is an isomorphism. Moreover, since every open subset V of U containing Y is by construction not simply connected (since $\operatorname{codim}_{\mathbb{P}^n}(\mathbb{P}^n \setminus f^{-1}(V) \geq 2)$, the universal covering space of V is $f^{-1}(V) \to V$), the embeddings $L \hookrightarrow X'$ and $Y \hookrightarrow Z$ cannot be Zariski equivalent.

Now look at the canonical commutative diagram

$$\begin{array}{ccc} K(Z) & \xrightarrow{\alpha_{Z,Y}} & K(Z_{/Y}) \\ \Big\downarrow f^* & & \Big\downarrow \tilde{f}^* \\ K(X') & \xrightarrow[\alpha_{X',L}]{} & K(X'_{/L}) \end{array}$$

in which the map $\alpha_{X',L}$ is an isomorphism by Theorem 9.14, and the map $\tilde{f}^*$ is an isomorphism by what we have said above. In particular, $K(Z_{/Y})$ is a field. Since the field extension $f^* : K(X) \to K(X')$ has degree $n+1$, it follows that Y is $G2$ (but not $G3$) in X.

Note also that in this example the normal bundle $N_{Y|Z}$ of Y in Z is isomorphic to $(n-1)\mathcal{O}_{\mathbb{P}^1}(1)$ (direct sum of $n-1$ copies of $\mathcal{O}_{\mathbb{P}^1}(1)$).

On the other hand, $\mathcal{O}_{\mathbb{P}^n}(1)$ is a G-sheaf in the sense of [114], page 69 (since G acts linearly on $\mathbb{P}^n$), and since f' is étale, we can apply Proposition 2 of [114] (page 70) to deduce that there exists an $\mathcal{L} \in \operatorname{Pic}(U)$ such that $f'^*(\mathcal{L}) \cong \mathcal{O}_{U'}(1) := \mathcal{O}_{\mathbb{P}^n}(1)|U'$. In particular, restriction map $\operatorname{Pic}(U) \to \operatorname{Pic}(Y)$ is surjective. Let $\varphi\colon X \to Z$ be a desingularization of Z such that $\varphi^{-1}(U) \to U$ is an isomorphism (recall that $U = \operatorname{Reg}(Z)$). Then the embedding $Y \hookrightarrow Z$ lifts to an embedding $Y \hookrightarrow X$. Since X is smooth, the restriction map $\operatorname{Pic}(X) \to \operatorname{Pic}(\varphi^{-1}(U)) \cong \operatorname{Pic}(U)$ is surjective, whence the restriction map $\operatorname{Pic}(X) \to \operatorname{Pic}(Y)$ is also surjective. Recalling that $Y \cong \mathbb{P}^1$ and $N_{Y|X} \cong (n-1)\mathcal{O}_{\mathbb{P}^1}(1)$, this shows that Y is a quasi-line on X, in the terminology of [19] (see also Definition 14.1 of Chapter 14 below). $\square$

The next result is a basic tool in studying formal functions. It is a refinement of a result of Hartshorne–Gieseker (see [58], Theorem 4.3), and also an algebraic analogue of an analytic result of Chow [35].

Theorem 9.16 ([18]). *Let Y be a closed connected subvariety of a normal irreducible variety X, and let $\zeta \in K(X_{/Y})$ be a formal-rational function which is algebraic* *over $K(X)$. Then there exists a finite surjective morphism $f: X' \to X$, of degree equal to the degree of the minimal polynomial of ζ over $K(X)$, from a normal irreducible projective variety X', with the following properties:*

(i) *There exists a closed embedding $i' : Y \hookrightarrow X'$ such that $f \circ i' = i$, where $i: Y \hookrightarrow X$ is the inclusion of Y in X.*

(ii) *The morphism f is étale in an open neighbourhood U' of $i'(Y)$ in X'.*

(iii) *$\tilde{f}^*(\zeta) \in K(X')$ (more precisely, $\tilde{f}^*(\zeta) \in \alpha_{X',i'(Y)}(K(X'))$, see the commutative diagram of Corollary 9.8).*

Remark 9.17. Let $x \in X$ be a normal point of an irreducible variety X. Denote by $\hat{\mathcal{O}}_{X,x}$ the completion of the local ring $\mathcal{O}_{X,x}$ with respect to its maximal ideal $m_{X,x}$. Then Theorem 9.16 asserts in this case that a formal-rational function $\zeta \in K(X_{/\{x\}}) = [\hat{\mathcal{O}}_{X,x}]_0$ is algebraic over $K(X)$ if and only if there exists a finite morphism $f: X' \to X$ and a point $x' \in X$ such that $f(x') = x$, at which f is étale (i.e. an étale neighbourhood (X', x') of (X, x)) such that $\hat{f}^*(\zeta) \in K(X')$. On the other hand, let $\mathcal{O}^h_{X,x}$ be the henselization of the local ring $\mathcal{O}_{X,x}$. By the definition of $\mathcal{O}^h_{X,x}$ we have $\mathcal{O}^h_{X,x} = \operatorname{dir\,lim}_{(X',x')} \mathcal{O}_{X',x'}$, where (X', x') runs over the set of all étale neighbourhoods of (X, x). Therefore Theorem 9.16 implies the following well-known facts:

i) $\mathcal{O}^h_{X,x} = \{\zeta \in \hat{\mathcal{O}}_{X,x} \mid \zeta \text{ algebraic over } K(X)\}$, and

ii) $[\mathcal{O}^h_{X,x}]_0 = \{\zeta \in K(X'_{/\{x'\}}) \mid \zeta \text{ algebraic over } K(X)\}$.

For the proof of Theorem 9.16 we need two lemmae, taken from [58].

Lemma 9.18 ([58]). *Suppose A and B are commutative Noetherian rings and I and J are ideals of A and B such that A and B are complete with respect to the I-adic and J-adic topologies. Let $f: A \to B$ be a homomorphism of unitary rings such that the map* $\operatorname{Spec}(f) : \operatorname{Spec}(B) \to \operatorname{Spec}(A)$ *is étale at every closed point of* $\operatorname{Spec}(B/J)$, *$f(I) \subseteq J$ (in particular, f is continuous), and the induced map $\overline{f}: A/I \to B/J$ is an isomorphism. Then f is an isomorphism.*

Proof. We first show that f is injective. Let $a \neq 0$ in A, and consider a maximal ideal m of A which contains the ideal $\{c \in A \mid ca = 0\}$. Since all maximal ideals of A contain I, and $\operatorname{Spec}(f)$ is étale at every point of $\operatorname{Spec}(B/J)$ the map $\hat{A}_m \to \hat{B}_m$ between the completions with respect to maximal ideals is an isomorphism. It follows that the map $A_m \to B_m$ is injective (because the maps $A_m \to \hat{A}_m$ and $B_m \to \hat{B}_m$ are injective). Now, a does not go to zero in A_m (by the choice of m), whence $f(a)$ is not zero in B_m. Hence $f(a) \neq 0$, fo f is injective.

We have $f(I)B \subseteq J$. We first show that $f(I)B = J$. Suppose not. Let L be the B-module $J/f(I)B$. Then for every maximal ideal m of A (which automatically contains I) the map $\hat{I}_m \to \hat{J}_m$ is an isomorphism since $A/I \cong B/J$ and $\hat{A}_m \cong \hat{B}_m$. On the other hand, there is an exact sequence

$$\hat{I}_m \to \hat{J}_m \to \hat{L}_m \to 0.$$

Hence $\hat{L}_m = 0$, which implies $L_m = 0$ for every maximal ideal m of A, i.e. $J = f(I)B$. From this we get $J^n = f(I^n)B$, $\forall n \geq 1$, therefore the map $I^n/I^{n+1} \to J^n/J^{n+1}$ is surjective. Now, let $b \in B$. Suppose we have constructed $a_1, \dots, a_n \in A$ so that $a_p - a_{p+1} \in I^p$ for every $p \leq n-1$ and $f(a_p) - b \in J^p$ for every $p \leq n$. Then we can find $c \in I^n$ so that $f(c) \equiv f(a_n) - b \mod J^{n+1}$. Then set $a_{n+1} = a_n - c$. Therefore we have constructed by induction a sequence $\{a_n\}_{n\geq 1}$ of elements of A such that $a_n - a_{n+1} \in I^n$ and $f(a_n) - b \in J^n$. Then the sequence $\{a_n\}_{n\geq 1}$ converges to an element $a \in A$ and $f(a) = b$ (since A is complete in the I-adic topology, and f is continuous). □

Lemma 9.19 ([58]). *Let $f\colon X' \to X$ be a finite morphism of algebraic varieties such that there is a closed subvariety Y of X, and the inclusion $Y \subset X$ lifts to an embedding $Y \hookrightarrow X'$ (in the sense of condition* (i) *of Theorem* 9.16*). Then f is étale in a Zariski open neighbourhood of Y in X' if and only if the morphism of formal schemes $\tilde{f} : X'_{/Y} \to X_{/Y}$ is an isomorphism.*

Proof. Let $y \in Y$. If $X'_{/Y} \to X_{/Y}$ is an isomorphism then the map $\mathcal{O}_{X,y} \to \mathcal{O}_{X',y}$ becomes an isomorphism after we take completions with respect to the ideals of Y in X and in X' respectively. Hence the map $\hat{\mathcal{O}}_{X,y} \to \hat{\mathcal{O}}_{X',y}$ between the completions with respect to the maximal ideals is also an isomorphism. Since the map $\mathcal{O}_{X,y} \to \mathcal{O}_{X',y}$ is étale if and only if the map $\hat{\mathcal{O}}_{X,y} \to \hat{\mathcal{O}}_{X',y}$ is an isomorphism, we are done. Conversely, assume that f is étale in a neighbourhood of Y in X'; then it is sufficient to apply Lemma 9.18 to see that the map $\tilde{f}\colon X'_{/Y} \to X_{/Y}$ is an isomorphism. □

Now we are ready to prove Theorem 9.16.

Proof of Theorem 9.16. We shall adapt the proof of Theorem 4.3 of [58] to our slightly different situation. Since X is normal and Y connected, $K(X_{/Y})$ is a field by Proposition 9.2. Since ζ is algebraic over $K(X)$, we get the finite field extension $K(X) \subseteq K(X)(\zeta)$. Then take as $f\colon X' \to X$ the normalization of X in $K(X)(\zeta)$. By construction, $\tilde{f}^*(\zeta) \in K(X')$. It remains to show that the conditions (i) and (ii) are fulfilled. Choose an open affine covering $\{\mathrm{Spec}(A_i)\}_i$ of X such that $Y \cap \mathrm{Spec}(A_i)$ is connected for every i. For any fixed index set $A = A_i$, and let $\hat{A}$ be the completion of A with respect to the ideal $I = I_i$ of $Y \cap \mathrm{Spec}(A)$ in $\mathrm{Spec}(A)$. By Lemma 9.1 and Proposition 9.2, $K(\mathrm{Spf}(\hat{A})) = [\hat{A}]_0$ is a field. It follows that $\hat{A}$ is an integral domain. Let $L = [\hat{A}]_0$ be the quotient field of $\hat{A}$. We claim that $\hat{A}$ is normal. Indeed, it is sufficient to check normality at every point of $x \in \mathrm{Spf}(\hat{A})$. But x is also a point of $A/I = \hat{A}/\hat{I}$, and the completions of $(\hat{A})_x$ and of A_x at their maximal ideals are isomorphic. Since the completion of A_x at its maximal ideal is integrally closed (by the analytic normality of normal rings, see [154], Vol. II, pages 313–320, or [116], 37.5), we infer that $(\hat{A})_x$ is also integrally closed. So, the claim that $\hat{A}$ is normal is proved. Set $K := [A]_0 = K(X)$ and $K' := K(X') = K(X)(\zeta)$. Let A' be the integral closure of A in K'. Since $\hat{A}$ is normal and $K' \subseteq L$, $A' \subseteq \hat{A}$. So we have the following canonical inclusions:

$$\begin{array}{ccccc} A & \longrightarrow & A' & \longrightarrow & \hat{A} \\ \downarrow & & \downarrow & & \downarrow \\ K & \longrightarrow & K' & \longrightarrow & L \end{array}$$

Let $\hat{I} := I\hat{A}$ and $I' := A' \cap \hat{I}$. With this definition, clearly $I' \cap A = I$, whence we get inclusions

$$A/I \hookrightarrow A'/I' \hookrightarrow \hat{A}/\hat{I} = A/I,$$

whose composite is the identity. Thus the first inclusion is an isomorphism $A/I \cong A'/I'$. Moreover, $I'\hat{A} = I\hat{A} = \hat{I}$. Set $\dim(A) = n$. Then the completion $\hat{A}'$ of A' with respect to I' is also of dimension n, and the inclusion $A' \hookrightarrow \hat{A}$ extends to a ring homomorphism $\varphi : \hat{A}' \to \hat{A}$. Therefore we get maps

$$\hat{A} \longrightarrow \hat{A}' \xrightarrow{\ \varphi\ } \hat{A},$$

whose composite is the identity of $\hat{A}$. It follows that φ is a surjection of integral domains of the same dimension, whence φ is an isomorphism. Now, reinstalling the indices i, we see that the ideals I_i fit together to form a coherent sheaf of ideals I, and so, a lifting of Y to X'. Thus we verified (i). Clearly the morphism $\hat{f} : X'_{/Y} \to X_{/Y}$ is an isomorphism. By Lemma 9.19, f is étale in a neighbourhood of Y in X'. □

As a consequence one immediately gets the following result of Hartshorne–Gieseker (see [58], Theorem 4.3).

Corollary 9.20 (Hartshorne–Gieseker). *Let Y be a closed subvariety of a normal projective variety X which is G2 in X. Then there exists a finite surjective morphism $f\colon X' \to X$ of degree $[K(X_{/Y}) : K(X)]$ from a normal projective variety X' such that the inclusion $Y \subset X$ lifts to an inclusion $i\colon Y \hookrightarrow X'$, f is étale in a neighbourhood of $i(Y)$, and $i(Y)$ is G3 in X'.*

Now we are ready to prove the following useful algebraicity criterion.

Theorem 9.21 ([18]). *Let $\zeta \in K(X_{/Y})$ be a formal-rational function of an irreducible variety X along a closed subvariety Y of X such that $K(X_{/Y})$ is a field (e.g. if X is normal and Y is connected, see Proposition 9.2). Then the following conditions are equivalent:*

(i) *ζ is algebraic over $K(X)$.*

(ii) *There is a proper surjective morphism $f\colon X' \to X$ from an irreducible variety X' and a closed subvariety Y' of X' such that $f(Y') \subseteq Y$ and $\tilde{f}^*(\zeta)$ is algebraic over $K(X')$.*

(iii) *There is a proper surjective morphism* $f\colon X' \to X$ *from an irreducible variety* X' *and a closed subvariety* Y' *of* X' *such that* $f(Y') \subseteq Y$ *and* $\tilde{f}^*(\zeta) \in K(X')$ *(more precisely, there exists a rational function* $t \in K(X')$ *such that* $\tilde{f}^*(\zeta) = \alpha_{X',Y'}(t)$*).*

If moreover X *is normal and* Y *is connected, the above conditions are also equivalent to the following one:*

(iv) *There is a finite surjective morphism* $f\colon X' \to X$*, with* X' *normal and irreducible and* $\deg(f) = \deg_{K(X)}(\zeta)$*, and a closed subvariety* Y' *of* X' *such that:* $f(Y') \subseteq Y$*,* $f|Y' : Y' \to Y$ *is an isomorphism,* f *is étale at every point of* Y'*, and* $\tilde{f}^*(\zeta) \in K(X')$*.*

Proof. Let us first prove that (iii)$\Longrightarrow$(i). By hypothesis, $\tilde{f}^*(\zeta) = \alpha_{X',Y'}(t)$ for some rational function $t \in K(X')$. Assume that ζ is transcendental over $K(X)$. Since $K(X_{/Y})$ is a field, the map $\tilde{f}^*$ is injective. It follows that $\tilde{f}^*(\zeta)$ is transcendental over $\alpha_{X',Y'}(f^*(K(X)))$, whence t is transcendental over $K(X)$ (more precisely, over $f^*(K(X))$). Thus we get the purely transcendental field subextension $K(X) \subset E := K(X)(t)$ of $K(X')$. Set $V := X \times \mathbb{P}^1$. Then V is identified with a variety whose field of rational functions is E such that the map $g^* : K(X) \to K(V) = E$ induced by the projection $g\colon V = X \times \mathbb{P}^1 \to X$ is just the subextension $K(X) \subset E$. Let $v : \tilde{X} \to X$ be the (birational) normalization morphism. By Corollary 9.10 the hypothesis that $K(X_{/Y})$ is a field is equivalent to the fact that $v^{-1}(Y)$ is connected. It follows that $v^{-1}(Y) \times \mathbb{P}^1$ is also connected. Since $v \times \mathrm{id}_{\mathbb{P}^1} : \tilde{X} \times \mathbb{P}^1 \to V = X \times \mathbb{P}^1$ is the normalization of V and $(v \times \mathrm{id}_{\mathbb{P}^1})^{-1}(g^{-1}(Y)) = v^{-1}(Y) \times \mathbb{P}^1$ is connected, Corollary 9.10 again implies that $K(V_{/g^{-1}(Y)})$ is a field.

Thus the field extension $f^* : K(X) \to K(X')$ factors as $K(X) \subset K(X)(t) = K(V) \subseteq K(X')$. It follows that there exists a unique dominant rational map $h\colon X' \dashrightarrow V$ such that $f = g \circ h$ (so that the extension $K(V) \subseteq K(X')$ coincides with $h^* : K(V) \to K(X')$). Let X'' be the normalization of the closure (in $X' \times V$) of the graph of h, let $h' : X'' \to V$ be the corresponding projection onto V (which is a proper surjective morphism), and let $u\colon X'' \to X'$ be the projection onto X'. Clearly, u is a birational proper morphism. Therefore, if we write $Y'' := u^{-1}(Y')$, then $h'(Y'') \subseteq g^{-1}(Y)$ and the induced maps $u^* : K(X') \to K(X'')$ and $\hat{u}^* : K(X'_{/Y'}) \to K(X''_{/Y''})$ are both isomorphisms. The fact that u^* is an isomorphism is obvious, while the fact that $\hat{u}^*$ is an isomorphism follows from Theorem 9.9. Thus, without loss of generality we may assume that the rational dominant map h is a proper surjective morphism. In other words, we may assume that f factors as $f = g \circ h$, with $g\colon V \to X$ and $h\colon X' \to V$ proper surjective morphisms such that $K(V) = K(X)(t)$, with $t \in K(X')$ transcendental over $K(X)$

(and $\hat{f}^*(\zeta) = \alpha_{X',Y'}(t)$). Thus we get the commutative diagram

$$\begin{array}{ccccc} K(X) & \xrightarrow{g^*} & K(V) & \xrightarrow{h^*} & K(X') \\ \Big\downarrow{\scriptstyle \alpha_{X,Y}} & & \Big\downarrow{\scriptstyle \alpha_{V,g^{-1}(Y)}} & & \Big\downarrow{\scriptstyle \alpha_{X',Y'}} \\ K(X_{/Y}) & \xrightarrow{\hat{g}^*} & K(V_{/g^{-1}(Y)}) & \xrightarrow{\tilde{h}^*} & K(X'_{/Y'}) \end{array}$$

At this point we can apply Theorem 9.11 to deduce that $K(V_{/g^{-1}(Y)})$ is canonically isomorphic to $[K(X_{/Y}) \otimes_{K(X)} K(V)]_0$. In particular, $\hat{g}^*(\zeta)$ is transcendental over $K(V)$ (because ζ is transcendental over $K(X)$). Since we saw above that $K(V_{/g^{-1}(Y)})$ is a field, the map $\tilde{h}^*$ is injective. It follows that $\tilde{h}^*(\hat{g}^*(\zeta)) = \tilde{f}^*(\zeta)$ is transcendental over $h^*(K(V))$. However, this is not possible, because $\tilde{f}^*(\zeta) = \alpha_{X',Y'}(t)$ and by the construction of V, $t \in h^*(K(V))$. This proves that the assumption that ζ is transcendental over $K(X)$ is absurd, whence (iii)$\Longrightarrow$(i).

Obviously, (iv)$\Longrightarrow$(iii) and (iii)$\Longrightarrow$(ii). If X is normal the implication (i)$\Longrightarrow$(iv) follows from Theorem 9.16.

Moreover, since we just proved that (iii)$\Longrightarrow$(i), it follows that (i)$\Longleftrightarrow$(iv) if X is normal. Therefore the conditions (i), (iii) and (iv) are all equivalent (if X is normal), and (iii)$\Longrightarrow$(ii). However, if X is not normal and (i) holds, let $u : \tilde{X} \to X$ be the normalization morphism. By Theorem 9.9, the birational morphism u yields the $K(X_{/Y}) \cong K(\tilde{X}_{/u^{-1}(Y)})$. By the implication (i)$\Longrightarrow$(iv) (applied to $\zeta \in K(\tilde{X}_{/u^{-1}(Y)})$), there is a finite surjective morphism $g \colon X' \to \tilde{X}$ and a closed subvariety $Y' \subset X'$ such that $g(Y') \subseteq u^{-1}(Y)$ and $\tilde{g}^*(\zeta) \in K(X')$. Then taking $f := u \circ g$, we get a finite surjective morphism $f \colon X' \to X$ such that $f(Y') \subseteq Y$ and $\tilde{f}^*(\zeta) \in K(X')$. In other words, we showed that (i)$\Longrightarrow$(iii) for an arbitrary irreducible (not necessarity normal) variety X. Therefore (i)$\Longleftrightarrow$(iii) and (iii)$\Longrightarrow$(ii).

It remains to prove that (ii)$\Longrightarrow$(iii). By the birational invariance again (Theorem 9.9) we may replace X' by its normalization X'', Y' by $v^{-1}(Y')$ and f by $f \circ v \colon X'' \to X$ (where $v \colon X'' \to X'$ is the normalization morphism), so that condition (ii) still holds for $f \circ u$. In other words, without loss of generality we may assume X' normal. Then the fact that $\tilde{f}^*(\zeta)$ is algebraic over $K(X')$ and Theorem 9.16 yield the existence of a finite surjective morphism $u \colon Z \to X'$ and a closed subset W of Z such that $u(W) = Y'$ and $\tilde{u}^*(\tilde{f}^*(\zeta)) \in K(Z)$. Then we get the proper surjective morphism $g := f \circ u \colon Z \to X$ such that $g(W) \subseteq Y$ and $\tilde{g}^*(\zeta) \in K(Z)$, whence (ii)$\Longrightarrow$(iii). □

Here is a consequence of Theorem 9.21 which shows that there is a deep relationship between Formal Geometry and connectivity in Projective Geometry:

Corollary 9.22 ([18]). *Let X be an irreducible variety, and let Y be a closed subvariety of X. The following conditions are equivalent:*

(i) *For every proper surjective morphism $f\colon X' \to X$ from an irreducible variety X', $f^{-1}(Y)$ is connected.*

(ii) *$K(X_{/Y})$ is a field and $K(X)$ is algebraically closed in $K(X_{/Y})$.*

(iii) *$K(X_{/Y})$ is a field and the algebraic closure of $K(X)$ in $K(X_{/Y})$ is purely inseparable over $K(X)$ (i.e. the field extension $K(X_{/Y})|K(X)$ is primary).*

Proof. Let $u : \tilde{X} \to X$ be the canonical morphism from the normalization of X in $K(X)$. Observe first that since u is birational and finite, by Theorem 9.9, $K(\tilde{X}) \cong K(X)$ and $K(\tilde{X}_{/u^{-1}(Y)}) \cong K(X_{/Y})$. Thus the conditions (ii) and (iii) are stable under passing to the normalization. On the other hand, if $f\colon X' \to X$ is an arbitrary proper surjective morphism, if $v\colon \tilde{X}' \to X'$ is the normalization morphism of X', there exists a commutative diagram, there exists a commutative diagram

$$\begin{array}{ccc} \tilde{X}' & \xrightarrow{\ v\ } & X' \\ {\scriptstyle \tilde{f}}\downarrow & & \downarrow{\scriptstyle f} \\ \tilde{X} & \xrightarrow[\ u\ ]{} & X \end{array}$$

where $\tilde{f} : \tilde{X}' \to X'$ is a proper surjective morphism. From this it clearly follows that the condition (i) holds for the pair (X, Y) if and only if it does for $(\tilde{X}, u^{-1}(Y))$. In other words, there is no loss of generality in assuming that X is normal.

Assume that (i) holds and we want to prove (ii). Since X is normal and Y is connected (just apply (i) to the identity morphism id_X), from Proposition 9.2 it follows that $K(X_{/Y})$ is a field. On the other hand, if there would exist an algebraic formal function $\zeta \in K(X_{/Y})$ over $K(X)$ which does not belong to $K(X)$, then by Theorem 9.21 and by normality of X, there is a finite morphism $f\colon X' \to X$ of degree > 1 (equal to the degree of the minimal polynomial of ζ over $K(X)$) and a closed subvariety Y' satisfying the condition (iv) of Theorem 9.21. Then we claim that Y' is a connected component of $f^{-1}(Y)$, and $Y' \neq f^{-1}(Y)$. Indeed, if we assume that $Y' = f^{-1}(Y)$ it would follow that $\deg(f) = 1$ because f is étale in an open neighbourhood U of Y' in X'. Thus $Y' \neq f^{-1}(Y)$. Write $f^{-1}(Y) = Y' \cup Y^*$, with $\varnothing \neq Y^* \not\subseteq Y'$. Since $Y' \subseteq U$, Y' is a section of $f|U$ and $f|U$ is étale, an elementary general fact (see [67], éxposé I, corollaire 5.3) implies that Y' is a connected component of $f^{-1}(Y)$, and since $\varnothing \neq Y^* \not\subseteq Y'$, it follows that $f^{-1}(Y)$ is not connected, contradicting (i). Thus (i)$\Longrightarrow$(ii).

Clearly (ii)$\Longrightarrow$(iii).

Now assume that (iii) holds true and let $f\colon X' \to X$ be an arbitrary proper surjective morphism. Using (iii) and a well-known result (see [66] IV, (4.3.1), or also [87], théorème 3.6), we infer that $\mathrm{Spec}(K(X_{/Y}) \otimes_{K(X)} K(X'))$ is irreducible, i.e. the ring $K(X_{/Y}) \otimes_{K(X)} K(X')$ has a unique minimal prime ideal p. Let S be the (multiplicative) set of all non-zero divisors of $K(X_{/Y}) \otimes_{K(X)} K(X')$. Then $S \cap p = \varnothing$, whence by Theorem 9.11 we have

$$K(X'_{/f^{-1}(Y)}) \cong [K(X_{/Y}) \otimes_{K(X)} K(X')]_0 = S^{-1}(K(X_{/Y}) \otimes_{K(X)} K(X')).$$

From this we infer that the ring $K(X'_{/f^{-1}(Y)})$ has also a unique minimal prime ideal (namely $pS^{-1}(K(X_{/Y}) \otimes_{K(X)} K(X'))$. Assume $f^{-1}(Y)$ not connected; then $K(X'_{/f^{-1}(Y)})$ is by Corollary 9.10 the product of at least two fields, whence $K(X'_{/f^{-1}(Y)})$ has more than one minimal prime ideal.

This proves that (iii)$\Longrightarrow$(i). □

In view of Theorem 9.11 one may ask the following question: let $f\colon X' \to X$ be a proper surjective morphism of irreducible algebraic varieties, and let $Y \subset X$ and $Y' \subset X'$ be closed subvarieties such that $f(Y') \subseteq Y$. Assume that $K(X_{/Y})$ and $K(X'_{/Y'})$ are both fields. Then a natural question is under which conditions the field extensions $K(X_{/Y})|K(X)$ and $K(X')|K(X)$ are linearly disjoint in $K(X'_{/Y'})$. Recall that, if $E|K$ and $F|K$ are two field subextensions of a field extension $\Omega|K$, $E|K$ and $F|K$ are said to be *linearly disjoint in* Ω if the canonical map $E \otimes_K F \to \Omega$, induced by the inclusions $E \subseteq \Omega$ and $F \subseteq \Omega$, is injective (see [31], Chap. V, §2, 3). In this sense we have the following simple positive answer:

Proposition 9.23. *Under the above hypotheses, assume that $K(X'_{/f^{-1}(Y)})$ is a field. Then the field extensions $K(X_{/Y})|K(X)$ and $K(X')|K(X)$ are linearly disjoint in $K(X'_{/Y'})$. In particular, $K(X_{/Y}) \cap K(X') = K(X)$. If moreover Y' is G3 in X' then Y is G3 in X.*

Proof. By Theorem 9.11 the field extensions $K(X_{/Y})$ and $K(X')$ of $K(X)$ are linearly disjoint in $K(X'_{/f^{-1}(Y)})$. On the other hand, from the proof of Corollary 9.8 it follows that the inclusions $K(X_{/Y}) \subseteq K(X'_{/Y'})$ and $K(X') \subseteq K(X'_{/Y'})$ are gotten from $K(X_{/Y}) \subseteq K(X'_{/f^{-1}(Y)})$ and $K(X') \subseteq K(X'_{/f^{-1}(Y)})$ by composing with the natural restriction map $K(X'_{/f^{-1}(Y)}) \to K(X'_{/Y'})$. Then the conclusion follows from the hypotheses, which imply that this restriction map is a field extension (in particular, it is injective).

For the last part, assume that Y' is G3 in X', i.e. the map $\alpha_{X',Y'} : K(X') \to K(X'_{/Y'})$ is an isomorphism. Since $K(X'_{/f^{-1}(Y)})$ is a field, the canonical map $K(X'_{/f^{-1}(Y)}) \to K(X'_{/Y'})$ (induced by the inclusion $Y' \subseteq f^{-1}(Y)$) is injective. Therefore the map

$$\alpha_{X',f^{-1}(Y)} : K(X') \to K(X'_{/f^{-1}(Y)})$$

is also an isomorphism, i.e. $f^{-1}(Y)$ is G3 in X'. Finally, by Theorem 9.11,

$$K(X'_{/f^{-1}(Y)}) \cong [K(X') \otimes_{K(Y)} K(X_{/Y})]_0.$$

This isomorphism and the fact that $\alpha_{X',f^{-1}(Y)}$ is an isomorphism implies that the map $\alpha_{X,Y} : K(X) \to K(X_{/Y})$ is also an isomorphism, i.e. Y is G3 in X. □

Remark 9.24. Let Y be a closed subvariety of a projective irreducible variety X, and assume that there exists a surjective morphism $f\colon X \to X'$, with $\dim(X') \geq 1$ such that $f(Y)$ is a point in X'. Then, assuming $K(X_{/Y})$ a field, the field extension $\alpha_{X,Y} : K(X) \to K(X_{/Y})$ is transcendental. In particular, Y cannot be $G2$ in X.

Indeed, since $y = f(Y)$ is a point of X', there exists a formal function $\xi \in \hat{\mathcal{O}}_{X',y} \subseteq K(X'_{/y})$ which is transcendental over $K(X')$ (in fact, there are plenty of such formal functions which are regular at y). Then by Theorem 9.21, $\tilde{f}^*(\xi)$ is transcendental over $K(X)$ (and in particular, $\tilde{f}^*(\xi) \in H^0(X_{/Y}, \mathcal{O}_{X_{/Y}})$ is a non-constant formal regular function along Y). Alternatively, we could have avoided using Theorem 9.21, by arguing as in [81], pages 64-65, Remark (2.10).

Another important result on formal-rational functions is the following theorem of Faltings:

Theorem 9.25 (Faltings [48]). *Let Y be a closed subvariety of a projective irreducible variety X. Assume that $X \subset \mathbb{P}^n$, $\dim(X) = d \geq 2$ and Y is a set-theoretic intersection of X with r hyperplanes of $\mathbb{P}^n$, with $r \leq d-1$. Then Y is $G3$ in X.*

Proof of Theorem 9.25 *in a special case.* Faltings' proof of Theorem 9.25 relies on a local result which requires rather delicate arguments. However, in the special case when $\dim(Y) = d - r$ (i.e. when Y is a set-theoretic complete intersection of X with r hyperplanes) Theorem 9.25 can be proved using Theorem 9.14 and Corollary 9.13 above, as follows (see [81], (4.3)). Since Y is the set-theoretic complete intersection of X with r hyperplanes $H_1, \ldots, H_r$, every irreducible component of Y is of dimension $d - r \geq 1$. Then $H := H_1 \cap \cdots \cap H_r$ is a linear subspace of $\mathbb{P}^n$ of dimesnion $n - r$. Pick a linear subspace L of $H = \mathbb{P}^{n-r}$ of dimension $n - d - 1$ such that $L \cap Y = \varnothing$ (this is possible because Y is of pure dimension $d - r$), and consider the projection $\pi : \mathbb{P}^n \setminus L \to \mathbb{P}^d$ of center L. Since $Y = X \cap H$ and $L \subseteq H$, $X \cap L = \varnothing$. Therefore $f := \pi|X\colon X \to \mathbb{P}^d$ is a finite surjective morphism such that $Y = f^{-1}(M)$, with M a linear subspace of P^d of dimension $d - r \geq 1$. By Theorem 9.14, M is $G3$ in $\mathbb{P}^d$. Then by Corollary 9.13, (i) we deduce that Y is $G3$ in X. □

Corollary 9.26. *Let Y be an effective ample Cartier divisor on a projective irreducible variety X of dimension ≥ 2. Then Y (or the support of Y) is $G3$ in X.*

Proof. By assumption the divisor mY is very ample on X for $m \gg 0$. This means that there exists a projective embedding $i\colon X \hookrightarrow \mathbb{P}^n$ and a hyperplane H of $\mathbb{P}^n$ such that $nY = X \cap H$, where the intersection is taken in the scheme-theoretic sense. This implies that $\mathrm{Supp}(Y)$ is the set-theoretic complete intersection of X with H. Then the above proof of Theorem 9.25 (in the case when Y is a complete intersection) applies to deduce that $\mathrm{Supp}(Y)$ is $G3$ in X. □

Corollary 9.26 can be improved in the following way:

Corollary 9.27. *Let Y be an effective Cartier divisor on a projective irreducible variety of dimension ≥ 2. Assume that the normal bundle $N_{Y|X} = \mathcal{O}_X(Y) \otimes \mathcal{O}_Y$*

is ample. Then Y is $G3$ in X. In particular, if Y is an irreducible curve which is an effective Cartier divisor on the projective irreducible surface X such that $(Y^2) > 0$, then Y is $G3$ in X.

Proof. By Theorem 3.6 there exist a birational projective morphism $f\colon X \to Z$ and a Zariski open neighbourhood U of Y in X with the property that $f|U\colon U \to f(U)$ is an isomorphism and $Y' := f(Y)$ is an ample Cartier divisor on Z. Then by Corollary 9.26, Y' is $G3$ in Z, and since $Y = f^{-1}(Y')$, from Corollary 9.13, (i) it follows that Y is also $G3$ in X. □

We close this chapter by stating another important result (due to Hartshorne [75], or also [74]) in connection with the extension problem of formal functions. It is a higher codimensional analogue of Corollary 9.27, where the situation is more delicate. Precisely we have the following:

Theorem 9.28 (Hartshorne [75]). *Let X be a complete variety of dimension ≥ 2, and let Y be a connected closed subvariety of X of dimension ≥ 1. Assume that $Y \subset \mathrm{Reg}(X)$ (i.e. Y is contained in the smooth locus of X), Y is a local complete intersection in X and the normal bundle $N_{Y|X}$ is ample in the sense of Hartshorne (see [74]). Then Y is $G2$ in X.*

We are not going to prove this result here. For the proof we refer the reader to [75], or also to [74] (where the proof is explained under an extra hypothesis, in which case it becomes more transparent).

Remark 9.29. Unlike the one-codimensional case (Corollary 9.27), under the hypotheses of Theorem 9.28 with $\mathrm{codim}_X(Y) \geq 2$, Y need not be $G3$ in X in general. For instance in Example 9.1, $N_{Y|X} \cong (n-1)\mathcal{O}_{\mathbb{P}^1}(1)$, whence $N_{Y|X}$ is ample, and Y is $G2$, but not $G3$ in X. However, Hartshorne conjectured in [74] (page 208, problem 4.8) that if Y is smooth, $N_{Y|X}$ is ample and $\dim(Y) \geq \frac{1}{2}\dim(X)$, then Y should be $G3$ in X. In Chapter 13 we shall prove this conjecture in the case when X is a homogeneous space (see Theorem 13.10 below).

Chapter 10

Lefschetz Theory and Meromorphic Functions

Grothendieck–Lefschetz Conditions

Definition 10.1 (Grothendieck [68]). Let Y be a closed subvariety of a projective variety X. We say that the pair (X, Y) satisfies the *Grothendieck-Lefschetz condition* $\mathrm{Lef}(X, Y)$ if for every open subset U of X containing Y and for every vector bundle E on U the natural map $H^0(U, E) \to H^0(X_{/Y}, \hat{E})$ is an isomorphism, where $\hat{E} = \pi^*(E)$, with $\pi \colon X_{/Y} \to U$ the canonical morphism. We also say that (X, Y) satisfies the *effective Grothendieck–Lefschetz condition* $\mathrm{Leff}(X, Y)$ if the Grothendieck–Lefschetz condition $\mathrm{Lef}(X, Y)$ holds and, moreover, for every formal vector bundle $\mathcal{E}$ on $X_{/Y}$ there exists an open subset U of X and a vector bundle E on U such that $\mathcal{E} \cong \hat{E}$.

Roughly speaking the Grothendieck–Lefschetz conditions express a certain kind of "positivity" of the closed subvariety Y in the ambient variety X. They are crucial in Grothendieck's algebraic approach to prove Lefschetz-type results in algebraic geometry (see [68]). For instance one of the main results proved by Grothendieck is the following (see [68], éxposé XII, cf. also [74], Theorem 1.5, page 172).

Theorem 10.2 (Grothendieck). *Let X be a smooth closed irreducible subvariety of $\mathbb{P}^n$. Let Y be a complete intersection of X with r hyperplanes of $\mathbb{P}^n$, and assume* $\dim(Y) \geq 2$. *Then the effective Grothendieck–Lefschetz condition* $\mathrm{Leff}(X, Y)$ *holds.*

This result is crucial in order to prove algebraically generalized Lefschetz theorems for the algebraic fundamental group, or for the Picard group (see loc. cit.). We shall not prove this result here because we shall practically not use it in this book (but we refer the reader to loc. cit. for the proof; cf. also [74] for a simplified proof). We also refer the reader to the recent monograph of Lazarsfeld [100] for the various aspects of positivity in algebraic geometry.

Theorem 10.3 (Faltings [49]). *Let Y be a closed irreducible subvariety of $\mathbb{P}^n$ of dimension $> \frac{n}{2}$. Then the effective Grothendieck–Lefschetz condition* $\mathrm{Leff}(\mathbb{P}^n, Y)$ *holds.*

Proof. By Theorem 9.14, Y is $G3$ in X. Clearly, Y intersects every hypersurface of $\mathbb{P}^n$. Therefore by Theorem 10.7 below, the Grothendieck–Lefschetz condition $\mathrm{Lef}(\mathbb{P}^n, Y)$ holds. Note that to get $\mathrm{Lef}(\mathbb{P}^n, Y)$ we used only the fact that Y is irreducible of positive dimension. To get $\mathrm{Leff}(\mathbb{P}^n, Y)$ when $\dim(Y) > \frac{n}{2}$ (which is the more difficult part of the proof) we refer the reader to [49], Corollary 5. □

We shall use Theorem 10.3 (via Proposition 10.10) only to prove Corollary 10.13 below, which asserts that for every closed irreducible subvariety Y of the complex projective space $\mathbb{P}^n$ of dimension $> \frac{n}{2}$, $\mathcal{O}_Y(1)$ is not divisible in $\mathrm{Pic}(Y)$.

We shall make use of the following:

Lemma 10.4. *Let $A \to B$ be a flat homomorphism of local Noetherian rings. Then $A = B \cap [A]_0$, where the intersection is taken in the total ring of fractions $[B]_0$.*

Proof. Clearly $A \subseteq B \cap [A]_0$. Conversely, let $b = \frac{u}{v} \in B$, with $u, v \in A$, and v a non-zero divisor of A. If v is invertible in A there is nothing to prove. If v belongs to the maximal ideal of A then $u = bv \in Bv \cap A = Av$, the latter equality coming from the fact that the map $A \to B$ is faithfully flat (as a flat homomorphism of local rings). Therefore $b \in A$, as desired. □

Corollary 10.5. *Let Y be a closed subvariety of an algebraic variety X. Then $\mathcal{O}_{X_{/Y},x} \cap K(X) = \mathcal{O}_{X,x}$, for every point $x \in Y$.*

Proof. Since the canonical morphism $X_{/Y} \to X$ is flat, the induced homomorphism of local rings $\mathcal{O}_{X,x} \to \mathcal{O}_{X_{/Y},x}$ is faithfully flat. Then Lemma 10.4 yields the conclusion. □

Corollary 10.6 (Gieseker [58]). *Let (X_i, Y_i), $i = 1, 2$, be two pairs each consisting of an irreducible variety X_i and a closed subvariety Y_i. Assume that Y_i is $G3$ in X_i, $i = 1, 2$, and there is an isomorphism of formal schemes $u : X_{1/Y_1} \to X_{2/Y_2}$. Then there are Zariski open subsets U_i of X_i containing Y_i, $i = 1, 2$, and an isomorphism $f \colon U_1 \to U_2$ such that $f(Y_1) = Y_2$ and $f_{/Y_1} = u$.*

Proof. The isomorphism u yields a k-isomorphism of fields $\hat{u}^* : K(X_{2/Y_2}) \cong K(X_{1/Y_1})$, whence a k-isomorphism of $K(X_2) \cong K(X_1)$ of fields of rational functions because Y_i is $G3$ in X_i, $i = 1, 2$. This yields a birational isomorphism $f \colon X_1 \to X_2$ such that $f^* = \hat{u}^*$. We will show that f is well defined at all points of Y_1. Indeed, if y_1 is an arbitrary point of Y_1, set $y_2 := u(y_1)$. Since u is an isomorphism of formal schemes, the induced map $\mathcal{O}_{X_{2/Y_2},y_2} \to \mathcal{O}_{X_{1/Y_1},y_1}$ is an isomorphism of local rings. Then by Corollary 10.5,

$$\mathcal{O}_{X_{i/Y_i},y_i} \cap K(X_i) = \mathcal{O}_{X_i,y_i}, \quad i = 1, 2.$$

Hence the birational isomorphism f yields an isomorphism of local rings $\mathcal{O}_{X_1,y_1} \cong \mathcal{O}_{X_2,y_2}$. So f is defined at all points of Y_1 and f^{-1} is defined at all points of Y_2. □

In the next chapter we shall need the following slight generalization of a result of Hartshorne–Speiser, see [74], Proposition 2.1, page 200.

Theorem 10.7. *Let X be a projective irreducible variety (of dimension ≥ 2) that satisfies locally the condition S_2 of Serre, and let Y be a closed subvariety of X. Assume that Y is G3 in X and that Y intersects every hypersurface of X. Then (X, Y) satisfies the Grothendieck–Lefschetz condition* $\mathrm{Lef}(X, Y)$. *In particular, for every vector bundle E on X, $H^0(X_{/Y}, \hat{E})$ is a finite dimensional vector space.*

Proof. Let L be a line bundle on an open subset U of X with $Y \subset U$. It is well known that there exists a Cartier divisor D on X such that $L \cong \mathcal{O}_X(D)$. This implies that L is isomorphic to a subsheaf of the constant sheaf $K(X)$. Then for every $L \in \mathrm{Pic}(U)$ we have a commutative diagram

$$\begin{array}{ccc} H^0(U, L) & \longrightarrow & K(X) \\ \Big\downarrow{\scriptstyle\beta} & & \Big\downarrow{\scriptstyle\alpha_{X,Y}} \\ H^0(X_{/Y}, \hat{L}) & \longrightarrow & K(X_{/Y}) \end{array}$$

in which the map $\alpha_{X,Y}$ is an isomorphism because Y is $G3$ in X. Notice that since $L \subset K(X) = \mathcal{M}_X$, the flatness of the morphism $X_{/Y} \to X$ implies $\hat{L} \subseteq \mathcal{M}_{X_{/Y}}$, whence $H^0(X_{/Y}, \hat{L}) \subseteq \Gamma(X_{/Y}, \mathcal{M}_{X_{/Y}}) = K(X_{/Y})$. In other words, the bottom horizontal map in the above diagram is injective. If follows that under our hypotheses the map β is also injective.

By Corollary 10.5, $\mathcal{O}_{X_{/Y},x} \cap K(X) = \mathcal{O}_{X,x}$ for every $x \in Y$. Recalling that in the above commutative diagram the right vertical map is an isomorphism, it follows that every section $s \in H^0(X_{/Y}, \hat{L})$ extends to a section $s' \in H^0(V, L)$, with V an open subset of U containing Y. Since Y meets every hypersurface of X, $\mathrm{codim}_U(U \setminus V) \geq 2$. On the other hand, since X satisfies S_2, $\dim(X) \geq 2$ and L is a line bundle on U, it follows that the restriction map $H^0(U, L) \to H^0(V, L)$ is an isomorphism (see [65]). In other words, we proved that the map β is an isomorphism for every line bundle L on U.

Now let E be an arbitrary vector bundle of rank $r \geq 2$ on U. Let $\mathcal{O}_X(1)$ be an ample line bundle on the projective variety X and set $\mathcal{O}_U(1) := \mathcal{O}_X(1)|U$. Since $\mathcal{O}_U(1)$ is ample on U, for every coherent sheaf F on U, $F \otimes \mathcal{O}_U(m)$ is generated by its global sections for $m \gg 0$. This implies that there is a surjection of the form $n\mathcal{O}_U \to F \otimes \mathcal{O}_U(m)$ $(n \geq 1)$, whence a surjection $n\mathcal{O}_U(-m) \to F$, where $n\mathcal{O}_U$ denotes the direct sum of n copies of $\mathcal{O}_U$. In other words we showed that every coherent sheaf F on U is a quotient of a direct sum of line bundles.

Applying this fact twice we get in particular that there is an exact sequence of the form

$$\bigoplus_{j=1}^{p} M_j^{-1} \to \bigoplus_{i=1}^{n} L_i^{-1} \to E^* \to 0,$$

with L_i and M_j line bundles on U, and E^* the dual of E. Dualizing we get an exact sequence

$$0 \to E \to \bigoplus_{i=1}^{n} L_i \to \bigoplus_{j=1}^{p} M_j.$$

Since the functor $F \to \hat{F}$ is exact ([66] I, (10.8.8)), we also get the exact sequence

$$0 \to \hat{E} \to \bigoplus_{i=1}^{n} \hat{L}_i \to \bigoplus_{j=1}^{p} \hat{M}_j.$$

Therefore we get the commutative diagram with exact rows

$$\begin{array}{ccccccc}
0 & \longrightarrow & H^0(U,E) & \longrightarrow & H^0(U,\bigoplus_{i=1}^{n} L_i) & \longrightarrow & H^0(U,\bigoplus_{j=1}^{p} M_j) \\
 & & \downarrow & & \downarrow & & \downarrow \\
0 & \longrightarrow & H^0(X_{/Y},\hat{E}) & \longrightarrow & H^0(X_{/Y},\bigoplus_{i=1}^{n} \hat{L}_i) & \longrightarrow & H^0(X_{/Y},\bigoplus_{j=1}^{p} \hat{M}_j)
\end{array}$$

Using the result just proved for line bundles, the last two vertical maps are isomorphisms, whence the first vertical map is also an isomorphism by the five lemma. The last statement also follows because $H^0(X_{/Y},\hat{E}) \cong H^0(X,E)$, and $H^0(X,E)$ is finite dimensional because X is projective. □

Corollary 10.8. *Let X be a closed irreducible subvariety of $\mathbb{P}^n$ of dimension ≥ 2 which satisfies locally the condition S_2 of Serre. Let Y be a closed subvariety of X which is the set-theoretic intersection of X with r hyperplanes of $\mathbb{P}^n$, where $r < \dim(X)$. Then the Grothendieck–Lefschetz condition* $\mathrm{Lef}(X,Y)$ *holds.*

Proof. By Theorem 9.25 of Faltings Y is $G3$ in X. Since $\dim(Y) \geq \dim(X) - r \geq 1$, Y meets every hypersurface of X. Therefore we can apply Theorem 10.7 to get the conclusion. □

Proposition 10.9. *Let Y be a closed subvariety of a smooth projective variety X such that the effective Grothendieck–Lefschetz condition* $\mathrm{Leff}(X,Y)$ *holds. Then Y is $G3$ in X.*

Proof. The proof is completely similar to (and in fact even easier than) the proof of Proposition 9.4. Indeed, since X is smooth, the formal scheme $X_{/Y}$ is regular, hence by a remark made before Lemma 9.1, for every non-zero formal-rational function $\xi \in K(X_{/Y})$ the sheaves P_ξ and Q_ξ are invertible. Then the arguments of the proof of Proposition 9.4 apply in our situation by using $\mathrm{Leff}(X,Y)$ instead of Theorem 9.3, and the fact that the sheaves P_ξ and Q_ξ are invertible. □

For further use let us first prove the following general statement:

Proposition 10.10. *Let Y be a closed subvariety of the projective irreducible variety X over an algebraically closed field k of characteristic zero. Then for every formal line bundle $\mathcal{L} \in \mathrm{Pic}(X_{/Y})$ such that $\mathcal{L}|Y \cong M^s$, with $s \geq 2$ an integer and $M \in \mathrm{Pic}(Y)$, there exists a formal line bundle $\mathcal{M} \in \mathrm{Pic}(X_{/Y})$ such that $\mathcal{L} \cong \mathcal{M}^s$ and $\mathcal{M}|Y \cong M$. The same statement holds if the characteristic of k is $p > 0$, provided that s is prime to p.*

Proof. For every $n \geq 0$ consider the infinitesimal neighbourhood $Y(n)$ of order n of Y in X. We have the inclusions of subschemes

$$Y(0) \subset Y(1) \subset Y(2) \subset \cdots \subset X.$$

Then giving a formal line bundle $\mathcal{L}$ on $X_{/Y}$ amounts to giving a sequence $\{L_n\}_{n\geq 0}$, with $L_n \in \mathrm{Pic}(Y(n))$ such that $L_{n+1}|Y(n) \cong L_n$ for every $n \geq 0$. The hypothesis says that $L_0 \cong M^s$ for some M in $\mathrm{Pic}(Y(0)) = \mathrm{Pic}(Y)$. We shall construct by induction a formal line bundle $\mathcal{M} = \{M_n\}_{n\geq 0}$ in $\mathrm{Pic}(X_{/Y})$ with the desired properties. Starting with $M_0 = M$, the induction step is the following:

Claim. Assume that for a fixed integer $n \geq 0$ there exists $M_n \in \mathrm{Pic}(Y(n))$ such that $L_n \cong M_n^s$. Then there exists $M_{n+1} \in \mathrm{Pic}(Y(n+1))$ such that $L_{n+1} \cong M_{n+1}^s$ and $M_{n+1}|Y(n) \cong M_n$.

Indeed, consider the exact sequence of cohomology

$$H^1(Y, \mathcal{I}^{n+1}/\mathcal{I}^{n+2}) \to \mathrm{Pic}(Y(n+1)) \to \mathrm{Pic}(Y(n)) \to H^2(Y, \mathcal{I}^{n+1}/\mathcal{I}^{n+2})$$

associated to the truncated exponential exact sequence

$$0 \to \mathcal{I}^{n+1}/\mathcal{I}^{n+2} \to \mathcal{O}^*_{Y(n+1)} \to \mathcal{O}^*_{Y(n)} \to 0\,,$$

where $\mathcal{I}$ is the sheaf of ideals of Y in X. To prove the claim, observe that in this cohomology sequence the extreme terms are vector spaces over k; in particular $H^2(Y, \mathcal{I}^{n+1}/\mathcal{I}^{n+2})$ has no torsion because $\mathrm{char}(k) = 0$. Then the class of M_n in

$$\mathrm{Pic}(Y(n))/\,\mathrm{Im}(\mathrm{Pic}(Y(n+1)) \to \mathrm{Pic}(Y(n))) \subseteq H^2(Y, \mathcal{I}^{n+1}/\mathcal{I}^{n+2})$$

is a torsion element of order dividing s. Since $H^2(Y, \mathcal{I}^{n+1}/\mathcal{I}^{n+2})$ has no torsion we infer that $M_n \in \mathrm{Im}(\mathrm{Pic}(Y(n+1)) \to \mathrm{Pic}(Y(n)))$, i.e. there exists $N \in \mathrm{Pic}(Y(n+1))$ such that $N|Y(n) \cong M_n$. Now

$$(L_{n+1} \otimes N^{-s})|Y(n) \cong L_n \otimes M_n^{-s} \cong \mathcal{O}_{Y(n)}.$$

Therefore $L_{n+1} \otimes N^{-s}$ is a line bundle on $Y(n+1)$ coming from the k-vector space $H^1(Y, \mathcal{I}^{n+1}/\mathcal{I}^{n+2})$. Since $\mathrm{char}(k) = 0$ every element of such a k-vector space is divisible by s, whence

$$L_{n+1} \otimes N^{-s} \cong P^s, \quad \text{with } P \in \mathrm{Pic}(Y(n+1)) \text{ such that } P|Y(n) \cong \mathcal{O}_{Y(n)}.$$

If we take $M_{n+1} = N \otimes P$ we get that $L_{n+1} \cong M^s_{n+1}$ and $M_{n+1}|Y(n) \cong M_n$, and the claim is proved.

The last assertion of the proposition comes from the above argument plus the observation that the k-vector space $H^2(Y, \mathcal{I}^{n+1}/\mathcal{I}^{n+2})$ over a field k of characteristic $p>0$ has no s-torsion and every element of the k-vector space $H^1(Y, \mathcal{I}^{n+1}/\mathcal{I}^{n+2})$ is divisible by s, for every $s > 0$ prime to p. □

Remarks 10.11. 1. For every abelian group E let us denote by $\mathrm{Tors}(E)$ the torsion subgroup of E. If in Proposition 10.10 we take $\mathcal{L} = \mathcal{O}_{X_{/Y}}$ we get the following consequence (assuming $\mathrm{char}(k) = 0$): the map $\mathrm{Tors}(\mathrm{Pic}(X_{/Y})) \to \mathrm{Tors}(\mathrm{Pic}(Y))$, induced by the natural restriction map $\mathrm{Pic}(X_{/Y}) \to \mathrm{Pic}(Y)$, is surjective.

2. For every pair of integers $r, s \geq 2$ consider the s-fold Veronese embedding $i : \mathbb{P}^r \hookrightarrow \mathbb{P}^{n(r,s)}$ over an algebraically closed field of characteristic zero, with $n(r,s) = \binom{r+s}{s} - 1$. Using Proposition 10.10 and Theorem 10.2 we can immediately deduce that $Y := i(\mathbb{P}^r)$ cannot be a set-theoretic complete intersection in $\mathbb{P}^{n(r,s)}$. Indeed, assume that there were homogeneous forms $f_1, \ldots, f_{n(r,s)-r} \in \mathbb{C}[T_0, T_1, \ldots, T_{n(r,s)}]$ defining Y in $\mathbb{P}^{n(r,s)}$ set-theoretically. If Y' is the subscheme of $\mathbb{P}^{n(r,s)}$ defined by the ideal generated by $f_1, \ldots, f_{n(r,s)-r}$, then Y' is a scheme-theoretic complete intersection of $\mathbb{P}^{n(r,s)}$ of dimension $r \geq 2$, and hence by Theorem 10.2, $\mathrm{Leff}(\mathbb{P}^{n(r,s)}, Y')$ holds. Since $Y'_{\mathrm{red}} = Y$, we infer that $X_{/Y} = X_{/Y'}$; in particular, $\mathrm{Leff}(\mathbb{P}^{n(r,s)}, Y)$ also holds. Set $\mathcal{L} := \widehat{\mathcal{O}_{\mathbb{P}^{n(r,s)}}(1)}$. Since $\mathcal{O}_{\mathbb{P}^{n(r,s)}}(1)|Y \cong \mathcal{O}_{\mathbb{P}^r}(s)$, $\mathcal{L}|Y \cong \mathcal{O}_{\mathbb{P}^r}(s)$, and in particular, $\mathcal{L}|Y$ is divisible by s in $\mathrm{Pic}(Y)$. By Proposition 10.10, $\mathcal{L}$ is also divisible by s in $\mathrm{Pic}(\mathbb{P}^{n(r,s)}_{/Y})$. Since $\mathrm{Leff}(\mathbb{P}^{n(r,s)}, Y)$ holds, $\mathbb{P}^{n(r,s)}$ is smooth and Y meets every hypersurface of $\mathbb{P}^{n(r,s)}$, we infer that the natural map $\mathrm{Pic}(\mathbb{P}^{n(r,s)}) \to \mathrm{Pic}(\mathbb{P}^{n(r,s)}_{/Y})$ is an isomorphism. It follows that $\mathcal{O}_{\mathbb{P}^{n(r,s)}}(1)$ is divisible by s in $\mathrm{Pic}(\mathbb{P}^{n(r,s)})$ (because $\mathcal{L}$ is divisible by s in $\mathrm{Pic}(\mathbb{P}^{n(r,s)}_{/Y})$), and since $s \geq 2$, this is a contradiction. Actually the arguments above prove that the effective Grothendieck–Lefschetz condition $\mathrm{Leff}(\mathbb{P}^{n(r,s)}, v_r(\mathbb{P}^r))$ does not hold if $r \geq 2$ and $s \geq 2$.

Notice on the other hand that the image of the n-fold Veronese embedding of $\mathbb{P}^1$ in $\mathbb{P}^n$ (the rational normal curve of degree n in $\mathbb{P}^n$) is known to be a set-theoretic complete intersection in $\mathbb{P}^n$ (this is completely elementary, see [145]).

Here is an important application of Faltings' Theorem 10.3 together with Proposition 10.10:

Corollary 10.12. *If Y is a closed irreducible subvariety of $\mathbb{P}^n$ $(n \geq 3)$ over an algebraically closed field of characteristic zero, such that* $\dim(Y) > \frac{n}{2}$, *then the class of $\mathcal{O}_Y(1)$ is not divisible in* $\mathrm{Pic}(Y)$.

Proof. Indeed, assume that $\mathcal{O}_Y(1) \cong M^s$ for some $M \in \mathrm{Pic}(Y)$ and $s \geq 1$. Then by Proposition 10.10, there exists a formal line bundle $\mathcal{M} \in \mathrm{Pic}(\mathbb{P}^n_{/Y})$ and an isomorphism $\widehat{\mathcal{O}_{\mathbb{P}^n}(1)} \cong \mathcal{M}^s$ and $\mathcal{M}|Y \cong M$, where $\widehat{\mathcal{O}_{\mathbb{P}^n}(1)}$ is the pull back of $\mathcal{O}_{\mathbb{P}^n}(1)$ on $\mathbb{P}^n_{/Y}$.

On the other hand, by Theorem 10.3, the effective Grothendieck–Lefschetz condition $\mathrm{Leff}(\mathbb{P}^n, Y)$ holds. Since Y meets every hypersurface of $\mathbb{P}^n$, this immediately implies that the restriction map $\mathrm{Pic}(\mathbb{P}^n) \to \mathrm{Pic}(\mathbb{P}^n_{/Y})$ is an isomorphism. In particular, $\mathcal{M}$ is of the form $\mathcal{M} \cong \widehat{\mathcal{O}_{\mathbb{P}^n}(t)}$ for some $t \in \mathbb{Z}$. Therefore $\widehat{\mathcal{O}_{\mathbb{P}^n}(1)} \cong \widehat{\mathcal{O}_{\mathbb{P}^n}(st)}$. Using the injectivity of the map $\mathrm{Pic}(\mathbb{P}^n) \to \mathrm{Pic}(\mathbb{P}^n_{/Y})$, we get $\mathcal{O}_{\mathbb{P}^n}(1) \cong \mathcal{O}_{\mathbb{P}^n}(st)$, whence $st = 1$, i.e. $s = t = 1$. □

Corollary 10.12 is particularly interesting in view of Theorem 8.7 above. Specifically, let Y be an arbitrary closed irreducible smooth subvariety of $\mathbb{P}^n$ of dimension $\geq \frac{n+2}{2}$. Then by Theorem 8.7, $\mathrm{Pic}(Y) \cong \mathbb{Z}$. From this fact and from Corollary 10.12 we can derive the following famous result of Barth–Larsen [24], or also [22]. The original proof of this theorem is topological. Subsequently Ogus gave in [117] an algebraic proof of it.

Corollary 10.13 (Barth–Larsen [24]). *For every closed irreducible smooth subvariety Y of $\mathbb{P}^n$ of dimension $\geq \frac{n+2}{2}$ over $\mathbb{C}$, $\mathrm{Pic}(Y) \cong \mathbb{Z}$, generated by the class of $\mathcal{O}_Y(1)$.*

Formal and Meromorphic Functions

Let X be a complex projective algebraic variety and let Y be a closed irreducible subvariety of X. For an arbitrary complex algebraic variety Z we shall denote by Z^{an} the corresponding analytic space. Then in the analytic category it makes sense to consider the formal completion $X^{\mathrm{an}}{}_{/Y^{\mathrm{an}}}$ (see e.g. [20]). By projectivity assumptions we can apply GAGA-type results (see [135]) to get a canonical isomorphism

$$K(X_{/Y}) \cong K(X^{\mathrm{an}}{}_{/Y^{\mathrm{an}}}),$$

where the second ring is the ring of formal-meromorphic functions in the analytic category (defined in a completely analogous way as in the algebraic setting, see Chapter 9). Let U be a connected complex neighbourhood of Y^{an} in X^{an}. Clearly $\mathcal{M}(U) \subseteq K(X^{\mathrm{an}}{}_{/Y^{\mathrm{an}}})$, and therefore we get natural inclusions

$$K(X) \subseteq \mathcal{M}(U) \subseteq K(X_{/Y}). \tag{10.1}$$

Using (10.1), Theorem 9.14 immediately yields the following analytic result of Barth:

Theorem 10.14 (Barth [21]). *Let Y be a connected closed subvariety of dimension ≥ 1 of the complex projective space $\mathbb{P}^n_{\mathbb{C}}$ $(n \geq 2)$. Then every meromorphic function defined in a connected complex neighbourhood U of Y in $\mathbb{P}^n_{\mathbb{C}}$ can be (uniquely) extended to a meromorphic function on $\mathbb{P}^n_{\mathbb{C}}$, and hence to a rational function of $\mathbb{P}^n_{\mathbb{C}}$.*

Remark 10.15. Barth proved this result independently of Hironaka and Matsumura, using analytic methods. In fact, Theorem 10.14 extends earlier analytic results due to Severi [137] and Remmert and Van de Ven [123]. In fact, as far as I know the first result concerning the extension problem of meromorphic functions on complex projective varieties is due to Severi [137], who proved Theorem 10.14 in 1932 in the case when Y is a smooth hypersurface in $\mathbb{P}^n$, using methods of complex variables à la Weierstrass.

The Gi conditions ($i = 2, 3$) introduced in Definition 9.12 have the following analytic counterparts:

Definition 10.16. Let Y be a connected closed subvariety of an irreducible complete algebraic variety X over $\mathbb{C}$.

1. Y is said to be *analytically* $G2$ if for any connected complex neighbourhood U of Y in X the restriction map $\mathcal{M}(X^{\mathrm{an}}) \to \mathcal{M}(U)$ defines a finite field extension.

2. Y is said to be *analytically* $G3$ if for any connected complex neighbourhood U of Y in X the restriction map $\mathcal{M}(X^{\mathrm{an}}) \to \mathcal{M}(U)$ is an isomorphism.

Of course, in this definition we have $\mathcal{M}(X^{\mathrm{an}}) = K(X)$ by GAGA results (see [135]). Clearly "analytic $G3$" implies "analytic $G2$". On the other hand by our remarks above "algebraic Gi" implies "analytic Gi" for $i = 2, 3$. In general the converse is not true. We illustrate this by the following example due to Serre.

Example 10.1. (Serre, see [74], page 231, Example 3.2) Let B be an elliptic curve over $\mathbb{C}$, and consider a non-trivial extension of vector bundles

$$0 \longrightarrow \mathcal{O}_B \longrightarrow E \xrightarrow{\ \varepsilon\ } \mathcal{O}_B \longrightarrow 0. \tag{10.2}$$

Indeed, the isomorphism classes of such extensions are classified by $H^1(B, \mathcal{O}_B)$, which is a one-dimensional vector space because B is an elliptic curve. Thus non-trivial extensions of type (10.2) exists. Consider the geometrically ruled surface $X := \mathbb{P}(E)$ associated to the rank 2 vector bundle E, and denote by $p\colon X \to B$ the canonical projection. Let Y be the section of p corresponding to the surjection ε. Since the extension (10.2) is non-trivial one can show (loc. cit.) that the following properties hold:

(i) $Y^2 = 0$ (and in particular, $X \setminus Y$ is not affine).

(ii) $(X \setminus Y)^{\mathrm{an}}$ is analytically isomorphic to $\mathbb{C}^* \times \mathbb{C}^*$, and therefore is a Stein manifold.

This example is interesting because one can show that Y is analytically $G3$ in X, although the field extension $K(X) \to K(X_{/Y})$ is huge.

To sketch the proof of this fact, consider a more general situation. Let Y be a closed irreducible subvariety of a smooth complex projective variety X of dimension ≥ 2 such that $(X \setminus Y)^{\mathrm{an}}$ is a Stein manifold. Then it is well known that there exists a strictly plurisubharmonic exhaustion function $\varphi : (X \setminus Y)^{\mathrm{an}} \to \mathbb{R}$, whence $\forall a \in \mathbb{R}$ the open subsets $\{\varphi < a\}$ are relatively compact in $(X \setminus Y)^{\mathrm{an}}$. The unions of Y^{an} and the complements of the adherence (in $(X \setminus Y)^{\mathrm{an}}$) of $\{\varphi < a\}$, $\forall a \in \mathbb{R}$, form a fundamental system of strictly pseudoconcave neighbourhoods of Y^{an} in X^{an}. Then one can apply well-known results of Andreotti [2] to deduce that for every such connected pseudoconcave neighbourhood U of Y^{an} in X^{an}, $\mathcal{M}(X) = K(X) \to \mathcal{M}(U)$ is a finite algebraic field extension, i.e. Y is analytically $G2$ in X.

In fact, as Mihnea Colţoiu kindly informed the authors of [18], one can say even more: every $\zeta \in \mathcal{M}(U)$ can be extended to a meromorphic function to the whole X^{an}. Indeed, on a Stein manifold Z of dimension ≥ 2 every meromorphic function defined in the complement of a compact set K such that $Z \setminus K$ is connected, can be extended to a unique meromorphic function on Z (see [88]).

Thus, coming back to Example 10.1 one deduces that the canonical map $\mathcal{M}(X^{\mathrm{an}}) = K(X) \to \mathcal{M}(U)$ is an isomorphism for every connected complex neighbourhood U of Y in X, i.e. Y is even analytically $G3$ in X.

Now we want to study the relationship between $\mathcal{M}(U)$ and $K(X_{/Y})$ more closely. Observe that the natural map $\alpha\colon K(X) \to K(X_{/Y})$ decomposes as $\alpha = \gamma \circ \beta$, where

$$\beta\colon K(X) = \mathcal{M}(X) \to \operatorname{dir\,lim}_U \mathcal{M}(U)$$

is the canonical map into the direct limit, and

$$\gamma : \operatorname{dir\,lim}_U \mathcal{M}(U) \to K(X_{/Y})$$

is induced by the inclusions $\mathcal{M}(U) \subseteq K(X_{/Y})$ (and where U runs over all connected complex neighbourhoods of Y in X). Clearly, the maps β and γ are both injective. With these observations we can state the following consequence of Theorem 9.16:

Proposition 10.17 ([18]). *Let Y be a closed connected subvariety of the normal projective variety X over $\mathbb{C}$. In the above notation one has*

$$\overline{K(X)} \subseteq \gamma(\operatorname{dir\,lim}_U \mathcal{M}(U)),$$

where $\overline{K(X)}$ denotes the algebraic closure of $K(X)$ in $K(X_{/Y})$.

Proof. Since X is normal and Y connected, $K(X_{/Y})$ is a field by Proposition 9.2. Let $\zeta \in K(X_{/Y})$ be an arbitrary formal-rational function which is algebraic over $K(X)$. By Theorem 9.16 there exists a finite surjective morphism $f\colon X' \to X$ and a closed subvariety Y' of X' with the following properties: $f(Y') \subseteq Y$, $f|Y' : Y' \to Y$ is an isomorphism, f is étale along Y', and $\tilde{f}^*(\zeta) \in K(X')$. In particular, for every connected open complex neighbourhood U' of Y' in X', $\tilde{f}^*(\zeta) \in \mathcal{M}(U')$. Since f is étale along Y' and $f|Y' : Y' \to Y$ is an isomorphism, for sufficiently small U' the restriction $f|U' : U' \to U := f(U')$ is an isomorphism. This implies that $\zeta \in \mathcal{M}(U)$. □

Corollary 10.18. *Under the hypotheses of Proposition* 10.17, *assume furthermore that Y is analytically G3 in X. Then $K(X)$ is algebraically closed in $K(X_{/Y})$. Moreover, for every proper surjective morphism $f\colon X' \to X$ from an irreducible variety X', $f^{-1}(Y)$ is connected.*

Proof. Indeed, since Y is analytically $G3$ in X, $K(X) = \mathcal{M}(U)$, $\forall U$, whence $K(X) = \operatorname{dir\,lim}_U \mathcal{M}(U)$. Then the first part of the corollary follows from Proposition 10.17. The last part follows from the first and from Corollary 9.22. □

Corollary 10.19. *Let X be the geometrically ruled surface of Example* 10.1 *associated to a non-trivial extension of vector bundles of type* (10.2), *and let Y be the section corresponding to the surjection ε. Then $K(X)$ is algebraically closed in $K(X_{/Y})$, $K(X) \neq K(X_{/Y})$, and for every proper surjective morphism $f\colon X' \to X$, $f^{-1}(Y)$ is connected.*

Proof. By what we have said above Y is analytically $G3$ in X. Then the result follows from Corollary 10.18, except for the fact that $K(X) \neq K(X_{/Y})$. In other words, we have prove that Y is not $G3$ in X. Observe first that Y meets every irreducible curve C of X. Indeed, if C is a closed irreducible curve of X such that $Y \cap C = \varnothing$, then $C \subset X \setminus Y$. But this is not possible because C is a projective curve and $(X \setminus Y)^{\mathrm{an}}$ is a Stein manifold.

Assume now that Y is $G3$ in X. Then since Y meets every curve of X, by Theorem 10.7 we get

$$\dim_{\mathbb{C}} H^0(X_{/Y}, \hat{E}) < \infty, \quad \text{for every vector bundle } E \text{ on } X. \tag{10.3}$$

On the other hand, since Y is the section of $p\colon X \to B$ corresponding to ε (in the exact sequence (10.2)), the normal bundle of Y in X is isomorphic to $\mathcal{O}_Y$.

Let $\mathcal{J} = \mathcal{O}_X(-Y)$ be the (invertible) ideal sheaf of Y in X. Thus for every $n \geq 0$ we get $\mathcal{J}^n/\mathcal{J}^{n+1} \cong \mathcal{O}_Y$, or else, the exact sequence

$$0 \to \mathcal{O}_Y \to \mathcal{O}_{Y(n+1)} \to \mathcal{O}_{Y(n)} \to 0,$$

where $Y(n) = (Y, \mathcal{O}_X/\mathcal{J}^{n+1})$ is the n-th infinitesimal neighbourhood of Y in X. If we tensor this exact sequence with a line bundle L on X such that $(L \cdot Y) = d > 0$ we get the exact sequence

$$0 \to L^{n+1} \otimes \mathcal{O}_Y \to L \otimes \mathcal{O}_{Y(n+1)} \to L \otimes \mathcal{O}_{Y(n)} \to 0,$$

which yields the cohomology exact sequence

$$0 \to H^0(Y, L^{n+1} \otimes \mathcal{O}_Y) \to H^0(Y(n+1), L \otimes \mathcal{O}_{Y(n+1)}) \to H^0(Y(n), L \otimes \mathcal{O}_{Y(n)}) \to$$
$$\to H^1(Y, L^{n+1} \otimes \mathcal{O}_Y).$$

Since Y is an elliptic curve and $\deg(L^{n+1} \otimes \mathcal{O}_Y) = d(n+1) > 0$, the last cohomology group is zero. Therefore the maps of the projective system

$$\{H^0(Y(n), L \otimes \mathcal{O}_{Y(n)})\}_{n \geq 0}$$

are all surjective. Moreover, by Riemann–Roch we have

$$\dim_{\mathbb{C}} H^0(Y, L^{n+1} \otimes \mathcal{O}_Y) = d(n+1),$$

and therefore the above cohomology exact sequence yields for every $n \geq 0$:

$$\dim_{\mathbb{C}} H^0(Y(n+1), L \otimes \mathcal{O}_{Y(n+1)}) = d(n+1) + \dim_{\mathbb{C}} H^0(Y(n), L \otimes \mathcal{O}_{Y(n)}). \quad (10.4)$$

Since by a general statement (see [66] III, **0**, (13.3.1))

$$H^0(X_{/Y}, \hat{L}) \cong \operatorname{inv\,lim}_n H^0(Y(n), L \otimes \mathcal{O}_{Y(n)}),$$

(10.4) yields the following

$$\dim_{\mathbb{C}} H^0(X_{/Y}, \hat{L}) = \dim_{\mathbb{C}}(\operatorname{inv\,lim}_n H^0(Y(n), L \otimes \mathcal{O}_{Y(n)})) = \infty. \quad (10.5)$$

Therefore, if we assume that Y is $G3$ in X, (10.5) contradicts (10.3). □

Remark 10.20. The arguments of the proof of the last part of Corollary 10.19, suitably refined, yield the following more general assertion. If Y is an irreducible curve on a smooth projective surface X such that $(Y^2) \leq 0$, then Y cannot be $G2$ in X. Therefore, in view of Corollary 9.27, an irreducible curve Y on a smooth projective surface X is $G2$ on X if and only if Y is $G3$ on X, if and only if $(Y^2) > 0$ (cf. Corollary 12.11 below for a more general result).

Chapter 11

Connectedness and Formal Functions

As is well known, Zariski's main reason for introducing and studying systematically the formal functions in algebraic geometry was to prove his famous connectedness theorem (see [151], or also [66] III, (4.3.1)). It basically asserts that for every proper morphism of Noetherian schemes $f\colon X \to Y$ such that the canonical map $\mathcal{O}_Y \to f_*(\mathcal{O}_X)$ is an isomorphism, the fiber $f^{-1}(y)$ is connected and non-empty for every $y \in Y$.

On the other hand, Corollary 9.22 above shows that there exists a close connection between formal functions and connectedness results. To illustrate this idea, let us first slightly reformulate the Fulton–Hansen connectedness theorem (Corollary 7.15 above). For every closed irreducible subvariety X of $\mathbb{P}^n(e) \times \mathbb{P}^n(e)$ of dimension $> n$, set $Y := X \cap \Delta$ (with Δ the diagonal of $\mathbb{P}^n(e) \times \mathbb{P}^n(e)$). Then for every proper surjective morphism $f\colon X' \to X$ from an irreducible variety X', $f^{-1}(Y)$ is connected. As in §1.7, $\mathbb{P}^n(e)$ is the weighted projective space of weights $e = (e_0, e_1, \dots, e_n)$. Thus using Corollary 9.22 and Corollary 7.15 we get:

Theorem 11.1. *Let X be a closed irreducible subvariety of $\mathbb{P}^n \times \mathbb{P}^n$ of dimension $> n$, and let Y be the intersection $X \cap \Delta$. Then $K(X_{/Y})$ is a field and $K(X)$ is algebraically closed in $K(X_{/Y})$ (via the map $\alpha_{X,Y} : K(X) \to K(X_{/Y})$).*

Our first objective in this chapter is to prove, following [11], that the map $\alpha_{X,Y} : K(X) \to K(X_{/Y})$ in Theorem 11.1 is actually an isomorphism. This is substantial strengthening of Theorem 7.14 and Corollary 7.15. Precisely we have the following:

Theorem 11.2 ([11]). *Let $f\colon X' \to \mathbb{P}^n(e) \times \mathbb{P}^n(e)$ be a proper morphism from an irreducible variety X' over an algebraically closed field of arbitrary characteristic, such that $\dim(f(X')) > n$, $f^{-1}(\Delta)$ is G3 in X'. Equivalently, for every closed irreducible subvariety X of $\mathbb{P}^n \times \mathbb{P}^n$ of dimension $> n$ (as in Theorem 11.1), $Y := X \cap \Delta$ is G3 in X.*

Proof. For simplicity we shall prove the theorem for ordinary projective spaces. We refer the reader to [11] for the general case, although the arguments are essentially the same. Let us briefly recall the notation of the proof of Theorem 7.14. In the projective space

$$P := \mathbb{P}^{2n+1} = \operatorname{Proj}(k[T_0, \dots, T_n; U_0, \dots, U_n])$$

consider the closed subschemes $L_1 = V_+(T_0, \dots, T_n)$ and $L_2 = V_+(U_0, \dots, U_n)$. Then $L_1 \cap L_2 = \varnothing$. Moreover, the closed subvariety $H = V_+(T_0 - U_0, \dots, T_n - U_n)$ of P is contained in $U := P \setminus (L_1 \cup L_2)$. Consider the morphism $g \colon U \to \mathbb{P}^n \times \mathbb{P}^n$ defined by

$$g([t_0, \dots, t_n; u_0, \dots, u_n]) = ([t_0, \dots, t_n], [u_0, \dots, u_n]),$$

which yields the isomorphism $H \cong \Delta$. Then we get the commutative diagram

$$\begin{array}{ccccc}
U_X := g^{-1}(X) & \xrightarrow{\varphi'} & U & \xrightarrow{\supseteq} & H \\
\Big\downarrow{\scriptstyle g'} & & \Big\downarrow{\scriptstyle g} & & \Big\downarrow{\scriptstyle \cong} \\
X & \xrightarrow{\varphi} & \mathbb{P}^n \times \mathbb{P}^n & \xrightarrow{\supseteq} & \Delta
\end{array}$$

with φ a closed immersion (the inclusion of X in $\mathbb{P}^n \times \mathbb{P}^n$). Set $Y := X \cap \Delta$, and denote by Z the closure of U_X in $\mathbb{P}^{2n+1}$.

Step 1. The ring $K(X_{/Y})$ is a field.

Indeed, if $u : \tilde{X} \to X$ is the normalization morphism, by Corollary 9.10 all we have to check is that $u^{-1}(Y)$ is connected. But this is a direct consequence of the Fulton–Hansen connectedness theorem (Corollary 7.15), because u is finite and surjective and $\dim(X) > n$. Step 1 is proved.

Now consider a proper morphism $h \colon X_1 \to Z$ with the following properties:

i) h is birational and the restriction $h|h^{-1}(U_X)$ defines an isomorphism between $h^{-1}(U_X)$ and U_X;

ii) the composite $f := g' \circ h \colon X_1 \to X$ is a proper surjective morphism.

The existence of (X_1, h) is obvious because one can take as X_1 the closure of the graph $\Gamma_{g'} \subset U_X \times X$ in $Z \times X$, and as h the restriction to X_1 of the projection $Z \times X \to Z$. Then the restriction to X_1 of the projection $Z \times X \to X$ is a proper surjective morphism $f = g' \circ h$. In particular, it makes sense to speak about the field extension $f^* : K(X) \to K(X_1)$.

Now we apply Theorem 9.11 to f to get the canonical isomorphism

$$K(X_{1/f^{-1}(Y)}) \cong [K(X_{/Y}) \otimes_{K(X)} K(X_1)]_0. \tag{11.1}$$

Then the crucial point in the proof of our theorem is the following:

Step 2. The canonical map $\alpha_{X_1,f^{-1}(Y)} : K(X_1) \to K(X_{1/f^{-1}(Y)})$ is an isomorphism.

Accepting step 2 for the time being we see that via the above isomorphism we get the fact that the map $K(X) \to K(X_{/Y})$ is an isomorphism, i.e. the conclusion of our theorem. Indeed, since X_1 is birationally isomorphic to Z, and hence to U_X, and since the morphism $g: U \to \mathbb{P}^n \times \mathbb{P}^n$ is a locally trivial $\mathbb{G}_m$-bundle (and hence $g' : U_X \to X$ is also a locally trivial $\mathbb{G}_m$-bundle), we infer that $K(X_1) = K(X)(t)$, with $t \in K(X_1)$ transcendental over $K(X)$. Then the isomorphism (11.1) becomes $K(X_{1/f^{-1}(Y)}) \cong K(X_{/Y})(t)$, with t transcendental over $K(X_{/Y})$. Therefore the map $\alpha_{X_1,f^{-1}(Y)}$ becomes, via these identifications,

$$\alpha_{X_1,f^{-1}(Y)} : K(X)(t) \to K(X_{/Y})(t),$$

with $\alpha_{X_1,f^{-1}(Y)}(t) = t$ and $\alpha_{X_1,f^{-1}(Y)}|K(X) = \alpha_{X,Y}$. Using this, the assertion that the map $\alpha_{X,Y} : K(X) \to K(X_{/Y})$ is an isomorphism follows from step 2. (Note that in the case of weighted projective spaces it is not true in general that $g': U_X \to X$ is a locally trivial $\mathbb{G}_m$-bundle; however, U_X is still birationally isomorphic to $X \times \mathbb{G}_m$ in this more general situation, which is enough to derive the conclusion of the theorem from step 2 in general.)

Now we proceed to the proof of step 2. Since $g|H$ defines an isomorphism $H \cong \Delta$, $g'|U_X \cap H$ defines an isomorphism $Z \cap H = U_X \cap H \cong X \cap \Delta = Y$. By the construction of h, the closed subvariety $W := h^{-1}(Z \cap H)$ is isomorphic via $f|W$ to $Z \cap H \cong Y$. Since $W \subset f^{-1}(Y)$ we get the canonical flat morphisms

$$X_{1/W} \longrightarrow X_{1/f^{-1}(Y)} \longrightarrow X_1,$$

which yield the homomorphisms of k-algebras

$$K(X_1) \xrightarrow{a} K(X_{1/f^{-1}(Y)}) \xrightarrow{b} K(X_{1/W}),$$

where $a = \alpha_{X_1,f^{-1}(Y)}$ and $b \circ a = \alpha_{X_1,W}$. Now step 2 follows from the following two steps:

Step 3. The composite $b \circ a$ is an isomorphism of k-algebras.

Step 4. $K(X_{1/f^{-1}(Y)})$ is a field.

Indeed, by step 4 the map b is injective, whence by step 3, b is an isomorphism. By step 3 again the map a is an isomorphism.

It remains therefore to prove steps 3 and 4.

To prove step 3 observe that since $H \subset U$ and $h|h^{-1}(U_X) : h^{-1}(U_X) \to U_X = Z \cap U$ is an isomorphism, we get

$$K(X_1) \cong K(Z) \text{ and } K(X_{1/W}) \cong K(Z_{/Z\cap H}).$$

So, we are reduced to prove that the canonical map $\alpha_{Z,Z\cap H} : K(Z) \to K(Z_{/Z\cap H})$ is an isomorphism, i.e. $Z \cap H$ is $G3$ in Z. Since H is given by $n+1$ homogeneous equations $T_i - U_i = 0$, $i = 0, 1, \ldots, n$, and since $\dim(Z) = \dim(U_X) = \dim(X)+1 > n+1$, this assertion follows from Theorem 9.25 of Faltings.

Finally, it remains to prove step 4. Let $v : \tilde{X}_1 \to X_1$ be the normalization morphism of X_1, and set $\tilde{f} := f \circ v : \tilde{X}_1 \to X \subset \mathbb{P}^n \times \mathbb{P}^n$. By Corollary 9.10 all we have to check is that $v^{-1}(f^{-1}(Y)) = \tilde{f}^{-1}(\Delta)$ is connected. But this follows again from the Fulton–Hansen connectedness theorem (Corollary 7.15) because $\dim(\tilde{f}(\tilde{X}_1)) = \dim(X) > n$.

Thus step 4, and thereby Theorem 11.2, is proved. □

Corollary 11.3 ([11]). *Let Y and Z be two closed irreducible subvarieties of $\mathbb{P}^n(e)$ such that $\dim(Y) + \dim(Z) > n$. Then $Y \cap Z \cong (Y \times Z) \cap \Delta$ is G3 in $Y \times Z$, where Δ is the diagonal of $\mathbb{P}^n(e) \times \mathbb{P}^n(e)$. In particular, if Y is a closed irreducible subvariety of $\mathbb{P}^n(e)$ $(n \geq 3)$ of dimension $> \frac{n}{2}$, then the diagonal Δ_Y is G3 in $Y \times Y$.*

Proof. Take $X = Y \times Z$ and apply Theorem 11.2 to the inclusion $X = Y \times Z \hookrightarrow \mathbb{P}^n(e) \times \mathbb{P}^n(e)$. □

Corollary 11.4 (Faltings [48]). *Let Y and Z be two closed irreducible subvarieties of $\mathbb{P}^n(e)$ such that $\dim(Y) + \dim(Z) > n$. Then $Y \cap Z$ is G3 in both Y and Z. In particular, Y is G3 in $\mathbb{P}^n(e)$, for every irreducible positive-dimensional closed subvariety Y of $\mathbb{P}^n(e)$.*

Proof. If $p \colon Y \times Z \to Y$ is the first projection, we clearly have $p((Y \times Z) \cap \Delta) = Y \cap Z$. Since $\dim(Y) + \dim(Z) > n$, by the Fulton–Hansen connectedness theorem (Corollary 7.15, (ii)) we infer that for every proper surjective morphism $f \colon V \to Y$ from an irreducible variety V, $f^{-1}(Y \cap Z)$ is connected. Therefore by Corollary 9.10, $K(Y_{/Y\cap Z})$ is a field. Moreover, Corollary 7.15, (ii) also implies that for every proper surjective morphism $g \colon U \to Y \times Z$ (from an irreducible variety U), $g^{-1}(p^{-1}(Y \cap Z))$ is connected, whence $K((Y \times Z)_{/p^{-1}(Y\cap Z)})$ is also a field, by Corollary 9.10 again.

Now, by Corollary 11.3, $(Y \times Z) \cap \Delta$ is $G3$ in $Y \times Z$. Then by Proposition 9.23, applied to $p \colon Y \times Z \to Y$, we infer that $Y \cap Z$ is $G3$ in Y. Similarly one proves that $Y \cap Z$ is $G3$ in Z. □

Remark 11.5. Let V be a closed subvariety of a projective irreducible variety Y. Assume that $Y \subset \mathbb{P}^n$. Then Corollary 11.4 produces a necessary condition (in terms of formal functions) for V to be the intersection of Y with another irreducible variety Z of $\mathbb{P}^n$ such that $\dim(Y) + \dim(Z) > n$. From the point of view of the classical problem of contact of surfaces along curves (see [56]) we get in particular a necessary condition for a curve C lying on an irreducible surface Y of $\mathbb{P}^3$ to be the intersection with another irreducible surface Z of $\mathbb{P}^3$ (cf. also the remark following Theorem 8.4).

Let now Y be a smooth projective subvariety of $\mathbb{P}^n$ of dimension $> \frac{n}{2}$, and assume that G is a finite abelian group of order d acting freely on Y, with $d \geq 2$

and d prime to char(k) if char$(k) > 0$. Denote by X the quotient Y/G. Since the action of G on Y is free and Y is smooth, X is also smooth.

Proposition 11.6 ([11]). *Under the above hypotheses the diagonal Δ_X is G2 (but not G3) in $X \times X$. Moreover, $[K(X \times X_{/\Delta_X}) : K(X \times X)] = d$.*

Proof. The fact that $K(X \times X_{/\Delta_X})$ is a field follows from Proposition 9.2, because $X \times X$ is smooth and Δ_X is connected. Denote by $u\colon Y \to X$ the canonical étale morphism. Then

$$(u \times u)^{-1}(\Delta_X) = (G \times G)\Delta_Y,$$

where $G \times G$ acts on $Y \times Y$ in the canonical way. A simple computation shows that if $g, g', h, h' \in G$ then

$$(g, gh)\Delta_Y \cap (g', g'h')\Delta_Y \neq \varnothing \iff (g, gh)\Delta_Y = (g', g'h')\Delta_Y.$$

This shows that

$$(u \times u)^{-1}(\Delta_X) = \Delta_1 \cup \cdots \cup \Delta_d, \tag{11.2}$$

where $\Delta_i := (e, h_i)\Delta_Y$, $\forall i = 1, \dots, d$, where $h_1 = e, h_2, \dots, h_d$ is a full system of representatives for the quotient group $(G \times G)/\Delta_G$, and e is the unit of G. It follows that $\Delta_i \cap \Delta_j = \varnothing$ for $i \neq j$, and for every i and j there is an element $(e, h_{ij}) \in G \times G$ such that $(e, h_{ij})\Delta_i = \Delta_j$. From (11.2) it follows

$$K(Y \times Y_{/(u\times u)^{-1}(\Delta_X)}) \cong \prod_i K(Y \times Y_{/\Delta_i}) \cong \prod_{d \text{ times}} K(Y \times Y_{/\Delta_Y}), \tag{11.3}$$

where the last product has d factors. By Corollary 11.3,

$$K(Y \times Y_{/\Delta_Y}) \cong K(Y \times Y),$$

whence (11.3) becomes

$$K(Y \times Y_{/(u\times u)^{-1}(\Delta_X)}) \cong \prod_{d \text{ times}} K(Y \times Y). \tag{11.4}$$

On the other hand, by Theorem 9.11 there is a canonical isomorphism

$$K(Y \times Y_{/(u\times u)^{-1}(\Delta_X)}) \cong [K(X \times X_{/\Delta_X}) \otimes_{K(X\times X)} K(Y \times Y)]_0. \tag{11.5}$$

From (11.4) and (11.5) we get

$$[K(X \times X_{/\Delta_X}) \otimes_{K(X\times X)} K(Y \times Y)]_0 \cong \prod_{d \text{ times}} K(Y \times Y),$$

or else,

$$K(X \times X_{/\Delta_X}) \otimes_{K(X\times X)} K(Y \times Y) \cong \prod_{d \text{ times}} K(Y \times Y),$$

because $K(X \times X_{/\Delta_X}) \otimes_{K(X\times X)} K(Y \times Y)$ coincides with its total ring of fractions (the extension $K(X \times X) \subset K(Y \times Y)$ is finite and $K(X \times X_{/\Delta_X})$ is a field). This proves our proposition. □

Example 11.1. If in Proposition 11.6 we take Y to be the Fermat surface in $\mathbb{P}^3$ of equation $x_0^5 + x_1^5 + x_2^5 + x_3^5 = 0$ with the action of the multiplicative group G of roots of order 5 of 1 ($\mathrm{char}(k) \neq 5$) on $\mathbb{P}^3$ given by

$$g \cdot [t_0, t_1, t_2, t_3] = [t_0, gt_1, g^2t_2, g^3t_3], \ \ \forall g \in G, \ \ \forall [t_0, t_1, t_2, t_3] \in \mathbb{P}^3,$$

this action yields a free action of G on Y (cf. the example preceding Corollary 7.17). Then the Godeaux surface $X := Y/G$ has the property that $K(X \times X_{/\Delta_X})$ is a field which is an extension of degree 5 of $K(X \times X)$. In particular, Δ_X is $G2$ (but not $G3$) in $X \times X$.

Now we want to deduce some further consequences of Corollary 11.3.

Theorem 11.7 ([11]). *Let Y be a closed irreducible subvariety of $\mathbb{P}^n(e)$ of dimension $> \frac{n}{2}$ ($n \geq 3$). Then the Grothendieck–Lefschetz condition* $\mathrm{Lef}(Y \times Y, \Delta_Y)$ *holds, and in particular, for every vector bundle E on $Y \times Y$ the natural map $H^0(Y \times Y, E) \to H^0(Y \times Y_{/\Delta_Y}, \hat{E})$ is an isomorphism.*

Proof. By Corollary 11.3, Δ_Y is $G3$ in $X := Y \times Y$. Let V be an arbitrary open subset of X containing Δ_Y. We claim that $\mathrm{depth}(\mathcal{O}_{X,x}) \geq 2$ for every (not necessarily closed) point $x \in X \setminus V$.

To see this, let p_1 and p_2 be the canonical projections of $X = Y \times Y$, and set $y_i = p_i(x)$, $i = 1, 2$. Observe first that neither y_1 nor y_2 is the generic point ξ of Y. Indeed assuming for instance $y_1 = \xi$, the closure $\overline{\{x\}}$ would intersect Δ_Y because $\overline{\{x\}} \supseteq p_2^{-1}(y_2) = Y \times \{y_2\}$, and $Y \times \{y_2\}$ does intersect Δ_Y. This is impossible because on the other hand $\overline{\{x\}} \subseteq X \setminus V$.

Thus the local rings $A_i := \mathcal{O}_{Y,y_i}$ ($i = 1, 2$) are domains of dimension ≥ 1. Therefore there are non-zero elements t_i in the maximal ideal of A_i, $i = 1, 2$. Observe that it is sufficient to prove that $t_1 \otimes 1$ and $1 \otimes t_2$ form an $A_1 \otimes_k A_2$-sequence, because $\mathcal{O}_{X,x}$ being a localization of $A_1 \otimes_k A_2$, this will imply $\mathrm{depth}(\mathcal{O}_{X,x}) \geq 2$. To check this, consider the exact sequence

$$0 \longrightarrow A_1 \xrightarrow{\ t_1\ } A_1 \longrightarrow A_1/tA_1 \longrightarrow 0$$

(where the map t_1 is the multiplication by t_1), which yields the exact sequence

$$0 \longrightarrow A_1 \otimes_k A_2 \xrightarrow{\ t_1 \otimes \mathrm{id}\ } A_1 \otimes_k A_2 \longrightarrow (A_1/tA_1) \otimes_k A_2 \longrightarrow 0.$$

Notice that the map $t_1 \otimes \mathrm{id}$ is just the multiplication by $t_1 \otimes 1$. Similarly the exact sequence

$$0 \longrightarrow A_2 \xrightarrow{\ t_2\ } A_2 \longrightarrow A_2/t_2A_2 \longrightarrow 0$$

yields by tensoring with A_1/t_1A_1 the exact sequence

$$0 \to (A_1/t_1A_1) \otimes_k A_2 \to (A_1/t_1A_2) \otimes_k A_2 \to (A_1/t_1A_1) \otimes_k (A_2/t_2A_2) \to 0,$$

whence the claim.

Then the conclusion of our theorem follows from the proof of Theorem 10.7 as soon as we check that Δ_Y meets every hypersurface of X. Assume therefore that D is a hypersurface of X such that $D \cap \Delta_Y = \varnothing$. In the notation of the proof of Theorem 11.2, $g'^{-1}(D)$ is a hypersurface of U_X which does not intersect $U_X \cap H$ (because $U_X \cap H \subseteq g'^{-1}(\Delta_Y)$). Taking the closure Z of U_X in $\mathbb{P}^{2n+1}(e,e)$ and the closure D' of $g'^{-1}(D)$ in Z, we get a hypersurface D' of Z which does not intersect $Z \cap H$. Recalling that Z is a closed irreducible subvariety of $\mathbb{P}^{2n+1}(e,e)$ of dimension $\geq n+2$ and $H = V_+(T_0 - U_0, \dots, T_n - U_n)$, this fact is impossible. □

Remarks 11.8. (i) If Y is locally S_2 then $Y \times Y$ has the same property, and therefore we can apply directly Theorem 10.7 to prove Theorem 11.7. In fact, Theorem 11.7 was stated in [11] under the extra hypothesis that Y is locally S_2.

(ii) Theorem 11.7 was proved for the first time (with completely different methods) by Speiser [142] in the case when Y is locally Cohen–Macaulay over a field of positive characteristic.

(iii) For some recent interesting results on the topology of small codimensional closed subvarieties in $\mathbb{P}^n$ we refer the reader to [83] and [104].

Theorem 10.7 asserts basically that the Grothendieck–Lefschetz condition $\mathrm{Lef}(X,Y)$ holds as soon as Y is $G3$ in X. On the other hand, by Proposition 10.9, the effective Grothendieck–Lefschetz condition $\mathrm{Leff}(X,Y)$ implies the $G3$ condition, provided X is smooth. Therefore one may ask whether under the hypotheses of Theorem 11.7 the effective Lefschetz condition $\mathrm{Leff}(Y \times Y, \Delta_Y)$ also holds. Proposition 10.10 above implies that the answer is in general negative. Indeed, we have the following:

Proposition 11.9. *If Y is a smooth closed subvariety of $\mathbb{P}^n$ $(n \geq 3)$ of dimension $> \frac{n}{2}$ over $\mathbb{C}$, the effective Grothendieck–Lefschetz condition* $\mathrm{Leff}(Y \times Y, \Delta_Y)$ *does not hold.*

Proof. Suppose that $\mathrm{Leff}(Y \times Y, \Delta_Y)$ holds. By the last part of the proof of Theorem 11.7, Δ_Y meets every hypersurface of $X = Y \times Y$, because $\dim(Y) > \frac{n}{2}$. Moreover, $Y \times Y$ is smooth because Y is so. Under these circumstances $\mathrm{Leff}(Y \times Y, \Delta_Y)$ immediately implies that the natural map $\mathrm{Pic}(X) \to \mathrm{Pic}(X_{/\Delta_Y})$ is an isomorphism.

Then Proposition 10.10 implies the following: if $L \in \mathrm{Pic}(Y \times Y)$ is a line bundle such that $L|\Delta_Y$ is divisible by a positive integer $s \geq 2$ in $\mathrm{Pic}(\Delta_Y) \cong \mathrm{Pic}(Y)$, then L is also divisible by s in $\mathrm{Pic}(Y \times Y)$. On the other hand, since $\dim(Y) > \frac{n}{2}$, by Remark 8.8, $H^1(Y, \mathcal{O}_Y) = 0$, whence by [73], exercise 12.6, page 292, the map $(L, M) \to p_1^*(L) \otimes p_2^*(M)$ defines an isomorphism $\mathrm{Pic}(Y) \times \mathrm{Pic}(Y) \cong \mathrm{Pic}(Y \times Y)$, where p_1 and p_2 are the canonical projections of $Y \times Y$. Via this identification the map $\mathrm{Pic}(Y \times Y) \to \mathrm{Pic}(\Delta_Y) \cong \mathrm{Pic}(Y)$ becomes $\mathrm{Pic}(Y) \times \mathrm{Pic}(Y) \to \mathrm{Pic}(Y)$ given by $(L_1, L_2) \to L_1 \otimes L_2$. Since by Remark 8.8 $\mathrm{Pic}(Y)$ is a free abelian group of finite rank, there exists a non-divisible line bundle $L \in \mathrm{Pic}(Y)$. Then (L, L) is non-divisible in $\mathrm{Pic}(Y) \times \mathrm{Pic}(Y) \cong \mathrm{Pic}(X) \cong \mathrm{Pic}(X_{/\Delta_Y})$, although its restriction L^2 to $\mathrm{Pic}(\Delta_Y) \cong \mathrm{Pic}(Y)$ is divisible by 2. This contradicts Proposition 10.10. □

As another application of Theorem 11.7 (via an idea of Speiser [142]), we shall derive a result on triviality of stratified vector bundles, generalizing some results of Ogus [117] (in characteristic zero) and Speiser [142] (in char. $p > 0$).

To give another application of Theorem 11.2 we need to recall some definitions and basic facts from Grothendieck's theory of stratified vector bundles and descent theory of faithfully flat morphisms of schemes (see [71], [67], or [142]).

Definition 11.10. Let Y be an algebraic variety over k, and consider the canonical projections

$$Y \xleftarrow{\ p_1\ } Y \times Y \xrightarrow{\ p_2\ } Y$$

and

$$\begin{array}{ccccc} Y \times Y & \xleftarrow{p_{21}} & Y \times Y \times Y & \xrightarrow{p_{31}} & Y \times Y \\ & & \Big\downarrow{\scriptstyle p_{32}} & & \\ & & Y \times Y & & \end{array}$$

Denote by $\Delta = \Delta_Y$ the diagonal of $Y \times Y$ and by Δ' the diagonal of $Y \times Y \times Y$. For every $r \geq 0$ denote by Δ_r (resp. by Δ'_r) the infinitesimal neighbourhood of order r of Δ in $Y \times Y$ (resp. of Δ' in $Y \times Y \times Y$). Then from the previous diagrams we get projections

$$Y \xleftarrow{\ q_1^r\ } \Delta_r \xrightarrow{\ q_2^r\ } Y$$

and

$$\begin{array}{ccccc} \Delta_r & \xleftarrow{q_{21}^r} & \Delta'_r & \xrightarrow{q_{31}^r} & \Delta_r \\ & & \Big\downarrow{\scriptstyle q_{32}^r} & & \\ & & \Delta_r & & \end{array}$$

Let F be a vector bundle on Y. By *descent data* on F we mean an isomorphism $\varphi\colon p_1^*(F) \to p_2^*(F)$ over $Y \times Y$ such that the cocycle condition

$$p_{31}^*(\varphi) = p_{32}^*(\varphi) \circ p_{21}^*(\varphi) \tag{11.6}$$

holds over $Y \times Y \times Y$. From Grothendieck's faithfully flat descent theory (see [67], VIII) applied to the structural morphism $p\colon Y \to \operatorname{Spec}(k)$ we know that there exist descent data on F if and only if F is the pull-back via p of a vector bundle over $\operatorname{Spec}(k)$, i.e. if and only if F is a trivial vector bundle. This fact is a very useful general triviality criterion for a vector bundle.

A *stratification* on F is a compatible system of isomorphisms

$$\varphi_r : (q_1^r)^*(F) \to (q_2^r)^*(F)$$

over Δ_r, $\forall r \geq 0$, such that $\varphi_0 = \mathrm{id}$ and the cocycle condition

$$(q_{31}^r)^*(\varphi_r) = (q_{32}^r)^*(\varphi_r) \circ (q_{21}^r)^*(\varphi_r)$$

holds on Δ_r', $\forall r \geq 0$.

A *stratified vector bundle* on Y is a vector bundle F on Y endowed with a stratification. In other words, giving a stratification on F amounts to giving "formal descent data" on F. That is to say, if $\mathcal{X}$ is the formal completion on $Y \times Y$ along Δ, and $\mathcal{X}'$ — the formal completion of $Y \times Y \times Y$ along Δ', we get projections

$$Y \xleftarrow{\ q_1\ } \mathcal{X} \xrightarrow{\ q_2\ } Y$$

and

$$\begin{array}{ccccc} \mathcal{X} & \xleftarrow{\ q_{21}\ } & \mathcal{X}' & \xrightarrow{\ q_{31}\ } & \mathcal{X} \\ & & \Big\downarrow{\scriptstyle q_{32}} & & \\ & & \mathcal{X} & & \end{array}$$

Similarly one defines "formal descent data" on F, which consist of an isomorphism $\psi\colon q_1^*(F) \to q_2^*(F)$ together with the cocycle condition

$$q_{31}^*(\psi) = q_{32}^*(\psi) \circ q_{21}^*(\psi) \tag{11.7}$$

on $\mathcal{X}'$. It is known that in characteristic zero, giving a stratification on F is equivalent to giving an integrable connection on F (see [71], or [117]).

Now, following Speiser [142], we prove:

Proposition 11.11 ([142]). *Let Y be an irreducible algebraic variety over a field k for which the Lefschetz condition* $\mathrm{Lef}(Y \times Y, \Delta)$ *holds, where $\Delta = \Delta_Y$ is the diagonal of $Y \times Y$. Then giving a stratification on an arbitrary vector bundle F on Y is equivalent to giving descent data on F. In particular, every stratified vector bundle on Y is trivial.*

Proof. Let the vector bundle F on Y have a stratification. By what we said above, this yields an isomorphism $\psi\colon q_1^*(F) \to q_2^*(F)$ satisfying the cocycle condition (11.7). Of course, $\widehat{p_i^*(F)} \cong q_i^*(F)$, $i = 1, 2$. From $\mathrm{Lef}(Y \times Y, \Delta)$ it follows that there exists a unique map $\varphi\colon p_1^*(F) \to p_2^*(F)$ such that $\hat{\varphi} = \psi$. From the fact that ψ is an isomorphism and from $\mathrm{Lef}(Y \times Y, \Delta)$ we deduce that φ is also an isomorphism. Now we need to verify the cocycle condition (11.6). For any vector bundle G on $Y \times Y \times Y$ consider the natural map

$$H^0(Y \times Y \times Y, G) \to H^0(Y \times Y \times Y_{/\Delta'}, \hat{G}). \tag{11.8}$$

By [66] I, (10.8.11) the map (11.8) is injective. Hence the cocycle condition for φ, viewed as a relation between two sections of a vector bundle G, holds if its completion holds in $\hat{G}$. But clearly we have the cocycle condition for $\hat{\varphi} = \psi$, so that we have it for φ as well.

The last part follows from the first one and from Theorem (1.1) of [67], exposé VIII, as we explained above. □

Combining Theorem 11.7 with Proposition 11.11 we get the following:

Theorem 11.12 ([11]). *Let Y be a closed irreducible subvariety of $\mathbb{P}^n(e)$, with $n \geq 3$, of dimension $> \frac{n}{2}$. Then every stratified vector bundle on Y is trivial.*

Remark 11.13. Theorem 11.12 was proved in characteristic zero by Ogus [117] in the case when Y is a local complete intersection of small codimension in $\mathbb{P}^n$ (i.e. $\dim(Y) > \frac{n}{2}$), as an important step of his algebraic proof of Barth's theorem. Ogus' approach used a very careful study of the vector bundles with integrable connections and the De Rham cohomology associated to them. Subsequently, Speiser proved the last part of Theorem 11.7 and Theorem 11.12 for locally Cohen–Macaulay closed subvarieties Y in $\mathbb{P}^n$ of small codimension, over a field of positive characteristic. Our methods in [11] (mostly reproduced here), based on formal geometry (and using results of Grothendieck, Hironaka–Matsumura and Faltings), are completely different from those of Ogus (in characteristic zero) or of Speiser (in positive characteristic), and offer not only characteristic free proofs, but also substantial generalizations.

Definition 11.14 ([55]). Let X be a projective irreducible variety of dimension $n \geq 2$ over k, and let $\sigma : E \to F$ be a homomorphism of vector bundles of ranks e and f on X. For every $s \geq 0$ consider the degeneracy locus

$$D_s(\sigma) = \{x \in X \mid \operatorname{rank} \sigma(x) \leq s\}.$$

Then $D_s(\sigma)$ has a natural scheme-structure [55]. For instance, if $E = \mathcal{O}_X$ then $\sigma \in H^0(X, F)$ and $D_0(\sigma)$ is the zero locus $Z(\sigma)$ of σ. If non-empty, $D_s(\sigma)$ has codimension $\leq (e-s)(f-s)$ in X.

Then we have the following fundamental connectedness result:

Theorem 11.15 (Fulton–Lazarsfeld [55]). *In the above notation, assume that the vector bundle $E^* \otimes F$ is ample and $k = \mathbb{C}$. Then:*

(i) *$D_s(\sigma)$ is non-empty when $n \geq (e-s)(f-s)$, and*

(ii) *$D_s(\sigma)$ is connected when $n > (e-s)(f-s)$.*

The proof given in [55] is topological. Recently Steffen [143] gave an algebraic proof of a slightly more general form of Theorem 11.15 which works over any algebraically closed field of characteristic zero.

Corollary 11.16. *In the notation of Theorem* 11.15 *assume $E^* \otimes F$ ample,* $\operatorname{char}(k) = 0$ *and $n > (e-s)(f-s)$. Then $f^{-1}(D_s(\sigma))$ is connected for every proper surjective morphism $f \colon X' \to X$.*

Proof. Consider the Stein factorization $f = g \circ h$, with $h\colon X' \to Z$ a proper morphism such that the canonical map $\mathcal{O}_Z \to h_*(\mathcal{O}_{X'})$ an isomorphism and $g\colon Z \to X$ a finite surjective morphism. Since by Zariski's connectedness theorem h is a proper surjective morphism with connected fibers, it is enough to prove that $g^{-1}(D_s(\sigma))$ is connected. In other words, we may assume that f is finite and surjective. Then $\dim(X') = \dim(X) = n$ and $f^*(E^*\otimes F) \cong f^*(E)^*\otimes f^*(F)$ is ample because $E^*\otimes F$ is so and f is finite. Then applying Theorem 11.15 to $(X', f^*(E), f^*(F), f^*(\sigma))$ we get that $D_s(f^*(\sigma)) = f^{-1}(D_s(\sigma))$ is connected. □

Combining Corollary 11.16 with Corollary 9.22 we get the following reformulation of the Fulton–Lazarsfeld theorem in terms of Formal Geometry:

Corollary 11.17. *Under the hypotheses of Corollary* 11.16, $K(X_{/Y})$ *is a field and* $K(X)$ *is algebraically closed in* $K(X_{/Y})$, *where* $Y := D_s(\sigma)$.

Using Theorem 9.28 and Corollary 11.17 we get:

Corollary 11.18 ([16]). *Let* $Y = Z(s)$ *be the zero locus of a section* $s \in H^0(X,E)$ *of an ample vector bundle* E *of rank* $r \geq 1$ *on a projective irreducible variety* X *of dimension* $d > r$. *Assume that* $\dim(Y) = d - r$ *and that* X *is smooth at every point of* Y. *Then* Y *is* $G3$ *in* X.

Proof. In view of Corollary 11.17 it is enough to show that the field extension $K(X_{/Y})|K(X)$ is algebraic. We shall actually show that Y is $G2$ in X. Indeed, the section s can be viewed as a map $s\colon \mathcal{O}_X \to E$. Then the scheme structure of Y is given by the sheaf of ideals $\mathcal{I}_Y := \operatorname{Im}(s^*)$, where $s^*\colon E^* \to \mathcal{O}_X$ is the dual map of s. Restricting to Y we get a surjective map $\alpha\colon E^*|Y \to \mathcal{I}_Y/\mathcal{I}_Y^2$. On the other hand, under our hypotheses Y is a local complete intersection in X of codimension r, and in particular, $\mathcal{I}_Y/\mathcal{I}_Y^2$ is a vector bundle of rank r on Y. In other words, α is a surjective map between two vector bundles on Y of the same rank, whence α is an isomorphism. It follows that under our hypotheses the normal bundle $N_{Y|X}$ is isomorphic to $E|Y$, and in particular, is ample. Then we can apply Theorem 9.28 to deduce that Y is $G2$ in X. □

Zak's original proof of the finiteness of the Gauss map via Formal Geometry

It is interesting to note that, according to a letter of Zak to Fulton (dated December 26, 1979, see the references of [53]), Zak's original proof of Theorem 7.22 used methods of formal geometry, and only after Zak became aware of the Fulton–Hansen connectedness theorem (at the end of 1979) did he realize that one can get a more elementary proof (see [52], Note (1), at the end of §7). I think it is in the spirit of this presentation to include Zak's proof of finiteness of the Gauss map (Corollary 7.23) using formal geometry (see [150], page 24). I am indebted to Zak for helping me to understand his argument.

Under the hypotheses of Corollary 7.23, we have to prove that the fibers of the Gauss map $f\colon Y \to \operatorname{Grass}(\mathbb{P}^n, \mathbb{P}^d)$, $f(y) = T_y$, are all finite. Suppose this is not so; then there exists a d-dimensional linear subspace L and a positive-dimensional irreducible closed subvariety Z of Y such that $T_z = L$ for all points $z \in Z$, i.e. $f(Z)$ is the point L in $\mathrm{Grass}(\mathbb{P}^n, \mathbb{P}^d)$ (recall that T_z denotes the tangent space of Y at the point z). Then by Remark 9.24,

$$H^0(Y_{/Z}, \mathcal{O}_{Y_{/Z}}) \neq k. \tag{11.9}$$

On the other hand, let M be a general $(n-d-1)$-dimensional linear subspace of $\mathbb{P}^n$ (in particular, $Y \cap M = L \cap M = \varnothing$), and consider the restriction $\pi\colon Y \to \mathbb{P}^d$ of the projection $\pi' : \mathbb{P}^n \setminus M \to \mathbb{P}^d$ of center M. By construction the restriction $\pi'|L\colon L \to \mathbb{P}^d$ is an isomorphism. This implies that the tangent map $d\pi_z\colon T_zY \to T_{\pi(z)}\mathbb{P}^d$ is an isomorphism of vector spaces for every point $z \in Z$ (here by T_zY and $T_{\pi(z)}\mathbb{P}^d$ we mean the tangent spaces as vector spaces), or else, the morphism π is étale at every point of Z. Moreover, since $Z \subseteq L$, the restriction $\pi|Z\colon Z \to \pi(Z)$ is an isomorphism. Therefore we can apply Lemma 9.19 to deduce that the morphism $\tilde{\pi}_{/Z} : Y_{/Z} \to \mathbb{P}^n_{/\pi(Z)}$ of formal schemes is an isomorphism. In particular, the natural map

$$\tilde{\pi}^*_{/Z} : H^0(\mathbb{P}^n_{/\pi(Z)}, \mathcal{O}_{\mathbb{P}^n_{/\pi(Z)}}) \to H^0(Y_{/Z}, \mathcal{O}_{Y_{/Z}}) \tag{11.10}$$

is an isomorphism. Finally, by the result of Hironaka–Matsumura (Theorem 9.14 above), $\pi(Z)$ is $G3$ in $\mathbb{P}^d$ (recall that Z is irreducible of positive dimension and the morphism π is finite), which in particular implies

$$H^0(\mathbb{P}^n_{/\pi(Z)}, \mathcal{O}_{\mathbb{P}^n_{/\pi(Z)}}) = k. \tag{11.11}$$

Indeed, pick a formal-regular function ζ of $\mathbb{P}^d$ along $\pi(Z)$; by Theorem 1.8, ζ extends to a rational function on $\mathbb{P}^d$ which is regular in a neighbourhood U of $\pi(Z)$, and since $\pi(Z)$ intersects every hypersurface of $\mathbb{P}^d$, the complement of U in $\mathbb{P}^n$ is of codimension ≥ 2. It follows that ζ is a regular function on $\mathbb{P}^d$, i.e. $\zeta \in k$. Therefore (11.11) and the isomorphism (11.10) together contradict (11.9). □

Remark 11.19. The equality (11.11) follows easily in the case when $W := \pi(Z)$ is smooth and $\mathrm{char}(k) = 0$. Indeed, consider the m-th infinitesimal neighbourhood $W(m)$ of W in $\mathbb{P}^d$, and let $\mathcal{I}$ be the sheaf of ideals of W in $\mathbb{P}^d$. Then the canonical exact sequence

$$0 \to \mathcal{I}^{m+1}/\mathcal{I}^{m+2} = S^{m+1}(E) \to \mathcal{O}_{W(m+1)} \to \mathcal{O}_{W(m)} \to 0,$$

(where E is the dual of the normal bundle $N_{W|\mathbb{P}^d}$ of W in $\mathbb{P}^d$) yields the exact sequence for every $m \geq 0$:

$$0 \to H^0(W, S^{m+1}(E)) \to H^0(W(m+1), \mathcal{O}_{W(m+1)}) \to H^0(W(m), \mathcal{O}_{W(m)}).$$

Since $N_{W|\mathbb{P}^d}$ is an ample vector bundle (as a quotient of $T_{\mathbb{P}^d}|W$), since in characteristic zero the symmetric powers commute with the operation of taking duals (see [74], exercise 4.9, page 114) and since the positive symmetric powers of ample vector bundles are again ample (see [74], Proposition 1.10, page 85), we get

$H^0(W, S^{m+1}(E)) = 0$ for every $m \geq 0$. Therefore the last exact sequence yields by induction on m that $H^0(W(m), \mathcal{O}_{W(m)}) = k$ for every $m \geq 0$. Finally, since

$$H^0(\mathbb{P}^n_{/W}, \mathcal{O}_{\mathbb{P}^n_{/W}}) \cong \operatorname{inv}\lim_m H^0(W(m), \mathcal{O}_{W(m)}),$$

we get $H^0(\mathbb{P}^n_{/W}, \mathcal{O}_{\mathbb{P}^n_{/W}}) = k$. □

Remark 11.20. Fulton–Hansen's connectedness theorem has been subsequently generalized by Hansen [72] to flag manifolds in the following way. If F is a flag variety of subspaces of $\mathbb{P}^m$, and if $f\colon X \to F \times F$ is a proper morphism from an irreducible variety X, with $\operatorname{codim}_{F\times F}(f(X)) < m$, then $f^{-1}(\Delta_F)$ is connected. Hansen used the geometry of Grassmannians to reduce to the connectedness theorem for projective spaces (see e.g. Corollary 7.15 above).

On the other hand, independently and using completely different methods, in [47] Faltings proved in characteristic zero an even more general result, as follows. Let $Z = G/P$ a projective rational homogeneous space, with G a linear algebraic group and P a parabolic subgroup of G. Denote by $\operatorname{ca}(Z)$ the coampleness of Z (see [47], or [60], or also Definition 13.16 below). Let $f\colon X \to Z \times Z$ be a proper morphism from an irreducible variety X, with $\operatorname{codim}_{Z\times Z}(f(X)) < \operatorname{ca}(Z)$. Then $f^{-1}(\Delta_Z)$ is connected. Since the coampleness of a flag variety of subspaces in $\mathbb{P}^m$ is m, Faltings' connectedness theorem recovers Hansen's connectedness theorem for flag manifolds in characteristic zero.

In proving the Fulton–Hansen connectivity theorem we saw that the main ingredient was Theorem 7.6, stating that if $\operatorname{cd}(X \setminus Y) \leq \dim(X) - 2$ then Y is connected, where Y is a closed subvariety of a projective irreducible variety X. Therefore it is of interest to prove a result (due to Hartshorne–Speiser) giving sufficient conditions under which such an inequality holds. This involves in an essential way considering formal-rational functions. We shall close this chapter by proving the following result (which together with Theorem 9.14 implies Corollary 9.15 above):

Theorem 11.21 (Hartshorne–Speiser). *Let Y be a closed subvariety of a smooth projective irreducible variety X of dimension $n \geq 2$. Assume that Y is G3 in X and Y intersects every hypersurface of X. Then* $\operatorname{cd}(X \setminus Y) \leq n - 2$.

To prove Theorem 11.21 we shall follow [74]. The idea is to use Theorem 10.7 above. The first ingredient of this proof is the following duality result (see [74], page 94).

Theorem 11.22 (Hartshorne's formal duality). *Let Y be a closed subvariety of a smooth projective irreducible variety X of dimension $n \geq 2$. Then for every coherent sheaf F on X there exists a canonical isomorphism*

$$H^i(X_{/Y}, \hat{F}) \cong (H^{n-i}_Y(X, \underline{\operatorname{Hom}}_{\mathcal{O}_X}(F, \omega_X)))', \quad \textit{for every } i \geq 0,$$

where V' denotes the algebraic dual of a (not necessarily finite-dimensional) k-vector space V, and $\omega_X := \Omega^n_X$.

Proof of Theorem 11.22. We shall follow [74], pages 95-96. We shall make use of the following facts:

i) (Serre duality, see [73], Theorem 7.6, page 243.) For every coherent sheaf F on X,
$$H^i(X,F) \cong (\mathrm{Ext}^{n-i}_{\mathcal{O}_X}(F,\omega_X))', \quad \forall i \geq 0.$$

ii) For every closed subscheme Y of a Noetherian scheme X of ideal sheaf $\mathcal{I}$, there is a canonical isomorphism
$$H^i_Y(X,F) \cong \mathrm{dir\,lim}_m \, \mathrm{Ext}^i_{\mathcal{O}_X}(\mathcal{O}_X/\mathcal{I}^m, F),$$
for every coherent sheaf F on X and $i \geq 0$ (see [65], 2.8).

iii) For every coherent sheaf F on X and $i \geq 0$ there exists a canonical isomorphism
$$H^i(X_{/Y},\hat{F}) \cong \mathrm{inv\,lim}_m \, H^i(Y(m),F_m),$$
where $Y(m)$ is the m-th infinitesimal neighbourhood of Y in X and $F_m := F \otimes \mathcal{O}_{Y(m)}$ (see [66] III, **0**, (13.3.1)).

iv) For all coherent sheaves M and N on X and a locally free sheaf F of finite rank on X there is a canonical isomorphism
$$\mathrm{Ext}^i_{\mathcal{O}_X}(F\otimes M, N) \cong \mathrm{Ext}^i_{\mathcal{O}_X}(M, \underline{\mathrm{Hom}}_{\mathcal{O}_X}(F,N)).$$
(Hint: use the well-known canonical isomorphism
$$\mathrm{Hom}_{\mathcal{O}_X}(F\otimes M, N) \cong \mathrm{Hom}_{\mathcal{O}_X}(M, \underline{\mathrm{Hom}}_{\mathcal{O}_X}(F,N)),$$
and take a resolution of M with locally free sheaves of finite ranks (X is projective); the details are left to the reader.)

Now, the proof of Theorem 11.22 goes as follows. By iii) we have
$$H^i(X_{/Y},\hat{F}) \cong \mathrm{inv\,lim}_m \, H^i(Y(m),F_m).$$
But by i) and iv),
$$H^i(Y(m),F_m) = H^i(X,F_m) \cong (\mathrm{Ext}^{n-i}_{\mathcal{O}_X}(F_m,\omega_X))' \cong (\mathrm{Ext}^{n-i}_{\mathcal{O}_X}(\mathcal{O}_X/\mathcal{I}^{m+1},G))',$$
where $G := \underline{\mathrm{Hom}}_{\mathcal{O}_X}(F,\omega_X)$. Therefore using ii) we get
$$\begin{aligned} H^i(X_{/Y},\hat{F}) &\cong \mathrm{inv\,lim}_m (\mathrm{Ext}^{n-i}_{\mathcal{O}_X}(\mathcal{O}_X/\mathcal{I}^{m+1},G))' \\ &\cong (\mathrm{dir\,lim}_m \, \mathrm{Ext}^{n-i}_{\mathcal{O}_X}(\mathcal{O}_X/\mathcal{I}^{m+1},G))' \cong (H^{n-i}_Y(X,G))'. \end{aligned}$$
□

We will also need the following:

Lemma 11.23. *Let U be a quasi-projective irreducible variety, $\mathcal{O}_U(1)$ be an ample line bundle on U, and $r \geq 0$ a non-negative integer. Then the following conditions are equivalent:*

(i) $\mathrm{cd}(U) \leq r$.

(ii) $H^i(U, \mathcal{O}_U(-m)) = 0$, *for every $i > r$ and $m \gg 0$.*

Proof. Clearly (i)$\Longrightarrow$(ii). Conversely, assume that (ii) holds true. Let F be an arbitrary coherent sheaf on X. Since $\mathcal{O}_U(1)$ is ample, $F(m)$ is generated by its global sections for $m \gg 0$, whence we get a surjection $p\mathcal{O}_U(-m) \to F \to 0$, with $p \geq 1$. Thus we get an exact sequence

$$0 \longrightarrow G \longrightarrow p\mathcal{O}_U(-m) \longrightarrow F \longrightarrow 0,$$

yielding the cohomology sequence

$$H^i(U, p\mathcal{O}_U(-m)) \longrightarrow H^i(U, F) \longrightarrow H^{i+1}(U, G).$$

We proceed by descending induction on i (everything being clear if $i > \dim(U)$). Therefore we may assume that $H^{i+1}(U, G) = 0$. By (ii), $H^i(U, p\mathcal{O}_U(-m)) = 0$, because $m \gg 0$. Thus $H^i(U, F) = 0$. $\square$

Proof of Theorem 11.21. Choose an ample line bundle $\mathcal{O}_X(1)$, and set $\mathcal{O}_U(1) := \mathcal{O}_X(1)|U$, where $U := X \setminus Y$. By Lemma 11.23 it will be sufficient to show that $H^{n-1}(U, \mathcal{O}_U(-m)) = 0$ for $m \gg 0$. Set $L := \omega_X(m)$. By Theorem 10.7 the canonical map $H^0(X, L) \to H^0(X_{/Y}, \hat{L})$ is an isomorphism. By Theorem 11.22 and Serre duality this isomorphism is the dual of the map

$$a \colon H^n_Y(X, \mathcal{O}_X(-m)) \to H^n(X, \mathcal{O}_X(-m)),$$

whence a is also an isomorphism.

Finally, consider the local cohomology exact sequence

$$H^{n-1}(X, \mathcal{O}_X(-m)) \to H^{n-1}(U, \mathcal{O}_U(-m)) \to$$

$$\to H^n_Y(X, \mathcal{O}_X(-m)) \xrightarrow{a} H^n(X, \mathcal{O}_X(-m)).$$

By Serre duality, $H^{n-1}(X, \mathcal{O}_X(-m)) \cong (H^1(X, \omega_X(m))'$, and this latter space is zero for $m \gg 0$ by a result of Serre (see [134], or also [66] II, or [73]). Recalling that the map a is an isomorphism, it follows that $H^{n-1}(U, \mathcal{O}_U(-m)) = 0$ for $m \gg 0$, as required. $\square$

Remark 11.24. A small extra argument proves the following result of Speiser [141]: if Y is a closed subvariety of the smooth projective (irreducible) variety X such that Y is $G2$ in X and $\mathrm{cd}(X \setminus Y) \leq n-2$, then Y is actually $G3$ in X (see [141], or also [74], pages 200-202). We shall omit the details because we shall not use this statement.

We close this chapter by proving Theorem 11.25 below which generalizes Fujita's Theorem 3.4 to higher codimension. Let Y be a projective non-singular variety of dimension $d \geq 2$ over the field $\mathbb{C}$ of complex numbers, and let N be an ample vector bundle on Y. Our problem is to decide whether there exists a smooth projective variety X of dimension $n = r + d$, an ample vector bundle E on X of rank r, a closed embedding $Y \hookrightarrow X$ and a global section $s \in H^0(X, E)$, such that the normal bundle of Y in X is isomorphic to N and Y coincides (schematically) with the zero locus $Z(s)$ of s. In this case it follows that $E|Y \cong N$. If $r = 1$, Theorem 3.4 gives an answer to this problem. Specifically, it follows that if the condition

$$H^1(Y, T_Y \otimes N^{-m}) = 0, \ \forall m \geq 1, \tag{11.12}$$

holds, then Y cannot occur as an ample divisor on a smooth projective variety X with normal bundle N, unless $X \cong \mathbb{P}^{d+1}$ and $N \cong \mathcal{O}_{\mathbb{P}^d}(1)$.

Let us fix a projective non-singular variety Y of dimension $d \geq 2$, and an ample vector bundle N on Y of rank $r \geq 1$ such that

$$H^1(Y, T_Y \otimes \mathbf{S}^m(N^*)) = H^1(Y, N \otimes \mathbf{S}^{m+1}(N^*)) = 0, \ \forall m \geq 1, \tag{11.13}$$

where $\mathbf{S}(N^*) = \bigoplus_{m=0}^{\infty} \mathbf{S}^m(N^*)$ is the symmetric $\mathcal{O}_Y$-algebra of the dual N^* of N.

Note that in any case the two vanishings in (11.13) hold for every m sufficiently large. This follows using Serre duality, the ampleness of N and $\dim(Y) \geq 2$. Note also that the vanishings $H^1(Y, N \otimes \mathbf{S}^{m+1}(N^*)) = 0$ for all $m \geq 1$ are consequences of Kodaira's vanishing theorem when $r = 1$. Thus (11.13) may be considered as a natural generalization of (11.12) to the case $r \geq 2$.

Theorem 11.25. *Let Y be a projective non-singular variety of dimension $d \geq 2$ over $\mathbb{C}$, and let N be an ample vector bundle on Y of rank $r \geq 1$ such that condition* (11.13) *above holds. Assume that there exists a line bundle L on Y and two integers $a, b > 0$ such that $L^a \cong \det(N)^b$ and $H^0(Y, N \otimes L^{-1}) \neq 0$. Then Y cannot be embedded in any smooth projective variety X of dimension $n = d + r$ as the zero locus $Z(s)$ of a section $s \in H^0(X, E)$ of an ample vector bundle E of rank r on X such that $E|Y \cong N$ (i.e. with normal bundle of Y in X isomorphic to N), unless $X \cong \mathbb{P}^n$ and $E \cong r\mathcal{O}_{\mathbb{P}^n}(1)$ (direct sum of r copies of $\mathcal{O}_{\mathbb{P}^n}(1)$).*

Notice that the hypothesis that $H^0(Y, N \otimes L^{-1}) \neq 0$ in Theorem 11.25 is automatically fulfilled if $r = 1$, because we can take $L = N$.

The proof of Theorem 11.25 will make use of the following consequence of Grothendieck's obstruction theory (see [67], éxp. III, cf. also [77]):

Theorem 11.26 (Grothendieck). *Let Y be a smooth projective variety of dimension $r \geq 1$, and let N be a vector bundle on Y such that condition* (11.13) *above holds. Then any non-singular formal scheme $\mathcal{X}$ with reduced scheme of definition Y, and with normal bundle N, is isomorphic to the formal completion of the geometric vector bundle $\mathbb{V}(N^*) := \operatorname{Spec}(\mathbf{S}(N^*))$ along its zero-section.*

Proof of Theorem 11.25. Assume by way of contradiction that a triple (X, E, s) as in Theorem 11.25, such that $Y = Z(s)$, does exist.

Step 1. The embedding $Y \hookrightarrow X$ is Zariski equivalent to the zero-section embedding $Y \hookrightarrow \mathbb{V}(N^*)$ of the geometric vector bundle $\mathbb{V}(N^*) := \operatorname{Spec}(\mathbf{S}(N^*))$ over Y (see [66], II). In particular, there exists a Zariski open neighbourhood U of Y in X and a morphism $\pi : U \to Y$ such that $\pi|Y = \mathrm{id}_Y$.

Proof of step 1. The normal bundles $N_{Y|X}$ and $N_{Y|\mathbb{V}(N^*)}$ are both isomorphic to N. Then the cohomological hypothesis (11.13) and Theorem 11.26 (applied to the formal scheme $\mathcal{X} = X_{/Y}$) imply that the embeddings $Y \hookrightarrow X$ and $Y \hookrightarrow \mathbb{V}(N^*)$ are formally equivalent, i.e. there exists an isomorphism of formal schemes $X_{/Y} \cong \mathbb{V}(N^*)_{/Y}$ inducing identity on Y.

On the other hand, since $Y = Z(s)$, by Corollary 11.18, Y is $G3$ in X. Moreover, by a result of Hartshorne [75], Lemma 7.1 and Corollary 6.8, the zero section $Y \subset \mathbb{V}(N^*)$ has the property that Y is also $G3$ in $\mathbb{V}(N^*)$. Note that Lemma 7.1 of [75] (whose proof is elementary) is stated in the case when N is a direct sum of r copies of an ample line bundle, but the same proof works in general (with only minor changes, using the fact that N is ample).

Since the embeddings $Y \hookrightarrow X$ and $Y \hookrightarrow \mathbb{V}(N^*)$ are formally equivalent and Y is $G3$ in both X and $\mathbb{V}(N^*)$ by Corollary 10.6, we deduce that they are also Zariski equivalent. In other words, there exist a Zariski open subset U of X containing Y, a Zariski open subset V of $\mathbb{V}(N^*)$ containing the zero section Y, and an isomorphism $U \cong V$ inducing identity on Y.

For the last part, take as π the composite of the isomorphism $U \cong V$ with the restriction to V of the canonical projection $\mathbb{V}(N^*) \to Y$. This proves step 1.

Step 2. If U is the Zariski open neighbourhood of Y in X from step 1, the restriction maps $\operatorname{Pic}(X) \to \operatorname{Pic}(U)$ and $\operatorname{Pic}(U) \to \operatorname{Pic}(Y)$ are isomorphisms.

Proof of step 2. First we claim that Y meets every closed subset W of X of dimension $\geq r$. Indeed, applying Theorem 11.15, (i) (or the more elementary Theorem 2.1 of [99]) to $s|W \in H^0(W, E|W)$ we get $W \cap Y = Z(s|W) \neq \varnothing$. In particular, Y meets every hypersurface of X, whence $\operatorname{codim}_X(X \setminus U) \geq 2$. Then the fact that the first map is an isomorphism follows because X is non-singular.

On the other hand, by step 1, there exists a retraction $\pi : U \to Y$ of the inclusion $Y \subset U$. This implies that the composite map

$$\operatorname{Pic}(Y) \xrightarrow{\pi^*} \operatorname{Pic}(U) \xrightarrow{\alpha} \operatorname{Pic}(Y)$$

is the identity, where α is the natural restriction map and π^* is the map induced by π. In particular, the map α is surjective. The injectivity of α follows from the fact that $\operatorname{Pic}(X) \cong \operatorname{Pic}(U)$ and from Lefschetz–Sommese theorem (see [140], Proposition (1.16), cf. also [99]); precisely, this result asserts that $H^i(X, Y; \mathbb{Z}) = 0$ for every $i \leq d$, and since $d \geq 2$, this implies (via the exponential sequence) that the restriction map $\operatorname{Pic}(X) \to \operatorname{Pic}(Y)$ is injective. Step 2 is proved.

From now on we shall denote by L' a line bundle on X such that $L'|Y \cong L$. Such a line bundle L' exists (and is unique up to isomorphism) in view of step 2. Since $L^a \cong \det(N)^b$ with $a, b > 0$ and $E|Y \cong N$ we infer that $L'^a \cong \det(E)^b$. Since E is ample, $\det(E)$ is also ample, so that L' is ample as well.

Step 3. Since $\operatorname{codim}_X(X \setminus U) \geq 2$, via the isomorphism $U \cong V$ of the last assertion of step 1, we get $\operatorname{codim}_{\mathbb{V}}(\mathbb{V} \setminus U) \geq 2$, where $\mathbb{V} := \mathbb{V}(N^*) = \operatorname{Spec}(\mathbf{S}(N^*))$. In particular, the restriction map $\operatorname{Pic}(\mathbb{V}) \to \operatorname{Pic}(U)$ is an isomorphism.

Proof of step 3. Consider the composite map

$$Y \overset{i}{\hookrightarrow} \mathbb{V} \overset{\pi'}{\longrightarrow} Y$$

(which is the identity), where, as above, π' is the canonical projection of the geometric vector bundle $\mathbb{V}$ and i is the zero section. By a well-known general assertion, the map $\pi'^* : \operatorname{Pic}(Y) \to \operatorname{Pic}(\mathbb{V})$ is an isomorphism. It follows that the restriction map $i^* : \operatorname{Pic}(\mathbb{V}) \to \operatorname{Pic}(Y)$ is also an isomorphism.

Now, assume by way of contradiction that there exists a hypersurface D of $\mathbb{V}$ such that $D \cap Y = \varnothing$. Then $\mathcal{O}_{\mathbb{V}}(D)|Y \cong \mathcal{O}_Y$, and the injectivity of i^* implies that $\mathcal{O}_{\mathbb{V}}(D) \cong \mathcal{O}_{\mathbb{V}}$. On the other hand, since $\mathbb{V} = \operatorname{Spec}(\mathbf{S}(N^*))$, we have

$$\begin{aligned} H^0(\mathbb{V}, \mathcal{O}_{\mathbb{V}}) &\cong H^0(Y, \pi'_*(\mathcal{O}_{\mathbb{V}})) \cong H^0(Y, \mathbf{S}(N^*)) \\ &\cong \bigoplus_{i=0}^{\infty} H^0(Y, \mathbf{S}^i(N^*)) \cong H^0(Y, \mathcal{O}_Y) = \mathbb{C}, \end{aligned}$$

because $\mathbf{S}^i(N^*) \cong \mathbf{S}^i(N)^*$ (we are in characteristic zero), $\mathbf{S}^i(N)$ are ample for every $i \geq 1$ (because N is ample), and the dual of an ample vector bundle has no non-zero global sections (exercise). Recalling that $\mathcal{O}_{\mathbb{V}}(D) \cong \mathcal{O}_{\mathbb{V}}$, $H^0(\mathbb{V}, \mathcal{O}_{\mathbb{V}}) \cong \mathbb{C}$ implies $D = 0$. This shows that $\operatorname{codim}_{\mathbb{V}}(\mathbb{V} \setminus U) \geq 2$.

The last assertion follows from the first using the fact that $\mathbb{V}$ is non-singular. Step 5 is proven.

Step 4. The open immersion $U \hookrightarrow X$ extends (uniquely) to an open immersion $\mathbb{V} \hookrightarrow X$.

Proof of step 4. By step 3, the line bundle $L_1 := \pi'^*(L)$ on $\mathbb{V}$ satisfies $L_1|U \cong L'|U$. Then I claim that L_1 is ample on $\mathbb{V}$. This follows from [66], II (4.6.13), (ii) and (5.1.6), taking into account that the morphism $\pi' : \mathbb{V} \to Y$ is affine (and in particular, $\mathcal{O}_{\mathbb{V}}$ is π'-ample), $L_1 \cong \pi'^*(L)$ and L is ample on Y.

On the other hand, by step 3 again,

$$H^0(\mathbb{V}, {L_1}^i) \cong H^0(U, {L'}^i) \cong H^0(X, {L'}^i), \quad \forall i \geq 0.$$

Set $S := \bigoplus_{i=0}^{\infty} H^0(\mathbb{V}, {L_1}^i)$. Using the definition of ampleness (see [66], II (4.5.2) and (4.5.3)) it follows that L_1 yields an open immersion $\mathbb{V} \hookrightarrow \operatorname{Proj}(S)$. Finally, since $S \cong \bigoplus_{i=0}^{\infty} H^0(X, L'^i)$ and L' is ample, $\mathrm{Proj}(S) \cong X$. To get the desired open immersion take the composite of $\mathbb{V} \hookrightarrow \mathrm{Proj}(S)$ with the isomorphism $\mathrm{Proj}(S) \cong X$. Step 4 is proven.

From now on, in view of step 4, we may assume $U = \mathbb{V}$ and $L_1 = \pi^*(L) = L'|\mathbb{V}$, where $\pi : \mathbb{V} = \mathbb{V}(N^*) \to Y$ is the canonical projection.

Step 5. Let $T_{\mathbb{V}|Y}$ be the tangent sheaf of $\mathbb{V}$ relative to the retraction $\pi : \mathbb{V} \to Y$. Then $H^0(\mathbb{V}, T_{\mathbb{V}|Y} \otimes {L_1}^{-1}) \neq 0$.

Proof of step 5. Consider the compactification $\mathbb{P} := \mathbb{P}(N^* \oplus \mathcal{O}_Y)$ of $\mathbb{V}$. Inside $\mathbb{P}$ we have the effective divisor $\mathbb{P}(N^*)$ such that $\mathbb{P} \setminus \mathbb{P}(N^*) = \mathbb{V}$ and $\mathcal{O}_{\mathbb{P}}(1) = \mathcal{O}_{\mathbb{P}}(\mathbb{P}(N^*))$. Now consider the Euler sequence relative to the canonical projection $p : \mathbb{P} = \mathbb{P}(N^* \oplus \mathcal{O}_Y) \to Y$:

$$0 \to \mathcal{O}_{\mathbb{P}} \to p^*(N \oplus \mathcal{O}_Y) \otimes \mathcal{O}_{\mathbb{P}}(1) \to T_{\mathbb{P}|Y} \to 0.$$

Restricting this to $\mathbb{V}$ and using $\mathcal{O}_{\mathbb{P}}(1) = \mathcal{O}_{\mathbb{P}}(\mathbb{P}(N^*))$ we get the exact sequence

$$0 \to \mathcal{O}_{\mathbb{V}} \to \pi^*(N \oplus \mathcal{O}_Y) \to T_{\mathbb{V}|Y} \to 0,$$

which upon tensoring with ${L_1}^{-1}$ becomes

$$0 \to {L_1}^{-1} \to \pi^*(N \oplus \mathcal{O}_Y) \otimes {L_1}^{-1} \to T_{\mathbb{V}|Y} \otimes {L_1}^{-1} \to 0.$$

Using the fact that the morphism $\pi : \mathbb{V} = \mathrm{Spec}(\mathbf{S}(N^*)) \to Y$ is affine we get $H^0(\mathbb{V}, L_1^{-1}) \cong H^0(Y, L^{-1} \otimes \mathbf{S}(N^*)) = 0$. Thus, to prove step 5 it will be sufficient to show that

$$H^0(\mathbb{V}, \pi^*(N \oplus \mathcal{O}_Y) \otimes {L_1}^{-1}) \neq 0.$$

To this extent we have

$$H^0(\mathbb{V}, \pi^*(N \oplus \mathcal{O}_Y) \otimes {L_1}^{-1}) \cong H^0(Y, N \otimes L^{-1} \otimes \mathbf{S}(N^*)) \oplus H^0(Y, L^{-1} \otimes \mathbf{S}(N^*)).$$

Thus $H^0(Y, N \otimes L^{-1} \otimes \mathbf{S}(N^*))$ contains the direct summand $H^0(Y, N \otimes L^{-1})$ which by hypothesis is not zero. Step 5 is proven.

Step 6. X is isomorphic to the projective cone over a polarized variety (Z, L'_Z), with $Z \in |L'|$ and $L'_Z := L'|Z$.

Proof of step 6. The canonical exact sequence of tangent bundles

$$0 \to T_{\mathbb{V}|Y} \to T_{\mathbb{V}} \to \pi^*(T_Y) \to 0$$

yields the exact sequence

$$0 \to H^0(\mathbb{V}, T_{\mathbb{V}|Y} \otimes {L_1}^{-1}) \to H^0(\mathbb{V}, T_{\mathbb{V}} \otimes {L_1}^{-1}),$$

which taking into account step 5, yields $H^0(\mathbb{V}, T_{\mathbb{V}} \otimes {L_1}^{-1}) \neq 0$. Therefore

$$H^0(X, T_X \otimes {L'}^{-1}) \cong H^0(\mathbb{V}, T_{\mathbb{V}} \otimes {L_1}^{-1}) \neq 0,$$

because $\operatorname{codim}_X(X \setminus \mathbb{V}) \geq 2$. Finally, by Theorem 1.3, $H^0(X, T_X \otimes {L'}^{-1}) \neq 0$ implies that X is isomorphic to the projective cone over (Z, L'_Z) for some $Z \in |L'|$. This proves step 6.

Step 7. (Conclusion.) By step 6, X is a projective cone. On the other hand, by hypotheses, X is smooth. It follows that X is isomorphic to $\mathbb{P}^n$, with $n = d + r$. Moreover, the embedding $Y \hookrightarrow \mathbb{P}^n$ admits a rational retraction $\pi : \mathbb{V} \to Y$. This implies that the normal sequence

$$0 \to T_Y \to T_{\mathbb{P}^n}|Y \to N \to 0 \tag{11.14}$$

splits. Then the splitting of (11.14) and Theorem 6.1 forces Y to be a linear subspace in $\mathbb{P}^n$. In particular, $X \cong \mathbb{P}^n$ and $E \cong r\mathcal{O}_{\mathbb{P}^n}(1)$. This completes the proof of Theorem 11.25. □

Remarks 11.27. i) Theorem 11.25 is new and answers a question raised by A. Lanteri in a private discussion. In recent years Lanteri and Maeda, and others, classified the triples (X, E, s) as above for which $Y = Z(s)$ is a special given variety (see e.g. [96], [97]). In particular, they produced examples of varieties Y that cannot be zero loci of sections of ample vector bundles.

ii) From the proof of Theorem 11.25 we see that in the hypothesis on L the fact that $L^a \cong \det(N)^b$ for some $a, b \geq 1$ was necessary only to deduce that its unique extension $L' \in \operatorname{Pic}(X)$ is ample. Thus this hypothesis may be replaced by the following one: there is a line bundle L on Y such that $H^0(Y, N \otimes L^{-1}) \neq 0$ and an *ample* line bundle L' on X such that $L'|Y \cong L$.

iii) Let us also remark that if $f : Y' \to Y$ is a finite étale morphism of projective non-singular varieties over $\mathbb{C}$ and if N is an ample vector bundle on Y, then the condition (11.13) holds for (Y, N) if it does for (Y', N'), with $N' := f^*(N)$. This follows easily taking into account that $T_{Y'} = f^*(T_Y)$ and using the fact that in characteristic zero the vanishing $H^1(Y', f^*(E)) = 0$ implies the vanishing $H^1(Y, E) = 0$ for every vector bundle E on Y.

iv) Let Y be a projective nonsingular variety of dimension $d \geq 2$ over $\mathbb{C}$, and L an ample line bundle on Y such that $H^1(Y, T_Y \otimes L^{-i}) = 0$ for every $i \geq 1$. Take $N = L^{a_1} \oplus \cdots \oplus L^{a_r}$, with $a_1 \geq \cdots \geq a_r \geq 1$. Thus, using also Kodaira's vanishing theorem, we see that condition (11.13) above is fulfilled. Moreover, since $\det(N) = L^{a_1 + \cdots + a_r}$, clearly $H^0(Y, N \otimes L^{-a_r}) \neq 0$. Thus the hypotheses of Theorem 11.25 are fulfilled in this case.

v) The cohomological condition (11.13) is stable under small deformations in the following sense. Let $f : \mathcal{Y} \to T$ be a smooth projective morphism (with T a smooth curve) and $\mathcal{N}$ a rank r vector bundle on $\mathcal{Y}$ such that the restriction $\mathcal{N}_t$ of

$\mathcal{N}$ to $Y_t = f^{-1}(t)$ is ample for every $t \in T$. Assume there exists a $t_0 \in T$ such that the pair $(Y, N) = (Y_{t_0}, \mathcal{N}_{t_0})$ satisfies (11.13). Then by the base-change theorems, $(Y_t, \mathcal{N}_t)$ also satisfies (11.13) for every t in a Zariski open neighbourhood U of t_0 in T.

On the other hand, assume that the pair (Y, N) satisfies the hypotheses of Theorem 11.25, i.e. the condition (11.13) is fulfilled and there exists a line bundle L on Y such that $L^a \cong \det(N)^b$ and $H^0(Y, N \otimes L^{-1}) \neq 0$ (for example, (Y, N) could be taken as in the example of the previous remark). Assume moreover

$$H^1(Y, \mathcal{O}_Y) = H^2(Y, \mathcal{O}_Y) = H^1(Y, N \otimes L^{-1}) = 0.$$

Then one can see that (Y_t, N_t) also satisfies the hypotheses of Theorem 11.25 for t in a Zariski open neighbourhood U of t_0 in T. In fact the condition $H^2(Y, \mathcal{O}_Y) = 0$ and the truncated exponential sequences

$$0 \to \mathcal{O}_Y \to \mathrm{Pic}(Y(m+1)) \to \mathrm{Pic}(Y(m)) \to 0, \quad m = 0, 1, 2, \ldots$$

(with $Y(m)$ the m-th infinitesimal neighbourhood of Y in $\mathcal{Y}$) show that L can be extended to a line bundle $\mathcal{L}$ on the formal completion $\mathcal{Y}_{/Y}$. Then by Artin's approximation theory [6] we can replace T by an étale neighbourhood T' of (T, t_0) to show that L can be extended to a line bundle L' on $\mathcal{Y} \times_T T'$. The conditions $H^1(Y, \mathcal{O}_Y) = 0$ and $L^a \cong \det(N)^b$ imply that $L'^a \cong \det(\mathcal{N})^b$ (in a neighbourhood of t_0). Moreover, the base-change theorems and $H^1(N \otimes L^{-1}) = 0$ shows that the function $t \to \dim_{\mathbb{C}} H^0(Y_t, N_t \otimes L_t^{-1})$ is locally constant near t_0. Therefore $H^0(Y_t, N_t \otimes L_t^2) \neq 0$ for every $t \in T$ near t_0.

Example 11.2. Take $Y = \mathbb{P}^d$, with $d \geq 2$, and let N be any ample vector bundle of rank r on $\mathbb{P}^d$. Assume that $Y = Z(s)$, with s a global section of an ample vector bundle E of rank $n = d + r$ on a projective non-singular variety X of dimension $n \geq 4$ such that $E|Y \cong N$. Then we claim that $N \cong r\mathcal{O}_{\mathbb{P}^d}(1)$. To check this, by the adjunction formula we get

$$\omega_X|Y \cong \omega_Y \otimes \det(N)^{-1} \cong \mathcal{O}_{\mathbb{P}^d}(-d-1) \otimes \det(N)^{-1}. \tag{11.15}$$

Let $\mathcal{L}$ be any line in $\mathbb{P}^d$. By a theorem of Grothendieck, $N|\mathcal{L} \cong \bigoplus_{i=1}^{r} \mathcal{O}_{\mathbb{P}^1}(a_i)$, and in particular,

$$\det(N)|\mathcal{L} \cong \mathcal{O}_{\mathbb{P}^1}(\sum_{i=1}^{r} a_i). \tag{11.16}$$

Since N is ample, $a_i \geq 1$, for every $i = 1, \ldots, r$. To prove the claim, according to [118], Theorem 3.2.1, page 51, it is enough to check that $a_1 = \cdots = a_r = 1$ for every line $\mathcal{L}$ of $\mathbb{P}^d$. From (11.15) and (11.16) we get

$$\omega_X^{-1}|\mathcal{L} \cong \mathcal{O}_{\mathbb{P}^1}(t), \quad \text{with } t := d + 1 + \sum_{i=1}^{r} a_i. \tag{11.17}$$

Since N is ample, $a_i \geq 1$, for every $i = 1, \ldots, r$. Consider the restriction maps

$$\mathrm{Pic}(X) \to \mathrm{Pic}(Y) \to \mathrm{Pic}(\mathcal{L}).$$

By a result of Lefschetz–Sommese (see [140], cf. also step 2 of the proof of Theorem 11.25) the first map is an isomorphism, and the second map is (obviously) also an isomorphism. Thus (11.17) implies that there exists a line bundle L on X such that $\omega_X^{-1} \cong L^t$. Since $\mathrm{Pic}(X) \cong \mathbb{Z}$ and $L|\mathcal{L} \cong \mathcal{O}_{\mathbb{P}^1}(1)$, L is ample, and in particular, X is a Fano n-fold of index t. Then a simple general result (see e.g. [29], page 81, Lemma 3.3.2) asserts that $t \leq n+1$, i.e. $\sum_{i=1}^{r} a_i + d + 1 \leq n + 1$. Since $n = d + r$, this implies $a_1 = \cdots = a_r = 1$, as required.

As a consequence we get that N satisfies the hypotheses of Theorem 11.25 if $Y = \mathbb{P}^d$ with $d \geq 3$, in which case we get $X \cong \mathbb{P}^n$ and Y is a linear subspace of X (the latter fact was proved before in [97]). However, the same conclusion can be also drawn directly from a theorem of Kobaiashi–Ochiai (see e.g. [29], Theorem 7.2.1). □

The next lemma (whose standard proof is left to the reader) provides "asymptotical" examples of pairs (Y, N) satisfying the hypotheses of Theorem 11.25.

Lemma 11.28. *Let N_1 be an arbitrary ample vector bundle of rank $r \geq 1$ and L an ample line bundle on a projective non-singular variety Y of dimension $d \geq 2$. Then $N := N_1 \otimes L^p$ satisfies the condition* (11.13) *above for every $p \gg 0$. Moreover, if we take $L = \det(N_1)$, then $H^0(Y, N \otimes L^{-1}) \neq 0$ for every $p \gg 0$, i.e. all the hypotheses of Theorem* 11.25 *are fulfilled in this case.*

Chapter 12

Further Results on Formal Functions

Definition 12.1 (Sommese [140], or also [29]). Let L be a line bundle on a projective variety X, and let $s \geq 0$ be a non-negative integer. L is said to be s-*ample* if there exists an $m > 0$ such that L^m is generated by its global sections and the fibers of the associated morphism $\varphi_{|L^m|} : X \to \mathbb{P}(H^0(X, L^m)^*)$ are all of dimension $\leq s$. For example L is 0-ample if and only if L is ample in the usual sense.

Let now E be a vector bundle of rank $r \geq 2$ on X. Then E is said to be s-*ample* if the tautological line bundle $\mathcal{O}_{\mathbb{P}(E)}(1)$ is s-ample on the projective bundle $\mathbb{P}(E)$ associated to E.

Our first objective in this chapter is to prove a refinement of Theorem 9.28 (see Theorem 12.6 below). For this we follow [17]. This refinement is going to be applied in Chapter 13 to homogeneous spaces. Let Y be a smooth connected projective variety of dimension d, and let E be a vector bundle on Y of rank r. Let $\mathcal{O}_Y(1)$ be a very ample line bundle on Y. First we prove a Bertini-type result which might also have an interest in itself:

Theorem 12.2 ([17]). *In the above situation, assume that* Y *is of dimension* $d \geq 2$, *and that* E *is* s*-ample for some integer* $s \geq 1$. *Then for every* $t \geq (d+r-1-s)/s$ *and for every general* $H \in |\mathcal{O}_Y(t)|$, *the restriction* $E_H := E \otimes \mathcal{O}_H$ *is* $(s-1)$*-ample.*

To prove Theorem 12.2 we need the following:

Lemma 12.3 ([17]). *Let* V *be a non-degenerate closed irreducible subvariety of* $\mathbb{P}^n$. *Let*

$$v_s : \mathbb{P}^n \to \mathbb{P}^m, \quad \textit{with } m := \binom{n+s}{s} - 1,$$

be the s*-fold Veronese embedding. Then the linear subspace* $\operatorname{Span}(v_s(V))$ *of* $\mathbb{P}^m$ *generated by* $v_s(V)$ *is of dimension* $\geq ns$.

Proof of Lemma 12.3. The proof consists of two steps. The first step asserts that for every positive integer $p \geq n$ there exist $p+1$ points $x_0, x_1, \ldots, x_p \in V$ such that each $n+1$ of them generate $\mathbb{P}^n$.

To see this use induction on p. Start by choosing $x_0, x_1, \dots, x_n \in V$ such that they generate $\mathbb{P}^n$. This is possible because V is non-degenerate in $\mathbb{P}^n$. So we may assume that the assertion holds for $x_0, \dots, x_p$ (with $p \geq n$) and we want to find a point x_{p+1} such that the assertion still holds for $x_0, x_1, \dots, x_{p+1}$. To do this, let $\{H_i\}_i$ be the (finite) family of hyperplanes of $\mathbb{P}^n$ generated by all subsets of n different points of $\{x_0, x_1, \dots, x_p\}$. Since V is irreducible and non-degenerate in $\mathbb{P}^n$, V cannot be contained in the union of all H_i's. Then picking an arbitrary $x_{p+1} \in V \setminus \bigcup_i H_i$, the points $x_0, x_1, \dots, x_{p+1}$ satisfy the requirements of the first step.

Now take points $x_0, x_1, \dots, x_{ns} \in V$ (i.e. $p = ns$) such that each $n+1$ of them generate $\mathbb{P}^n$. Fix $1 \leq i \leq ns$, and let $H_{i,1}, \dots, H_{i,s}$ be the s hyperplanes of $\mathbb{P}^n$, each generated by n points corresponding to a partition of $\{x_0, \dots, x_{ns}\} \setminus \{x_i\}$ with subsets of n points. If $h_{i,j}$ is the equation of the hyperplane $H_{i,j}$, $j = 1, \dots, s$, then $f_i := h_{i,1} \cdots h_{i,s}$ is the equation of a hypersurface of degree s of $\mathbb{P}^n$ such that $f_i(x_i) \neq 0$ and $f_i(x_j) = 0$ for every $j \neq i$. In other words, the points $x'_0, \dots, x'_{ns}$, with $x'_i := v_s(x_i)$, of the closed subvariety $v_s(V)$ of $\mathbb{P}^m$ have the following property: for every $0 \leq i \leq ns$ there is a hyperplane L_i of $\mathbb{P}^m$ (namely, the one which corresponds to the hypersurface of degree s of $\mathbb{P}^n$ of equation $f_i = 0$) such that L_i contains all x'_j for $j \neq i$, but $x'_i \notin L_i$. Therefore the linear subspace of $\mathbb{P}^m$ generated by $x'_0, \dots, x'_{ns}$ is of dimension ns. In conclusion, $\dim(\operatorname{Span}(v_s(V))) \geq ns$. □

Corollary 12.4. *The assumptions are as in Lemma* 12.3 *except that V might be degenerate in $\mathbb{P}^n$. Then* $\dim(\operatorname{Span}(v_s(V))) \geq s \cdot \dim(\operatorname{Span}(V))$.

Remark 12.5. The statement of Lemma 12.3 and its proof was communicated to the authors of [17] by Iustin Coandă. In fact, all that is needed is an estimate of the type given in Corollary 12.4. Our original estimate of the dimension of $\operatorname{Span}(v_s(V))$ was far weaker than the above one (but good enough to prove Theorem 12.6 below). However, more recently, Mark Green told me that Lemma 12.3 is essentially due to Castelnuovo.

Proof of Theorem 12.2. Set $P := \mathbb{P}(E)$, and denote by $\mathcal{O}_P(1)$ the tautological line bundle on P. For $m \gg 0$ let

$$\varphi' : P \to \mathbb{P}(H^0(P, \mathcal{O}_P(m))^*)$$

be the morphism associated to the linear system $|\mathcal{O}_P(m)|$ (which by the definition of s-ampleness of E has no base points fo $m \gg 0$). Let $\varphi\colon P \to Z$ be the morphism with connected fibers deduced from φ' by Stein factorization. By the definition of s-ampleness, the fibers of φ have all dimension $\leq s$. Moreover, $\mathcal{O}_P(m) = \varphi^*(H')$, with H' an ample line bundle on Z.

Consider the (reduced) closed subscheme C of Z defined by

$$C := \{z \in Z \mid \dim(\varphi^{-1}(z)) = s\}.$$

Then $\dim(C) \leq u := d + r - 1 - s$ (by the definition of C and because $\dim(P) = d + r - 1$). Set $V := \varphi^{-1}(C)_{\mathrm{red}}$, $\psi := \varphi|V\colon V \to C$, and $j = j_t\colon V \to \mathbb{P}^n$, where j

is the composite of the inclusion of V into P followed by the canonical projection $\pi\colon P \to Y$, followed by the embedding of Y into $\mathbb{P}^N$ associated to $|\mathcal{O}_Y(t)|$, with $t \geq \frac{u}{s}$.

Let $c \in C$ be an arbitrary point. By the definition of φ, $\mathcal{O}_P(m)|\psi^{-1}(c)$ is trivial, and therefore for every $y \in Y$, $\mathcal{O}_P(m)|\psi^{-1}(c) \cap \pi^{-1}(y)$ is also trivial. Recalling that $\mathcal{O}_P(m)|\pi^{-1}(y)$ is ample, this implies that $\psi^{-1}(c) \cap \pi^{-1}(y)$ is finite for every $c \in C$ and $y \in Y$. In other words, the morphism

$$\pi|\psi^{-1}(c)_{\mathrm{red}} : \psi^{-1}(c)_{\mathrm{red}} \to \pi(\psi^{-1}(c))_{\mathrm{red}}$$

is finite for every $c \in C$. It follows that

$$j|\psi^{-1}(c)_{\mathrm{red}} : \psi^{-1}(c)_{\mathrm{red}} \to j(\psi^{-1}(c)_{\mathrm{red}})$$

is also finite for every $c \in C$.

Now, let W be an arbitrary irreducible component of dimension s of $\varphi^{-1}(c)_{\mathrm{red}}$ (with $c \in C$). Then we get the finite surjective morphism $W \to W' := j(W)$. In particular, W' is irreducible of dimension s. By the choice of t (i.e. $t \geq \frac{u}{s}$) and by Corollary 12.4 we get

$$\dim(\mathrm{Span}(v_t(W'))) \geq t \cdot \dim(\mathrm{Span}(W')) \geq t \cdot s \geq u \geq \dim(C), \qquad (12.1)$$

for every irreducible component W of an arbitrary fiber of ψ. As long as (12.1) holds we can apply a theorem of Bertini–Hironaka (see [79], Lemma 2.1, or also Proposition 12.15 of Appendix C at the end of this chapter) to deduce that

$$\dim(\psi^{-1}(c) \cap \pi^{-1}(H)) < \dim(\psi^{-1}(c)) = s \qquad (12.2)$$

for every $c \in C$ and for every general $H \in |\mathcal{O}_Y(t)|$. Recalling that for every $z \in Z \setminus C$, $\dim(\varphi^{-1}(z)) \leq s-1$, we have proved the following. Setting $P' := \pi^{-1}(H) = \mathbb{P}(E|H)$ and $i\colon P' \hookrightarrow P$ the canonical closed embedding, and using (12.2), the morphism $\varphi_1 := \varphi \circ i$ has the following two properties:

(i) $\varphi_1^*(H') = \mathcal{O}_{P'}(m)$, with H' an ample line bundle on Z, and

(ii) The fibers of φ_1 are all of dimension $\leq s-1$.

But these two conditions imply that $E_H = E|H$ is $(s-1)$-ample. □

We are ready to prove the desired refinement of Theorem 9.28 of Hartshorne:

Theorem 12.6 ([17]). *Let X be a smooth projective connected variety, and let Y be a smooth connected closed subvariety of X of dimension $d > 0$. Assume that the normal bundle $N_{Y|X}$ of Y in X is $(d-1)$-ample. Then Y is G2 in X.*

Proof. If $d = 1$ then $E := N_{Y|X}$ is ample and the result follows from Theorem 9.28. So assume $d \geq 2$. Then applying Theorem 12.2 repeatedly, we find a positive integer $t > 0$ and $d-1$ general hypersurfaces $H_1, \dots, H_{d-1} \in |\mathcal{O}_X(t)|$ with the property that $E|(H_1 \cap \cdots \cap H_i)$ is $(d-i-1)$-ample for every $i = 1, \dots, d-1$, where $\mathcal{O}_X(1)$ is a a very ample line bundle on X. In particular, if we set $Y_1 := H_1 \cap \cdots \cap H_{d-1}$, then Y_1 is a smooth projective curve such that $E|Y_1$ is ample.

On the other hand, look at the canonical exact sequence of normal bundles

$$0 \to N_{Y_1|Y} \to N_{Y_1|X} \to N_{Y|X}|Y_1 = E|Y_1 \to 0.$$

Since Y_1 is a complete intersection in Y of $d-1$ hypersurfaces of degree t, $N_{Y|Y} \cong (d-1)\mathcal{O}_{Y_1}(t)$, and therefore is ample. Since on the other hand we just saw that $E|Y_1$ is ample, we infer from the above exact sequence that $N_{Y_1|X}$ is also ample. Applying Theorem 9.28 again we infer that Y_1 is $G2$ in X, i.e. the canonical map $K(X) \to K(X_{/Y_1})$ is a finite field extension.

Now, since $Y_1 \subset Y$ we get canonical maps

$$K(X) \to K(X_{/Y}) \to K(X_{/Y_1}).$$

Since Y is connected and X is smooth, $K(X_{/Y})$ is a field by Proposition 9.2. Therefore the map $K(X) \to K(X_{/Y})$ defines a finite field extension (since the above composite is), i.e. Y is $G2$ in X. □

The next result gives natural examples of d-dimensional closed subvarieties Y of an abelian variety X with $(d-1)$-ample normal bundle. Before stating it, let us first give a definition. Let Y be a closed irreducible subvariety of an abelian variety X. We shall say that Y *generates* X if there exists a point $y_0 \in Y$ such that the subgroup generated by all elements of the form $y - y_0$, $\forall y \in Y$ (in the group-theoretical sense) coincides with X. This is the same as saying that the subgroup generated by the closed subvariety $Y - Y$ coincides with X. If Y passes through the origin, Y generates X if and only if the subgroup generated by Y coincides with X (see Definition 13.1 below for a more general concept of generatedness in the framework of homogeneous spaces). If general, if Y is a closed irreducible subvariety of X passing through the origin, it is easy to see that the subgroup K of X generated by Y is an abelian subvariety of X (see the more general discussion at the beginning of the next chapter).

As an example, let Y be a smooth projective variety and let $X = \mathrm{Alb}(X)$ be the Albanese variety of X. If $\varphi\colon Y \to X$ is the canonical morphism then $\varphi(Y)$ generates X (we didn't make explicit the reference point $y_0 \in Y$ such that $\varphi(y_0) = 0$). Indeed, if K is the abelian subvariety generated by $\varphi(Y)$ then by the universal property of the Albanese variety, there exists a unique morphism of abelian varieties $f\colon X \to K$ such that $f \circ \varphi = \varphi$. If $g := i \circ f\colon X \to X$, with $i\colon K \to X$ the natural inclusion, it follows that $g \circ \varphi = \varphi$, whence by the universal property again, $g = \mathrm{id}_X$. This implies that i is surjective, i.e. $K = X$. In particular, if Y is a curve, $X = \mathrm{Jac}(Y)$ is the Jacobian variety of Y and φ is the canonical embedding of Y in its Jacobian, then $\varphi(Y)$ generates X.

Proposition 12.7 ([17]). *Let X be an abelian variety over an algebraically closed field of characteristic zero, and let Y be a smooth connected closed subvariety of*

X of dimension d, with $d \geq 1$. Then the normal bundle $N_{Y|X}$ is $(d-1)$-ample if and only if Y generates X.

Proof. There is no loss of generality in assuming that Y passes through the origin of X (just replace Y by $Y - y_0$, with $y_0 \in Y$, if $0 \notin Y$).

Assume first that $N_{Y|X}$ is $(d-1)$-ample. By Theorem 12.6, Y is $G2$ in X. Let K be the abelian subvariety generated by Y. If $K \neq X$ then Y is contained in the fiber over the origin of the (non-trivial) canonical morphism $X \to X/K$, and this is not possible because Y is $G2$ in X, by remark 9.24. So $K = X$, i.e. Y generates X.

Assume now that Y generates X. If $d = 1$ we have to show that $N_{Y|X}$ is ample. This follows from a result of Hartshorne (see [76], Proposition 4.1, or also Theorem 12.13 of Appendix C at the end of this chapter). So assume $d > 1$. Let $H_1, \ldots, H_{d-1}$ be general hyperplane sections of X (with respect to some projective embedding of X), and denote by Y' the intersection $Y \cap H_1 \cap H_2 \ldots \cap H_{d-1}$. We claim that Y' also generates X.

To see this, take an arbitrary abelian subvariety of K of X such that $K \neq X$, and consider the composite $f\colon Y \to X \to X/K$. By remark (2.1.26), Y' is not contained in any fiber of f (since Y' is a complete intersection in Y, and therefore $G2$ in Y). In particular, Y' is not contained in K, whence Y' generates X.

Since Y' is a smooth curve generating X, by Hartshorne's result quoted above (see [76], Proposition 4.1, or also Theorem 12.13 of Appendix C) the normal bundle $N_{Y'|X}$ is ample. Hence its quotient $N_{Y|X}|Y'$ is also ample.

Assume that $E := N_{Y|X}$ is not $(d-1)$-ample. This implies that, if $n := \dim(X)$, the morphism $f : \mathbb{P}(E) \to \mathbb{P}^{n-1}$, which is the composite of the closed immersion $\mathbb{P}(E) \to \mathbb{P}(T_X|Y) \cong Y \times \mathbb{P}^{n-1}$ with the projection onto $\mathbb{P}^{n-1}$, has a fiber of dimension d, i.e. containing $Z := Y \times p$, for some $p \in \mathbb{P}^{n-1}$. Here we use the fact that E is generated by its global sections. Restricting to $Y' = Y \cap H_1 \cap \ldots \cap H_{d-1}$, we get that the composite of f with the inclusion $\mathbb{P}(E|Y') \to \mathbb{P}(E)$ has a fiber of positive dimension, which contradicts the ampleness of $E|Y'$. □

Next, following [15], we prove another criterion for extending formal-rational functions. For this we need some definitions.

Definition 12.8. According to [61], an algebraic variety Z over k is said to be *semi-affine* if there exists a proper surjective morphism $f\colon Z \to V$, with V an affine variety. Alternatively, Z is semi-affine if and only if the canonical morphism of schemes $\varphi\colon Z \to \operatorname{Spec}(\Gamma(Z, \mathcal{O}_Z))$ is proper (in which case $\operatorname{Spec}(\Gamma(Z, \mathcal{O}_Z))$ is actually an algebraic variety, see [61], Corollary 3.6). We refer the reader to [61] for the basic properties of semi-affine varieties. For example, if Y is a closed subvariety of a complete irreducible variety X of dimension $d \geq 2$ such that $X \setminus Y$ is a semi-affine variety, then Y has pure codimension one in X (see [61], Corollary 6.8).

Theorem 12.9 ([15]). *Let X be a complete irreducible algebraic variety over k of dimension $d \geq 2$, and Y a closed (possibly reducible) subvariety of X. Assume* *that $U := X \setminus Y$ is semi-affine, U contains at most finitely many complete $(d-1)$-dimensional irreducible subvarieties, and Y is the support of an effective Cartier divisor D of X (the latter assumption is always fulfilled if X is smooth). Then Y is G3 in X, and in particular, Y is connected.*

Proof. According to Corollary 8.5 of [61], the hypothesis that U is semi-affine implies that there exists a blowup $f: X' \to X$ with center outside U such that $f^{-1}(U)$ is the complement of the support of an effective Cartier divisor D' on X' whose linear system $|D'|$ has no base points. Let $\varphi: X' \to P^n$ be the morphism associated to $|D'|$. Set $Z := \varphi(X')$. Then there exists a hyperplane section H of Z such that $\varphi^*(H) = D'$. In particular, the restriction $\psi: f^{-1}(U) \to Z \setminus H$ of φ is a proper, surjective morphism onto the affine variety $Z \setminus H$. Since $D' > 0$ and X' is complete, $\dim(Z) > 0$ (otherwise the divisor D' would be linearly equivalent to zero). If $\dim(Z) = 1$, by the theorem on dimension of fibers and the fact that ψ is proper, the reduced fibers $\varphi^{-1}(z)_{\text{red}}$ is a complete $(d-1)$-dimensional subvariety of $\varphi^{-1}(U) \cong U$ for every $z \in Z$. But this contradicts the hypothesis that U contains at most finitely many complete $(d-1)$-dimensional irreducible subvarieties. Therefore $\dim(Z) \geq 2$.

Now, since $\dim(Z) \geq 2$ and H is a hyperplane section on Z (in particular, a complete intersection in X), Corollary 9.26 shows that H is $G3$ in Z, i.e. the canonical map of k-algebras $\alpha_{Z,H} : K(Z) \to K(Z_{/H})$ is an isomorphism. Since $\varphi: X' \to Z$ is a proper surjective morphism such that $\varphi^*(H) = D'$ and $Y' := \text{Supp}(D')$, by Corollary 9.13, (i) we infer that the canonical map $\alpha_{X',Y'} : K(X') \to K(X'_{/Y'})$ is an isomorphism.

Finally, since f is a proper birational morphism such that $f^{-1}(Y) = Y'$, by Theorem 9.9 we get $K(X'_{/Y'}) \cong K(X_{/Y})$. In other words, the map $\alpha_{X,Y} : K(X) \to K(X_{/Y})$ is an isomorphism (because $\alpha_{X',Y'}$ is so), i.e. Y is $G3$ in X. □

Corollary 12.10. *Let X be a complete irreducible variety of dimension $d \geq 2$ and $U \subset X$ an affine open subset of X. Assume that $Y := X \setminus Y$ is the support of an effective Cartier divisor of X (e.g. when X is smooth). Then Y is G3 in X.*

Recall that a divisor D on a smooth projective variety of dimension $d \geq 2$ is said to be *nef* and *big* if $(D \cdot C) \geq 0$ for every irreducible curve $C \subset X$ and $(D^{\cdot d}) > 0$.

Corollary 12.11. *Let $D > 0$ be an effective nef and big divisor on a smooth projective surface X. Then $Y := \text{Supp}(D)$ is G3 in X.*

Proof. We first claim that $U := X \setminus Y$ is semi-affine. This statement was noticed by F. Russo (see [125], Corollary 5). For the sake of completeness we indicate a direct proof, which is a simple consequence of some fundamental results of Zariski [152]. For every $n \geq 1$ consider the fixed component B_n of the linear system $|nD|$. In other words, B_n is the maximal effective divisor such that $\dim |nD - B_n| = \dim |nD|$. By a result of Zariski [152], Theorem 10.1, the B_n's are bounded from above. This implies that $\text{Supp}(nD \setminus B_n) = \text{Supp}(nD)$ for n sufficiently large. Set

$D'_n := nD - B_n$, $n \gg 0$. Then the linear system $|D'_n|$ has no fixed components and $\mathrm{Supp}(D'_n) = \mathrm{Supp}(nD) = \mathrm{Supp}(D)$, $\forall n \gg 0$.

Since $|D'_n|$ has no fixed components, by another result of Zariski (see [152], Theorem 6.1, or also [12], Theorem 9.17), the linear system $|mD'_n|$ has no base points for m sufficiently large. Moreover,

$$U = X \setminus \mathrm{Supp}(D) = X \setminus \mathrm{Supp}(D'_n) = X \setminus \mathrm{Supp}(mD'_n), \quad \forall m \gg 0.$$

Since $|mD'_n|$ has no base points for $m \gg 0$, U is semi-affine by [61], Proposition 1.3.

Next we prove that U contains at most finitely many complete curves. Let C be a complete irreducible curve contained in U. Then $C \cap \mathrm{Supp}(D) = \varnothing$, and in particular, $(D \cdot C) = 0$. Let $C_1, \ldots, C_s$ be arbitrary complete irreducible curves contained in U. Since D is nef and big and $(D \cdot C_i) = 0$, $i = 1, \ldots, s$, by Hodge index theorem (see e.g. [12], Corollary (2.4)) the intersection matrix $\| \, (C_i \cdot C_j) \, \|_{i,j=1,\ldots,s}$ is negative definite. In particular, $C_1, \ldots, C_s$ define linearly independent elements in the finite-dimensional $\mathbb{Q}$-vector space $\mathrm{NS}(X) \otimes_{\mathbb{Z}} \mathbb{Q}$, where $\mathrm{NS}(X)$ denotes the Néron–Severi group of X, whence s is bounded by the dimension of this vector space.

Now the corollary follows from Theorem 12.9. □

Remarks 12.12. i) The hypothesis that U contains at most finitely many complete $(d-1)$-dimensional subvarieties is necessary in Theorem 12.9 because of the following example. Let $g\colon X \to B$ be a proper surjective morphism from a smooth projective variety X to a smooth projective curve B. Pick $m \geq 2$ arbitrary points $b_1, \ldots, b_m$ of B and set

$$Y := \bigcup_{i=1}^{m} g^{-1}(b_i).$$

Then Y supports effective Cartier divisors, $U := X \setminus Y$ is semi-affine (because U properly dominates the affine curve $B \setminus \{b_1, \ldots, b_m\}$), and Y is not connected (because $m \geq 2$).

ii) One may naturally ask whether Corollary 12.11 can be extended in dimension ≥ 3. We do not know the answer in general. However, we can only notice that the proof of Corollary 9.9 made essential use of the fact that the complement $X \setminus \mathrm{Supp}(D)$ was semi-affine if $\dim(X) = 2$. Unfortunately, if D is a nef and big effective divisor on a smooth projective variety of dimension ≥ 3 the complement $X \setminus \mathrm{Supp}(D)$ need not be semi-affine in general. In fact, by [152] (see also [12], chap. 14) there are pairs (S, Δ), with S a smooth complex projective surface and Δ an effective nef and big divisor on S such that

$$R(S, \Delta) := \bigoplus_{i=0}^{\infty} \Gamma(S, \mathcal{O}_S(i\Delta))$$

is not a finitely generated $\mathbb{C}$-algebra. By [152], Theorem 10.6, this means that $\mathcal{O}_S(i\Delta)$ is not generated by its global sections for every $i > 0$. Consider the $\mathbb{P}^1$-bundle $X := \mathbb{P}(\mathcal{O}_S(\Delta) \bigoplus \mathcal{O}_S)$ associated to the sheaf $\mathcal{O}_S(\Delta) \oplus \mathcal{O}_S$ (X coincides to the projective closure of the vector bundle $\mathbb{V}(\mathcal{O}_S(\Delta)))$. Let D be the "zero-section" of the canonical projection $\pi\colon X \to S$, so that $\mathcal{O}_X(1) = \mathcal{O}_X(D)$. Since Δ is nef and big on S, D is also nef and big on X. I claim that the complement $X \setminus \mathrm{Supp}(D)$ is not semi-affine; indeed, otherwise, by [61], Proposition 9.2, it would follow that $\mathcal{O}_S(i\Delta)$ is generated by its global sections for some $i > 0$.

Note that the above example was given by Russo in [125] to show that $\mathrm{Proj}(R(X, D))$ is not always an algebraic variety when $\dim(X) \geq 3$ (on the contrary, if D is a nef and big divisor on a smooth projective surface X, then $\mathrm{Proj}(R(X, D))$ is always a quasi-projective surface, see [125]).

iii) Under the hypotheses and notation of Corollary 12.11 and its proof, let $\varphi\colon U \to \mathrm{Spec}(\Gamma(U))$ be the canonical dominant morphism, where $\Gamma(U) := \Gamma(U, \mathcal{O}_U)$. The fact that U is semi-affine implies that φ is a proper morphism (in fact, this is one of the equivalent definitions of semi-affineness, see [61]), and since φ is dominant, φ is also surjective. Then we claim that $\Gamma(U)$ is a finitely generated k-algebra. This follows from [61], Corollary 3.6, since U is semi-affine. In particular, $\mathrm{Spec}(\Gamma(U))$ is an affine variety. However, one can also give the following direct argument. Since U is the complement of the support of the effective divisor $\Delta := mD'_n$, $m \gg 0$, such that the linear system $|\Delta|$ has no base points, by a result of Zariski (see [152], Corollary 10.3, see also [12], Proposition 9.2) the graded k-algebra

$$R(X, \Delta) := \bigoplus_{i=0}^{\infty} \Gamma(X, \mathcal{O}_X(i\Delta)),$$

is finitely generated; then using [66] I, (6.8.2) one has (in the notation of [66] I) $\Gamma(U) = R(X, \Delta)_{(\alpha)}$, where $\alpha \in \Gamma(X, \Delta)$ is a global equation of Δ (cf. also [153]).

Appendix C

The first aim of this appendix is to prove the following special case of a result of Hartshorne, which was used in the proof of Proposition 12.7 (see [76], Proposition 4.1, for a more general version):

Theorem 12.13 (Hartshorne [76]). (char$(k) = 0$.) *Let X be an abelian variety, and let Y be a smooth curve in X that generates X. Then the normal bundle $N_{Y|X}$ is ample.*

The proof of Theorem 12.13 makes use of the following ampleness criterion for vector bundles on curves:

Theorem 12.14 (Gieseker [59]). (char$(k) = 0$.) *Let E be a vector bundle over a smooth projective curve Y that is generated by its global sections. Then E is ample if and only if every quotient line bundle L of E is ample.*

Proof of Theorem 12.13. We may assume that Y passes through the origin of X. Let $\mathrm{Alb}(Y)$ be the Albanese variety of Y (i.e. the Jacobian variety of the curve Y). Let $Y \hookrightarrow \mathrm{Alb}(Y)$ be the canonical inclusion. By the universal property of the Albanese variety, the inclusion $Y \hookrightarrow X$ can be uniquely extended to a morphism of abelian varieties $f : \mathrm{Alb}(Y) \to X$. It is clear that saying that Y generates X is equivalent to saying that the morphism f is surjective (cf. also Corollary 13.6 below). Therefore f yields an injective map ($\mathrm{char}(k) = 0$)

$$H^0(X, \Omega^1_X) \hookrightarrow H^0(\mathrm{Alb}(Y), \Omega^1_{\mathrm{Alb}(Y)}).$$

Since the latter vector space can be canonically identified with $H^0(Y, \Omega^1_Y)$, we actually get that the canonical map $H^0(X, \Omega^1_X) \to H^0(Y, \Omega^1_Y)$ is injective. Since Ω^1_X is trivial and Y is projective, $H^0(X, \Omega^1_X) \cong H^0(Y, \Omega^1_X|Y)$, whence the canonical map (induced by the canonical map $\Omega^1_X|Y \to \Omega^1_Y$)

$$H^0(Y, \Omega^1_X|Y) \to H^0(Y, \Omega^1_Y) \tag{12.3}$$

is also injective.

On the other hand, the normal bundle $N_{Y|X}$ is a quotient of $T_X|Y$, and since T_X is trivial, $N_{Y|X}$ is generated by its global sections. Therefore we may apply Theorem 12.14 of Gieseker. Let L be the quotient line bundle of $N_{Y|X}$. We have to check that L is ample on Y, i.e. $\deg(L) > 0$. Suppose to the contrary: $\deg(L) \leq 0$. Now L is also generated by its global sections, so we would have $\deg(L) = 0$, and in fact, $L \cong \mathcal{O}_Y$. Therefore the dual $N^*_{Y|X}$ has a subbundle isomorphic to $\mathcal{O}_Y$, and in particular, $H^0(Y, N^*_{Y|X}) \neq 0$. Now the canonical exact sequence

$$0 \to N^*_{Y|X} \to \Omega^1_X|Y \to \Omega^1_Y \to 0,$$

implies that

$$H^0(Y, N^*_{Y|X}) = \mathrm{Ker}(H^0(Y, \Omega^1_X|Y) \to H^0(Y, \Omega^1_Y)),$$

which contradicts the injectivity of the map (12.3). □

Proof of Theorem 12.14. By a general property any quotient of an ample vector bundle is again ample (see e.g. [74]), so if E is ample, any quotient line bundle of E is also ample.

Conversely, assume that any quotient line bundle of E is ample. We shall proceed by induction on the rank of E. If E is a line bundle the statement is trivial. Assume therefore that E has rank $r \geq 2$. If F is an arbitrary quotient vector bundle of E of rank $\leq r - 1$ we may assume by induction that F is ample.

Now since E is generated by its global sections, by Proposition 1.5 a general section $s \in H^0(X, E)$ does not vanish at any point of Y. Such a section s yields an injective map $\mathcal{O}_Y \to E$ (sending 1 to s), and the fact that s does not vanish at any point of Y translates into the fact that the subsheaf $\mathcal{O}_Y$ of E is actually a vector subbundle of E. Therefore we obtain an exact sequence of vector bundles

$$0 \to \mathcal{O}_Y \to E \to F \to 0, \tag{12.4}$$

with F a quotient vector bundle of rank $r-1$. By induction hypothesis, F is ample. Moreover, the exact sequence (12.4) does not split, because otherwise $\mathcal{O}_Y$ would be a quotient of E, hence $\mathcal{O}_Y$ would be ample by hypothesis; but this is impossible because the curve Y is projective. Thus the proof of Theorem 12.14 is reduced to the following:

Claim 1. ($\mathrm{char}(k) = 0$) If we are given a non-split exact sequence (12.4) of vector bundles on a smooth projective curve Y, then E is ample if F is so.

The proof of claim 1 will make use of the following:

Claim 2. ($\mathrm{char}(k) = 0$) Let $f\colon Z \to Y$ be a finite flat morphism of irreducible varieties, and let E be a vector bundle on Y. Then for every $i \geq 0$ the natural maps $H^i(Y,E) \to H^i(Z, f^*(E))$ are injective.

We shall first prove claim 2. Since $\mathrm{char}(k) = 0$ the natural map $\mathcal{O}_Y \to f_*(\mathcal{O}_Z)$ has a retraction, namely the trace map $f_*(\mathcal{O}_Z) \to \mathcal{O}_Y$. (The existence of the trace map makes use of the fact that f is finite and flat.) In particular, $\mathcal{O}_Y$ is a direct summand of $f_*(\mathcal{O}_Z)$. It follows that E is a direct summand of $f_*(\mathcal{O}_Z)\otimes E$. Since f is finite, $H^i(Z, f^*(E))$ is canonically isomorphic to $H^i(Y, f_*(f^*(E)))$ and $f_*(f^*(E))$ is isomorphic to $E \otimes f_*(\mathcal{O}_Z)$. In particular, the map $H^i(Y,E) \to H^i(Z, f^*(E))$ is injective, and claim 2 is proved.

To prove claim 1, we have to show that $\mathcal{O}_{\mathbb{P}(E)}(1)$ is ample on $\mathbb{P}(E)$. We may regard $\mathbb{P}(F)$ as a closed subscheme of $\mathbb{P}(E)$. In fact $\mathbb{P}(F)$ can be described more explicitly as follows. Let $s \in H^0(Y,E)$ be the section given by the inclusion $\mathcal{O}_Y \hookrightarrow E$ of (**). Then s yields a section of $p^*(E)$ (where $p : \mathbb{P}(E) \to Y$ is the canonical projection), and hence a section s' of $\mathcal{O}_{\mathbb{P}(E)}(1)$ on $\mathbb{P}(E)$ (as $\mathcal{O}_{\mathbb{P}(E)}(1)$ is a quotient of $p^*(E)$). Then the zero locus $Z(s')$ is just $\mathbb{P}(F)$ and $\mathcal{O}_{\mathbb{P}(E)}(1)|\mathbb{P}(F) = \mathcal{O}_{\mathbb{P}(F)}(1)$. In particular, $\mathcal{O}_{\mathbb{P}(E)}(1)|\mathbb{P}(E) \setminus \mathbb{P}(F) = \mathcal{O}_{\mathbb{P}(E)\setminus\mathbb{P}(F)}$.

Now to prove that $\mathcal{O}(1) := \mathcal{O}_{\mathbb{P}(E)}(1)$ is ample we use Kleiman's form of the Nakai–Moishezon criterion for ampleness (see [89], Chap. III, Theorem 1). According to this criterion we have to show that the following condition holds true: if Z is an arbitrary integral subscheme of dimension > 0 of $\mathbb{P}(E)$, then there is a section $t \in H^0(Z, \mathcal{O}(n)|Z)$ for some $n > 0$ which vanishes at some point of Z, but does not vanish identically.

If Z is contained in $\mathbb{P}(F)$ this condition follows from the fact that $\mathcal{O}(1)|\mathbb{P}(F) = \mathcal{O}_{\mathbb{P}(F)}(1)$ is ample, whence $\mathcal{O}(1)|Z$ is also ample. Assume therefore that Z is not contained in $\mathbb{P}(F)$. If moreover Z meets $\mathbb{P}(F)$ the section $s'|Z$ does not vanish identically (because s' does not vanish at any point of $\mathbb{P}(E)\setminus\mathbb{P}(F)$), but does vanish on $\mathbb{P}(F) \cap Z \neq \varnothing$. So it will be sufficient to prove that every integral subscheme Z of $\mathbb{P}(E)$ of positive dimension meets $\mathbb{P}(F)$. Suppose it does not. Clearly we may assume that Z is a curve. If Z is contained in a fiber of p then certainly Z meets $\mathbb{P}(F)$. If Z is not contained in any fiber of p we get a finite flat morphism $f := p|Z\colon Z \to Y$ (f is surjective and Y is a smooth curve) such that $f^*(E)$ has $\mathcal{O}_Z$ as a quotient (because there is the canonical surjection $p^*(E) \to \mathcal{O}(1)$ and $Z \cap \mathbb{P}(F) = \varnothing$ implies $\mathcal{O}(1)|Z \cong \mathcal{O}_Z$). This implies that the extension induced by (12.4)

$$0 \to \mathcal{O}_Z \to f^*(E) \to f^*(F) \to 0$$

is trivial. Since $H^1(Y, F^*)$ classifies the isomorphism classes of all extensions of F by $\mathcal{O}_Y$, this implies that the natural map

$$H^1(Y, F^*) \to H^1(Z, f^*(F^*))$$

is not injective, contradicting claim 2. So Z meets $\mathbb{P}(F)$ and claim 1 (and thereby Theorem 12.14) is proved. □

The second aim of this appendix is to prove a result of Bertini type (see Proposition 12.15 below) that was used in the proof of Theorem 12.2. We shall follow [79] closely.

First we need some setup. We shall fix a universal domain Ω, which is an algebraically closed field containing k, of infinite transcendence degree over k. All the base fields of the algebraic schemes we are going to consider will be subfields of Ω containing k, over which Ω has infinite transcendence degrees. A geometric point of an algebraic k-scheme X is a k-morphism $\mathrm{Spec}(\Omega) \to X$. This amounts to giving a point x of the scheme X together with a k-isomorphism of the residue field $k(x)$ onto a subfield of Ω. These two isomorphic fields will be simply denoted by $k(x)$ when there will be no danger of confusion. Let Z be an algebraic k-scheme and X_i a closed subscheme of $Z_{K_i} := Z \otimes_k K_i$, $i = 1, 2$, where K_i are field subextensions of $\Omega|k$ of finite transcendence degree over k. We shall say that X_1 "contains" X_2 (and write $X_1 \supset X_2$) if $X_{1L} \supset X_{2L}$ in the scheme-theoretical sense (i.e. the ideal sheaf defining X_{1L} in Z_L is contained in the ideal sheaf defining X_{2L} in Z_L), where L is the subfield of Ω generated by K_1 and K_2.

For a field subextension $k \subset K$ of Ω, a K-subscheme Y of a k-scheme X is by definition a closed subscheme Y of X_K. In particular, it makes sense to speak of a K-linear subspace of the projective space $\mathbb{P}^n_k$ over k. If W is a K-subscheme of $\mathbb{P}^n_k$, the *linear dimension* of W in $\mathbb{P}^n_k$ is by definition the dimension of the smallest linear $\overline{K}$-subspace of $\mathbb{P}^n_k$ which contains W, where $\overline{K}$ is the algebraic closure of K. By a *k-generic hyperplane* we mean the closed subscheme of $\mathbb{P}^n_K = \mathrm{Proj}(K[T_0, \dots, T_n])$ defined by a polynomial of the form $\sum_{i=0}^{n} u_i T_i$, where the coefficients $u_0, \dots, u_n$ are algebraically independent elements of K over k, and $K = k(u)$ is the subfield of Ω generated by $u_0, \dots, u_n$ over k. Finally, if X is a k-subscheme of $\mathbb{P}^n_k$ and if H is a k-generic hyperplane as above, then $X \cap H$ denotes the subscheme of X_K, with $K := k(u)$, whose ideal sheaf is generated in $\mathcal{O}_{X_K}$ by the ideal sheaf of H in $\mathbb{P}^n_K$.

Proposition 12.15 (Bertini–Hironaka [79]). *Let $f\colon X \to Y$ be a morphism of algebraic k-schemes, and $j\colon X \to \mathbb{P}^n_k$ a proper morphism. Assume that for every $y \in Y$ and for every irreducible component W of $f^{-1}(y)_{\mathrm{red}}$ such that* $\dim(W) = \dim(f^{-1}(y))$*, the linear dimension of $j(W)$ in $\mathbb{P}^n_k$ is* $\geq \dim(Y)$*. Then for every general hyperplane H of $\mathbb{P}^n$,* $\dim(f^{-1}(y) \cap j^{-1}(H)) < \dim(f^{-1}(y))$*, $\forall y \in Y$.*

The key point in the proof of Proposition 12.15 is the following:

Lemma 12.16 ([79], Lemma 2.1). *Let $f\colon X \to Y$ be a morphism of algebraic k-schemes, and $j\colon X \to \mathbb{P}^n_k$ a proper morphism. Let H be a k-generic hyperplane in $\mathbb{P}^n_k$, and W an algebraic $k(y)$-subscheme of $f^{-1}(y)$ for some geometric point y of Y. If $j(W) \subseteq H$, then the linear dimension of $j(W)$ in $\mathbb{P}^n_k$ is* $\leq \operatorname{tr.deg}_k(k(y)) - 1$.

Proof. Since j is proper and $W \subseteq f^{-1}(y)$, $j(W)$ is a $k(y)$-subscheme of $\mathbb{P}^n_k$. Let r denote the linear dimension of $j(W)$ in $\mathbb{P}^n_k$. Then there are $r+1$ projectively independent geometric points $t_0, t_1, \ldots, t_r$ of the smallest linear $\overline{k(y)}$-subspace of $\mathbb{P}^n$ containing $j(W)$ such that $k(t_i)$ is algebraic over $k(y)$, $i = 0, 1, \ldots, r$. If the equation of H is $\sum_{i=0}^{n} u_i T_i$, $t_i \in H$ for $0 \leq i \leq r$, this implies $r+1$ independent $\overline{k(y)}$-linear dependences among the coefficients of u_i with $0 \leq i \leq n$. Hence

$$\operatorname{tr.deg}_k(k(y)) + (n+1) - (r+1) \geq \operatorname{tr.deg}_k(k(y,u)) \geq \operatorname{tr.deg}_k(k(u)) \geq n+1,$$

so that $\operatorname{tr.deg}_k(k(y)) - 1 \geq r$. □

Proof of Proposition 12.15. By hypothesis, for every irreducible component W of $f^{-1}(y)_{\mathrm{red}}$ such that $\dim(W) = \dim(f^{-1}(y))$, the linear dimension of $j(W)$ in $\mathbb{P}^n_k$ is $\geq \dim(Y)$. Since $\operatorname{tr.deg}(k(y)) \leq \dim(Y)$, we infer that this linear dimension is $\geq \operatorname{tr.deg}(k(y))$ for every $y \in Y$. Therefore by Lemma 12.16, a k-generic hyperplane H of $\mathbb{P}^n_k$ cannot contain $j(W)$ for any such W. In other words, if H is a k-generic hyperplane H of $\mathbb{P}^n_k$, then for every such W, $W \not\subseteq j^{-1}(H)$. This implies that for H k-generic, $\dim(f^{-1}(y) \cap j^{-1}(H)) < \dim(f^{-1}(y))$ for every $y \in Y$. Then the conclusion follows by the general philosophy concerning Bertini-type theorems (see e.g. [87]). □

Chapter 13

Formal Functions on Homogeneous Spaces

Unless otherwise specified, throughout this chapter k will be an algebraically closed field of characteristic zero.

This chapter is based upon [18], except for Example 13.1 and Proposition 13.20. Let X be a projective homogeneous space over k, acted on transitively by the connected algebraic group G, and let Y be a closed irreducible subvariety of X. Fix a point $p \in Y$ and consider the morphism $\varphi\colon G \to X$ defined by $\varphi(g) = g \cdot p$. Then the subset $G_{Y,p} := \varphi^{-1}(Y) = \{g \in G | g \cdot p \in Y\}$ is closed in G. Denote by G_Y the subgroup of G generated by $G_{Y,p}$. It is easily seen that G_Y depends only on Y, and not on the reference point $p \in Y$. A basic elementary fact is that G_Y is a closed subgroup of G (see [35], or also the discussion following Definition 13.1 below when G is a linear algebraic group).

Definition 13.1 (Chow [35]). One says that the closed subvariety Y *generates the homogeneous space* X if $G_Y = G$, i.e. if $G_{Y,p}$ generates G (in the group-theoretical sense).

Let us first remark that if Y generates X then $g \cdot Y$ also generates X for every $g \in G$. Indeed, if we fix a point $p \in Y$ then $G_{g\cdot Y, g\cdot p} = gG_{Y,p}g^{-1}$ as is easily seen, whence $G_{Y,p}$ generates G if and only if $G_{g\cdot Y,g\cdot p}$ generates G.

From now on we shall fix a smooth point p of Y as the reference point. Since for all $x \in X$ the fiber $\varphi^{-1}(x)$ is isomorphic to the isotropy group G_x of x, we infer that the morphism $\varphi|G_{Y,p} : G_{Y,p} \to Y$ is smooth, and in particular is open. Moreover, $G_{Y,p}$ is irreducible if G acts on X by connected isotropy groups, because the fibers of $\varphi|G_{Y,p}$ are all irreducible, Y is irreducible and the morphism $\varphi|G_{Y,p}$ is open. This is the case when G is a linear algebraic group. Indeed, the isotropy group $P = G_p$ of the reference point p is a parabolic subgroup of G (since $X \cong G/G_p$ is projective), whence P contains a Borel subgroup B. If P^0 is the connected component of P containing the unit element e of G, then P^0 still contains B (since B is connected). In particular, P^0 is also a parabolic subgroup of G and P is contained in the normalizer $N_G(P^0)$ of P^0 in G. But by a result of Chevalley (see [30], Theorem 11.15), $P^0 = N_G(P^0)$, whence $P = P^0$, i.e. P is connected. (Alternatively, the canonical morphism $\pi\colon G/P^0 \to G/P = X$ induced by the inclusion $P^0 \subseteq P$ is a finite étale morphism of degree equal to the index of P^0 in P. Since X is a rational variety, it is simply connected, whence π is an isomorphism, i.e. $P^0 = P$.)

If G is an arbitrary connected algebraic group, $p = \varphi(e)$ is a smooth point of Y and the morphism $\varphi|G_{Y,p}$ is smooth, whence e is a smooth point of $G_{Y,p}$. In particular, there is a unique irreducible component, call it H, of $G_{Y,p}$ passing through the point e. By using general dimensionality arguments, Chow observes in [35] that Y generates X if and only if H generates G (in the group-theoretical sense).

In the above notation, fix a natural number $i \geq 1$ and consider the morphism $\varphi_i\colon H^{2i} \to G$ (where $H^{2i} = H \times \cdots \times H$ ($2i$ times)) defined by

$$\varphi_i(g_1, \ldots, g_{2i}) = g_{2i}^{-1} g_{2i-1} \cdots g_t^{(-1)^{t+1}} \cdots g_2^{-1} g_1.$$

Since $\varphi_i(g_1, \ldots, g_{2i}) = \varphi_{i+1}(g_1, \ldots, g_{2i}, e, e)$, we infer that $\varphi_i(H^{2i}) \subseteq \varphi_{i+1}(H^{2(i+1)})$ for every $i \geq 1$. Moreover, because $g = \varphi_1(e, g)$, $\forall g \in G$, $H \subseteq \varphi_1(H^2)$. From the definition of φ_i we get that $g^{-1} \in \varphi_i(H^{2i})$, for every $g \in \varphi_i(H^{2i})$, and $\varphi_i(H^{2i}) \cdot \varphi_j(H^{2j}) \subseteq \varphi_{i+j}(H^{2(i+j)})$ for every $i, j \in \mathbb{N}$. It follows that the subgroup $\langle H \rangle$ generated by H coincides with $\bigcup_{i=1}^{\infty} \varphi_i(H^{2i})$.

On the other hand, the ascending sequence

$$\overline{\varphi_1(H^2)} \subseteq \overline{\varphi_2(H^4)} \subseteq \cdots \subseteq \overline{\varphi_i(H^{2i})} \subseteq \cdots$$

of closed irreducible subsets (where "overline" means taking closure) has to become stationary, i.e. there is an $n \geq 1$ such that $\overline{\varphi_n(H^{2n})} = \overline{\varphi_{n+p}(H^{2(n+p)})}$ for every $p \geq 1$. Set $F := \overline{\varphi_n(H^{2n})}$. Then F is the closure in G of the subgroup $\langle H \rangle$, and so F is a closed (connected) subgroup of G. Since by a general result of Chevalley $\varphi_n(H^{2n})$ contains a non-empty open subset U of F, we infer that

$$\langle H \rangle = \bigcup\nolimits_{g \in \langle H \rangle} gU,$$

and in particular, $\langle H \rangle$ is an open subgroup of F, whence $\langle H \rangle = F$. In other words, we proved the well-known fact that the subgroup $\langle H \rangle$ generated by a closed irreducible subvariety H passing through e is a closed connected group of G. Note that we also showed that there exists an $n \geq 1$ such that the morphism $\varphi_n\colon H^{2n} \to \langle H \rangle$ is dominant. In particular, $\varphi_n(H^{2n})$ generates the subgroup $\langle H \rangle$.

Proposition 13.2. *Let X be a projective homogeneous space, and let Y be a closed irreducible subvariety of X. Then Y generates X if and only if Y intersects every irreducible hypersurface of X.*

Proof. This result was proved by Speiser in the case when X is an abelian variety (see [141]), and by Debarre in general in [41], (2.1). The proof which follows (taken from [18]) is an adaptation of Speiser's proof to the general case. Assume that Y generates X and that there is an irreducible hypersurface D of X such that $Y \cap D = \varnothing$. If $\varphi\colon G \to X$ is the morphism $\varphi(g) = g \cdot p$ from the beginning of this chapter, set $D' := \varphi^{-1}(D)$.

Claim 1. For every $g \in G$ we have either $gG_{Y,p} \cap D' = \varnothing$, or $gG_{Y,p} \subseteq D'$.

Indeed, since $G_{Y,p} = \varphi^{-1}(Y)$ claim 1 is a consequence of the following statement: either $gY \cap D = \varnothing$, or $gY \subseteq D$. Assume that this statement does not hold, i.e. $gY \cap D \neq \varnothing$ and $gY \not\subseteq D$. Then $Y \cap g^{-1}D \neq \varnothing$ and $Y \not\subseteq g^{-1}D$. Hence the cycles $Y \cdot D = 0$ and $Y \cdot g^{-1}D > 0$ are numerically equivalent on Y, but this is impossible since Y is a projective variety. This proves claim 1.

Claim 2. For every $g, h \in G_{Y,p}$ one has $D'h^{-1}g \subseteq D'$.

To prove claim 2, let $g, h \in G_{Y,p}$ and $\delta \in D'$ be arbitrary elements. Then $\delta \in \delta h^{-1}G_{Y,p} \cap D'$ (because $h \in G_{Y,p}$), and so, by claim 1 we get $\delta h^{-1}G_{Y,p} \subseteq D'$. In particular, $\delta h^{-1}g \in D'$, and since δ was arbitrary in D', we get $D'h^{-1}g \subseteq D'$.

By applying claim 2 repeatedly we get that for any $n \geq 1$ and for every $g_1, g_2, \ldots, g_{2n} \in G_{Y,p}$ one has

$$D' g_{2n}^{-1} g_{2n-1} \cdots g_2^{-1} g_1 \subseteq D',$$

from which, using the discussion following Definition 13.1, we get $D' \cdot G_Y \subseteq D'$. So far we didn't use the hypothesis that Y generates X. This means that $G = G_Y$. Therefore $D' \cdot G \subseteq D'$, which is obviously impossible because D' is a hypersurface of G.

The converse is immediate. Indeed, if Y does not generate X, set $X = G/P$, with P the isotropy group of the reference point $p \in Y$, and denote by P' the (closed) subgroup of G_Y (generated by $G_{Y,p}$), and by $\pi\colon X = G/P \to X' := G/P'$ the canonical (surjective) morphism. Then $\pi(Y)$ is a point of X' and $\dim(X') \geq 1$ (because Y does not generate X), whence there exists an irreducible hypersurface H of X' such that $\pi(Y) \notin H$. Then any component of $\pi^{-1}(H)$ is an irreducible hypersurface of X which does not meet Y. □

Corollary 13.3. *Let X be a projective homogeneous space, and let Y be a closed irreducible subvariety of X of dimension ≥ 2. Let $\mathcal{O}_X(1)$ be a very ample line bundle on X. If Y generates X then $Y' := Y \cap H$ also generates X, for every general $H \in |\mathcal{O}_X(1)|$.*

Proof. Let Z be an arbitrary hypersurface. By Proposition 13.2 it will be sufficient to show that $Y' \cap Z \neq \varnothing$. Since Y generates X, by Proposition 13.2, $Y \cap Z \neq \varnothing$, and since $\dim(Y) \geq 2$, $\dim(Y \cap Z) \geq 1$ (because the ambient space is smooth). This implies that $H \cap (Y \cap Z) \neq \varnothing$ (since H is a hyperplane section), whence $Y' \cap Z \neq \varnothing$. □

Theorem 13.4 (Chow [36]). (i) *Let X be a projective homogeneous space over k, and let Y be a closed irreducible subvariety of X. Then the natural map $K(X) \to K(X_{/Y})$ is an algebraic field extension if and only if Y generates X in the sense of Definition* 13.1.
(ii) *Under the hypotheses of* (i), *assume that Y generates X. Let $\zeta \in K(X_{/Y})$ be a formal-rational function which is algebraic of degree d over $K(X)$. Then there exists a finite étale morphism $f\colon X' \to X$ of degree d from a smooth irreducible projective variety X' such that: $\tilde{f}^*(\zeta) \in K(X')$, the irreducible components of $f^{-1}(Y)$ are mutually disjoint, numerically equivalent to each other as cycles on X' and are mapped by f isomorphically onto Y.*

Proof of part (ii) *of Theorem* 13.4. We follow [18], using a beautifully simple idea of Chow [35]. By Theorem 9.16 there exists a finite surjective morphism $f\colon X' \to X$, with X' normal and irreducible, a closed subvariety Y' of X' with the following properties: $f(Y') \subseteq Y$, the restriction $f|Y' : Y' \to Y$ is an isomorphism, f is étale in a neighbourhood of Y', and $\tilde{f}^*(\zeta) \in K(X')$. We will show that f is everywhere étale.

Let $\Delta' \subset X'$ be the ramification locus of f, and let $\Delta := f(\Delta')$ be the branch locus of f. Since f is étale along Y', $Y' \cap \Delta' = \varnothing$. Let u be the generic point of the group G, and let v be the generic point of Y. If $s\colon G \times X \to X$ is the morphism arising from the action of G on X, then by transitivity of this action we get that $s(u, v) = u \cdot v$ is the generic point of X. It follows that $u \cdot Y = s(\{u\} \times Y)$ is dense in X, whence $f^{-1}(u \cdot Y))$ is also dense in X', and in particular, $f^{-1}(u \cdot Y)$ is an irreducible scheme over $k(u) = K(G)$ (because X' is irreducible). Consider the incidence variety $Z := \{(gy, g) \mid g \in G,\ y \in Y\} \subset X \times G$, and let $\psi\colon Z \to X$ and $\pi\colon Z \to G$ be the morphisms defined by $\psi(gy, g) = gy$ (ψ is surjective because G acts transitively on X) and $\pi(gy, g) = g$. Clearly $\psi|\pi^{-1}(u)$ defines a $k(u)$-isomorphism between $\pi^{-1}(u)$ and $u \cdot Y$. It follows that the $k(u)$-scheme of finite type $f^{-1}(u \cdot Y)$ is irreducible. If $\overline{k(u)}$ is an algebraic closure of $k(u)$ we infer that the irreducible components of the $\overline{k(u)}$-scheme $(\pi \circ \varphi)^{-1}(u) \otimes_{k(u)} \overline{k(u)} \cong f^{-1}(u \cdot Y) \otimes_{k(u)} \overline{k(u)}$ are all conjugate over $k(u)$ under the natural action of the Galois group $\mathrm{Gal}(\overline{k(u)}|k(u))$. Thus we get the commutative diagram with cartesian squares

$$\begin{array}{ccccc}
f^{-1}(u \cdot Y) \otimes_{k(u)} \overline{k(u)} & \longrightarrow & (u \cdot Y) \otimes_{k(u)} \overline{k(u)} & \longrightarrow & \mathrm{Spec}(\overline{k(u)}) \\
\downarrow & & \downarrow & & \downarrow \\
Z' = X' \times_X Z & \xrightarrow{\ \varphi\ } & Z & \xrightarrow{\ \pi\ } & G \\
\rho \downarrow & & \downarrow \psi & & \\
X' & \xrightarrow{\ f\ } & X & &
\end{array}$$

Clearly the ramification locus of φ is $\Delta_1 := \rho^{-1}(\Delta')$. On the other hand, by the proof of Corollary 9.22 we know that Y' is an isolated connected component of $f^{-1}(Y)$, which is isomorphic to Y. It follows that $(\pi \circ \varphi)^{-1}(e)$ has an isolated irreducible component (corresponding to Y') which is contained in the étale locus of φ. Since this component is a specialization of an isolated irreducible component T of the geometric generic fiber $f^{-1}(u \cdot Y) \otimes_{k(u)} \overline{k(u)}$, it follows that T does not intersect the ramification locus of the morphism $f^{-1}(u \cdot Y) \otimes_{k(u)} \overline{k(u)} \to (u \cdot Y) \otimes_{k(u)} \overline{k(u)}$. Since the irreducible components of $f^{-1}(u \cdot Y) \otimes_{k(u)} \overline{k(u)}$ are conjugate under the action of $\mathrm{Gal}(\overline{k(u)}|k(u))$, we infer that the morphism

$$f^{-1}(u \cdot Y) \otimes_{k(u)} \overline{k(u)} \to (u \cdot Y) \otimes_{k(u)} \overline{k(u)}$$

is étale and every irreducible component of $f^{-1}(u \cdot Y) \otimes_{k(u)} \overline{k(u)}$ is isolated and maps isomorphically onto $(u \cdot Y) \otimes_{k(u)} \overline{k(u)}$. In particular, $(\pi \circ \varphi)^{-1}(u) \cap \Delta_1 = \varnothing$. It follows that $(\pi \circ \varphi)^{-1}(u) \cap \Delta_1 = \varnothing$, which implies that $u \notin \pi(\varphi(\Delta_1))$. Set $U := G \setminus \pi(\varphi(\Delta_1))$. Since the morphisms π and φ are proper, $\pi(\varphi(\Delta_1))$ is a proper closed subset of G, whence U is a non-empty open subset of G. Therefore for all $g \in U$, $f^{-1}(g \cdot Y) \cap \Delta' = \varnothing$. In other words, $(g \cdot Y) \cap \Delta = \varnothing$, $\forall g \in U$.

Now, since Y generates X, $g \cdot Y$ also generates X, and since $(g \cdot Y) \cap \Delta = \varnothing$, $\forall g \in U$, $\mathrm{codim}_X(\Delta) \geq 2$, by Proposition 13.2. Finally, since X is smooth we can apply the purity of the branch locus (see e.g. [1], VI, Theorem (6.8), or [68], exposé X,Theorem 3.4) to deduce that $\mathrm{codim}_X(\Delta) \geq 2$ implies $\Delta = \varnothing$, i.e. f is everywhere étale. In particular, since X is smooth and f étale, X' is also smooth.

Actually we got even more: the irreducible components of $f^{-1}(g \cdot Y)$ are mutually disjoint, all isomorphic to $g \cdot Y$, and numerically equivalent to each other for every $g \in G$. □

Remark 13.5. The implication "Y generates X" $\Longrightarrow$ "the fields extension $K(X) \to K(X_{/Y})$ is algebraic" of (i) is the hard part of Theorem 13.4. Chow's proof given in [36] follows the idea of the proof of the analytic counterpart of it (see [35]) plus arguments involving non-archimedian function theory.

Note also that the proof of the reverse implication of (i) is easy. Indeed, assume that $K(X) \to K(X_{/Y})$ is an algebraic field extension. Then Y must generate X. Assume not. Since $X = G/P$ we can take as the reference point p just e mod P. Let $Q := G_Y$ be the (closed) subgroup of G generated by $G_{Y,p}$. Since $Y \subseteq Q/P$, Y is mapped to a point by the canonical proper surjective morphism $f\colon X = G/P \to G/Q$. Since $\dim(G/Q) \geq 1$, it follows from Remark 9.24 that the field extension $K(X) \to K(X_{/Y})$ is transcendental, a contradiction. □

Corollary 13.6 (Hironaka–Matsumura [81]). *Let Y be a closed irreducible subvariety of an abelian variety X, let $i\colon Y \to \mathrm{Alb}(Y)$ be the canonical morphism into the strict Albanese variety $\mathrm{Alb}(Y)$ of Y in the sense of [81] (if Y is smooth, the strict Albanese variety coincides with the usual one), and let $f : \mathrm{Alb}(Y) \to X$ be the morphism induced by the inclusion $Y \to X$ arising from the universal property of* *strict Albanese varieties (f is a homomorphism of abelian varieties if we assume that Y passes through the origin 0_X of X). Then:*

(i) *Y generates X if and only if f is surjective.*

(ii) *The morphism i is a closed embedding and $i(Y)$ is $G3$ in $\mathrm{Alb}(Y)$.*

(iii) *If Y generates X then Y is $G2$ in X. Moreover, Y is $G3$ in X if and only if Y generates X and the canonical morphism f has connected fibers.*

Proof. Part (i) is trivial, as well as the fact that i is a closed embedding (because $f \circ i$ coincides with the inclusion $Y \subset X$). Clearly $i(Y)$ generates $A := \mathrm{Alb}(Y)$. If Y is smooth, Proposition 12.7 implies that the normal bundle of $i(Y)$ in A is $(\dim(Y) - 1)$-ample, whence by Theorem 12.6, $i(Y)$ is $G2$ in A, i.e. the field extension $K(A) \to K(A_{/i(Y)})$ is finite. If Y is singular, by Theorem 13.4, (i), the field extension $K(A) \to K(A_{/i(Y)})$ is algebraic anyway. Assume that there is a formal-rational function $\zeta \in K(A_{/i(Y)}) \setminus K(A)$. Since in both cases ζ is algebraic over $K(A)$, by part (ii) of Theorem 13.4 there is a finite étale morphism $g\colon A' \to A$ such that $\tilde{g}^*(\zeta) \in K(A')$ and such that the embedding $i\colon Y \to A$ lifts to an embedding $i' : Y \to A'$. Then by a theorem of Serre–Lang (see [114], Chapter IV, §18) there is a unique structure of abelian variety on A' such that g becomes an isogeny of abelian varieties. Then by the universal property of $\mathrm{Alb}(Y)$, there is a unique morphism $h\colon A \to A'$ such that $h \circ i = i'$. It follows that $g \circ h = \mathrm{id}_A$, which would disconnect A' if $\deg(g) > 1$. Thus $\deg(g) = 1$. Recalling that $\deg(g) = \deg_{K(A)}(\zeta)$ we infer that $\zeta \in K(A)$, i.e. $i(Y)$ is $G3$ in A. So, part (ii) is proved.

Part (iii) follows now easily from (ii). Indeed, look at the map $\tilde{f}^* : K(X_{/Y}) \to K(A_{/i(Y)})$. By (ii), $K(A_{/i(Y)}) = K(A)$, whence $\tilde{f}^*(K(X_{/Y})) \subseteq K(A)$. But the extension $f^* : K(X) \to K(A)$ is finitely generated, so that the subextension $K(X) \to K(X_{/Y})$ is also finitely generated. Being also algebraic, it is finite, i.e. Y is $G2$ in X. To prove the last statement of (iii), we apply Corollary 9.7 according to which there exists a finite morphism $u\colon X' \to X$ with the following properties: X' is a normal projective variety, the inclusion $Y \subset X$ lifts to an inclusion $Y \subset X'$, u is étale along Y, and Y is $G3$ in X. Then the argument of the proof of part (ii) of Theorem 13.4 applies also in this situation to yield the fact that u is everywhere étale. Since X is an abelian variety, X' is also an abelian variety and u is a homomorphism of abelian varieties (by the same theorem of Serre–Lang quoted above). By the universal property of $\mathrm{Alb}(Y)$ the morphism f factors as $\mathrm{Alb}(Y) \to X' \to X$, and this yields immediately the conclusion. □

Remark 13.7. The above proof of the $G2$ condition in Corollary 13.6, taken from [18], is completely different from Hironaka–Matsumura's proof [81].

Corollary 13.8. *Let X be a projective homogeneous space over k, and let Y be a closed irreducible subvariety of X that generates X. Assume furthermore that one of the following two conditions is satisfied:*

(i) *X is a rational variety, or*

(ii) $Y \cdot Y \neq 0$ (*in the Chow ring of cycles modulo numerical equivalence*).

Then Y *is* $G3$ *in* X.

Proof. Since Y generates X, by Theorem 13.4, (i) the extension $K(X) \subseteq K(X_{/Y})$ is algebraic. Let $\zeta \in K(X_{/Y})$ be an arbitrary formal-rational function. Since ζ is algebraic over $K(X)$, by Theorem 13.4, (ii) we find a finite étale morphism $f\colon X' \to X$ of degree equal to $\deg_{K(X)}(\zeta)$, with X' normal, such that a closed subvariety Y' of X' with the following properties: $f(Y') \subseteq Y$, the restriction $f|Y' : Y' \to Y$ is an isomorphism, and $\tilde{f}^*(\zeta) \in K(X')$. Under hypothesis (i), X is algebraically simply connected (as any rational variety, see e.g. [67]), hence f is an isomorphism. This implies that $\deg_{K(X)}(\zeta) = 1$, i.e. $\zeta \in K(X)$. Thus the map $\alpha_{X,Y} : K(X) \to K(X_{/Y})$ is an isomorphism in case (i).

Assume now (ii), and that f is not an isomorphism. Since f is a finite étale morphism, it follows that $d := \deg(f) \geq 2$. Then $f^{-1}(Y)$ consists of n irreducible components $Y_1 = Y', Y_2, \ldots, Y_d$. By Theorem 13.4, (ii), Y_i is numerically equivalent to $Y_1 = Y'$, $i = 2, \ldots, d$, and these components do not meet each other. It follows that

$$Y_1 \cdot f^*(Y) = Y_1 \cdot (Y_1 + Y_2 + \cdots + Y_d) = dY_1 \cdot Y_2 = 0.$$

Then, by the projection formula, we get $0 = f_*(Y_1 \cdot f^*(Y)) = f_*(Y_1) \cdot Y = Y \cdot Y$, contradicting the hypothesis (ii). This completes the proof of Corollary 13.8. □

Remark 13.9. Corollary 13.8 is an algebraic analogue of an analytic result of Chow [35]. We are going to use it in connection with the following problem raised by Hartshorne [74], page 208, problem 4.8): let Y be a smooth closed subvariety of a smooth projective variety of dimension n such that $\dim(Y) \geq \frac{n}{2}$ and the normal bundle $N_{Y|X}$ is ample. Is then Y $G3$ in X?

The following result answers this problem affirmatively in the case when X is a projective homogeneous space.

Theorem 13.10 ([18]). *Let* Y *be a smooth closed subvariety of a projective homogeneous space* X *of dimension* n *such that* $\dim(Y) \geq \frac{n}{2}$ *and the normal bundle* $N_{Y|X}$ *is ample. Then* Y *is* $G3$ *in* X.

Proof. Since the normal bundle $N_{Y|X}$ is ample, Y is $G2$ in X by Theorem 9.28. (Alternatively, the ampleness of $N_{Y|X}$ implies that Y generates X, so that, instead of using Hartshorne's result, one could use Chow's Theorem 13.4, (i) above to get that the field extension $K(X_{/Y})|K(X)$ is algebraic.) So, to prove $G3$, in view of Corollary 13.8, (ii), it will be sufficient to check that $Y \cdot Y \neq 0$. To do this, we first apply a result of Kleiman [91], valid if $\mathrm{char}(k) = 0$ (see also [73], Theorem 10.8, page 273), which asserts that for general $g \in G$ (with G the connected algebraic group acting transitively on X) either $Y \cap gY = \varnothing$, or Y and gY have a non-empty proper intersection. Clearly, we have to rule out the first possibility. Since $N_{Y|X}$ is ample and $\dim(Y) + \dim(gY) = 2\dim(Y) \geq n$ (by hypothesis), we can conclude that $Y \cap gY \neq \varnothing$ by applying a result of Lübke (see [102], or also [54] for a slightly more general result, with a different proof). □

For further consequences of Theorem 13.4 we need to recall the following:

Definition 13.11 (Ran–Debarre [122], [42]). Let Y be a d-dimensional closed irreducible subvariety ($d > 0$) of an n-dimensional abelian variety X, and let $s \geq 0$ be a non-negative integer. Then Y is said to be *s-geometrically non-degenerate* in X if for every abelian subvariety K of X the image under the canonical morphism $X \to X/K$ is either X/K, or has dimension $\geq d - s$. If Y is 0-geometrically non-degenerate in X one simply says that Y is geometrically non-degenerate ([122]).

A closed irreducible subvariety Y is s-geometrically non-degenerate precisely if for every closed subvariety Z of X such that $\dim(Y) + \dim(Z) \geq \dim(X) + s$ one has $Y \cap Z \neq \varnothing$ (see [42], (1.5), or also Theorem 13.22 of Appendix D at the end of this chapter). Examples of s-geometrically non-degenerate closed subvarieties are smooth irreducible (positive dimensional) closed subvarieties Y of an abelian variety X whose normal bundle $N_{Y|X}$ is s-ample, (see [42], (1.3), or also Proposition 13.23 of Appendix D below). Every irreducible (positive-dimensional) closed subvariety Y of a simple abelian variety X is geometrically non-degenerate.

Corollary 13.12. *Let Y be a closed irreducible subvariety of dimension d of an abelian variety X of dimension n, and assume that $d \geq \frac{n}{2}$ and that Y is $(d - [\frac{n}{2}])$-geometrically non-degenerate in X. Then Y is G3 in X. In particular, every closed irreducible subvariety of dimension $d \geq \frac{n}{2}$ of a simple n-dimensional abelian variety X is G3 in X.*

Proof. Since $d - [\frac{n}{2}] \leq d - 1$, the hypothesis implies that for every proper abelian subvariety K of X (i.e. $K \neq X$), the image of Y in X/K is at least one-dimensional. This implies that Y cannot be contained in any proper abelian subvariety of X, hence Y generates X. Thus, by Corollary 13.6, (iii) of Hironaka–Matsumura, Y is $G2$ in X.

To get the result, it will be sufficient to check that condition (ii) of Corollary 13.8 is fulfilled. By [91], for general $x \in X$ the intersection of Y with the translate $t_x(Y) = Y + x$ is either empty, or smooth of dimension $2d - n$. In particular, Y and $t_x(Y)$ are in general position for $x \in X$ general. To show that $Y \cdot Y \neq 0$, it remains to show that for every $x \in X$, $Y \cap t_x(Y) \neq \varnothing$. This follows from the above mentioned intersection properties of s-geometrically non-degenerate closed subvarieties [42], (1.5) (see also Theorem 13.22 of Appendix D). □

The last part of Corollary 13.12 was first proved by Hironaka–Matsumura [81], and in the analytic context, by Barth [21].

Corollary 13.13 (Sommese [140]). *Let Y be a positive dimensional connected closed subvariety of an abelian variety X. Assume that* $\dim(Y) - \operatorname{codim}(Y) \geq m$, *where m is the maximum dimension of a proper abelian subvariety of X. Then Y is G3 in X.*

Proof. The inequality $\dim(Y) - \operatorname{codim}(Y) \geq m$ implies that Y cannot be contained in any proper abelian subvariety of X, whence Y generates X, and so, by Corollary 13.6, (iii), Y is $G2$ in X. Then, as in the proof of Corollary 13.12, it will be sufficient to show that $Y \cap t_x(Y) \neq \emptyset$, $\forall x \in X$. Let K be an arbitrary proper abelian subvariety of X, and denote by $\varphi\colon X \to X/K$ the canonical morphism. Clearly, $\dim(\varphi(Y)) \geq \dim(Y) - \dim(K) \geq \dim(Y) - m$. Therefore Y is m-geometrically non-degenerate, and since $\dim(Y) + \dim(t_x(Y)) = 2\dim(Y) \geq \dim(X) + m$ (by hypothesis), by [42], (1.5) (or also Theorem 13.22 of Appendix D), $Y \cap t_x(Y) \neq \emptyset$, $\forall x \in X$, as claimed. □

Corollary 13.14 (Babakarian–Hironaka [8]). *Let Y be a closed irreducible subvariety of positive dimension of a Grassmann variety X. Then Y is $G3$ in X.*

Proof. Indeed, if X is a Grassmann variety, then X is a homogeneous space $X = G/P$, with P a maximal parabolic subgroup of G, whence for every closed subvariety Y of positive dimension we have $G_Y = G$, i.e. Y generates X. Then the conclusion follows from Corollary 13.8, (i). □

Remark 13.15. If X is a projective space Corollary 13.14 is of course identical to Theorem 9.14. This generalization of Theorem 9.14 to Grassmann varieties had been obtained for the first time by Babakarian and Hironaka in [8] (even in arbitrary characteristic).

Before stating the last application let us recall Goldstein's notion of coampleness of a projective homogeneous space X (see [60]).

Definition 13.16. Let X be a projective homogeneous space X. Then the tangent bundle T_X is generated by its global sections. Consider the projective bundle $\mathbb{P}(T_X)$ associated to T_X. It follows that the tautological line bundle $\mathcal{O}_{\mathbb{P}(T_X)}(1)$ is also generated by its global sections. Consider then the morphism $\varphi : \mathbb{P}(T_X) \to \mathbb{P}^N$ associated to the complete linear system $|\mathcal{O}_{\mathbb{P}(T_X)}(1)|$. By definition the *ampleness* of X, denoted $\mathrm{amp}(X)$, is the maximum fiber dimension of the morphism φ. By definition the *coampleness* of X, denoted $\mathrm{ca}(X)$, is given by the formula $\mathrm{ca}(X) := \dim(X) - \mathrm{amp}(X)$.

Goldstein studied the coampleness of a rational projective homogeneous space $X = G/P$ over $\mathbb{C}$, showing that for G a semi-simple complex Lie group (which we can always assume), $\mathrm{ca}(X) \geq r$, where r is the minimum of ranks of the simple factors of G. In particular, for a complex rational projective homogeneous space X, $\mathrm{ca}(X) \geq 1$. More precisely, Goldstein computed $\mathrm{ca}(X)$ explicitly in each of the cases corresponding to the Dynkin diagrams (see [60]). For example, in the case $G = \mathrm{SL}_{n+1}(\mathbb{C})$ one has $\mathrm{ca}(X) = n$ for every homogeneous space of the form $X = G/P$, with P an arbitrary parabolic subgroup of G. For example, $\mathrm{ca}(G(n,r)) = n$ for every Grassmann variety $G(n,r)$ of r-planes in $\mathbb{P}^n$.

Next we will show that the closed subvarieties of a homogeneous space X of small codimension always generate X.

Lemma 13.17. *Let X be a projective homogeneous space and let Y be a closed irreducible subvariety of X. If $\mathrm{codim}_X(Y) < \mathrm{ca}(X)$ then Y generates X.*

Proof. The argument (taken from [18]) is completely formal. Let $X := G/P$. Indeed, assume that Y does not generate X. Taking $p = e \mod P$, let $Q := G_Y$ be the subgroup of G generated by $G_{Y,p}$. Q is a closed subgroup of G containing P. Moreover, Y is a closed subvariety of Q/P. Consider the canonical fibration $X = G/P \to G/Q$ with fiber Q/P. Then the normal bundle $N_{F|X}$ of $F := Q/P$ in $X = G/P$ is trivial, and since it is a quotient of $T_X|F$, we get

$$\mathrm{amp}(X) = \mathrm{amp}(T_X) \geq \mathrm{amp}(T_X|F) \geq \mathrm{amp}(N_{F|X}) = \dim(F).$$

It follows that $\mathrm{codim}_X(F) \geq \mathrm{ca}(X)$. Since $Y \subseteq F$ we have $\mathrm{codim}_X(Y) \geq \mathrm{codim}_X(F)$, whence $\mathrm{codim}_X(Y) \geq \mathrm{ca}(X)$, contradicting the hypothesis. □

Corollary 13.18 (Faltings [47]). *Let Y be a closed irreducible subvariety of a rational projective homogeneous space $X = G/P$ over $\mathbb{C}$. If $\mathrm{codim}_X(Y) < \mathrm{ca}(X)$ then Y is $G3$ in X.*

Proof. Combine Lemma 13.17 with Corollary 13.8, (i). □

Following [17] we can give an independent proof of Corollary 13.18 (in the case when Y is smooth) as an application of Theorem 12.6.

Proof. Since the normal bundle $N_{Y|X}$ is a quotient of $T_X|Y$, and since T_X is $(\dim(X) - \mathrm{ca}(X))$-ample, $N_{Y|X}$ is also $(\dim(X) - \mathrm{ca}(X))$-ample. The inequality assumed in Corollary 13.18 is equivalent to $\dim(Y) - 1 \geq \dim(X) - \mathrm{ca}(X)$. In particular, $N_{Y|X}$ is $(d-1)$-ample, with $d := \dim(Y)$. Therefore Y is $G2$ in X by Theorem 12.6 (in this way we avoid using the difficult part (i) of Chow's Theorem 13.4). To get that Y is $G3$ in X use the fact that Y is $G2$ in X and Corollary 13.8, (i). □

Another application of Corollary 13.4 is the following:

Theorem 13.19 ([17]). *If X is a homogeneous space over $\mathbb{C}$ as above, then the diagonal Δ_X is $G3$ in $X \times X$.*

Proof. The normal bundle of Δ_X in $X \times X$ is just the tangent bundle T_X of X. According to Goldstein [60], $\mathrm{ca}(X) > 1$. In particular, T_X is $(d-1)$-ample, with $d := \dim(X)$. Therefore we can apply Theorem 12.6 above to deduce that Δ_X is $G2$ in $X \times X$.

To get $G3$ from $G2$ in this case one proceeds exactly as in the proof of Corollary 13.18 (using Corollary 13.8, (i)), since $X \times X$ is a rational projective homogeneous space. □

Example 13.1. (See [82], [119], or also [85], if $n = 2$.) Consider the projective bundle $X := \mathbb{P}(T_{\mathbb{P}^n})$ associated to the tangent bundle $T_{\mathbb{P}^n}$ of $\mathbb{P}^n$, with $n \geq 2$. Assume that the ground field is $\mathbb{C}$. We have $\dim(X) = 2n - 1$. The Euler sequence

$$0 \to \mathcal{O}_{\mathbb{P}^n} \to (n+1)\mathcal{O}_{\mathbb{P}^n}(1) \to T_{\mathbb{P}^n} \to 0$$

implies that X is a closed subvariety of $\mathbb{P}((n+1)\mathcal{O}_{\mathbb{P}^n}(1)) \cong \mathbb{P}^n \times \mathbb{P}^n$. More precisely,

$$X \cong \{([x_0, \ldots, x_n], [y_0, \ldots, y_n]) \in \mathbb{P}^n \times \mathbb{P}^n \mid x_0y_0 + \cdots + x_ny_n = 0\}. \tag{13.1}$$

Now identifying $\mathbb{P}^n$ with its dual $(\mathbb{P}^n)^*$ via the map $a = [a_0, \ldots, a_n] \to H_a$ (where H_a is the hyperplane in $\mathbb{P}^n$ of equation $a_0x_0 + \cdots + a_nx_n = 0$), we also get the following description of X:

$$X \cong \{(x, H) \in \mathbb{P}^n \times (\mathbb{P}^n)^* \mid x \in H\}.$$

In particular, X is a flag variety, and hence, a rational projective homogeneous space (see [118], page 47 for another description of $X = \mathbb{P}(T_{\mathbb{P}^n})$). In case $n = 2$ the threefold X is sometimes called *Hitchin's flag manifold* (see [119]).

From now on we shall regard X as a submanifold of $\mathbb{P}^n \times \mathbb{P}^n$ via the isomorphism (13.1). Denote by $\pi_i \colon X \to \mathbb{P}^n$, $i = 1, 2$, the restrictions to X of the canonical projections of $\mathbb{P}^n \times \mathbb{P}^n$, so that $\pi_1 \colon X \to \mathbb{P}^n$ identifies with the canonical projection $\mathbb{P}(T_{\mathbb{P}^n}) \to \mathbb{P}^n$. Let L be a line in $\mathbb{P}^n$. Then $\pi_1^{-1}(L) = \mathbb{P}(T_{\mathbb{P}^n}|L)$ is a $\mathbb{P}^{n-1}$-bundle over $L = \mathbb{P}^1$. We claim that

$$\pi_1^{-1}(L) \cong \mathbb{P}((n-1)\mathcal{O}_{\mathbb{P}^1}(1) \oplus \mathcal{O}_{\mathbb{P}^1}(2)). \tag{13.2}$$

To see this observe that the normal sequence

$$0 \to T_L = \mathcal{O}_{\mathbb{P}^1}(2) \to T_{\mathbb{P}^n}|L \to N_{L|\mathbb{P}^n} = (n-1)\mathcal{O}_{\mathbb{P}^1}(1) \to 0$$

splits, whence $T_{\mathbb{P}^n}|L \cong (n-1)\mathcal{O}_{\mathbb{P}^1}(1) \oplus \mathcal{O}_{\mathbb{P}^1}(2)$.

Using (13.2) we get that $\pi_1^{-1}(L)$ can be identified to the blow up of $\mathbb{P}^n$ along a 2-codimensional linear subspace M of $\mathbb{P}^n$. In fact the blow up morphism $\pi_1^{-1}(L) \to \mathbb{P}^n$ is just the restriction $\varphi := \pi_2|\pi_1^{-1}(L)$. Let now $C \subset \mathbb{P}^n$ be a line in $\mathbb{P}^n$ such that $C \cap M = \varnothing$, and set $Y := \varphi^{-1}(C)$. Then Y is a section of the morphism $\rho := \pi_1|\pi_1^{-1}(L) : \pi_1^{-1}(L) \to L$ and the restriction of π_2 yields an isomorphism $Y \cong C = \mathbb{P}^1$ such that

$$N_{Y|\pi_1^{-1}(L)} \cong (n-1)\mathcal{O}_{\mathbb{P}^1}(1). \tag{13.3}$$

On the other hand,

$$N_{\pi_1^{-1}(L)|X} \cong \rho^*(N_{L|\mathbb{P}^n}) = (n-1)\mathcal{O}_X(1,0)|\pi_1^{-1}(L), \tag{13.4}$$

where $\mathcal{O}_X(a,b) := \pi_1^*(\mathcal{O}_{\mathbb{P}^n}(a)) \otimes \pi_2^*(\mathcal{O}_{\mathbb{P}^n}(b))$, $\forall a, b \in \mathbb{Z}$. Since Y is a section of ρ, from (13.4) it follows that

$$N_{\pi_1^{-1}(L)|X} \cong (n-1)\mathcal{O}_{\mathbb{P}^1}(1). \tag{13.5}$$

Consider now the canonical exact sequence

$$0 \to N_{Y,\pi_1^{-1}(L)} \cong (n-1)\mathcal{O}_{\mathbb{P}^1}(1) \to N_{Y|X} \to N_{\pi_1^{-1}(L)|X}|Y \cong (n-1)\mathcal{O}_{\mathbb{P}^1}(1) \to 0.$$

Since $H^1(\mathbb{P}^1, \mathcal{O}_{\mathbb{P}^1}) = 0$ this exact sequence splits, and therefore

$$N_{Y|X} \cong (2n-2)\mathcal{O}_{\mathbb{P}^1}(1). \tag{13.6}$$

Since $Y \cong \mathbb{P}^1$ then (13.6) shows that Y is a quasi-line in the sense of [19], see also Definition 14.1 of the next chapter. Moreover, $\mathcal{O}_X(1,0)|Y \cong \mathcal{O}_{\mathbb{P}^1}(1)$, and Y is a conic in $\mathbb{P}^{n^2+2n}$ with respect to the Segre embedding $X \hookrightarrow \mathbb{P}^n \times \mathbb{P}^n \hookrightarrow \mathbb{P}^{n^2+2n}$.

Now we have the following:

Proposition 13.20. *In the notation of Example* 13.1, *the formal completions* $X_{/Y}$ *and* $\mathbb{P}^{2n-1}_{/line}$ *are not isomorphic.*

Proof. By (13.6), $N_{Y|X}$ is ample, and hence by Theorem 9.28, Y is $G2$ in X. Since X is a rational projective homogeneous space by Theorem 13.4, (ii), we infer that Y is $G3$ in X. On the other hand, any line in $\mathbb{P}^{2n-1}$ is $G3$ in $\mathbb{P}^{2n-1}$ by Theorem 9.14. Assume now that there exists a formal isomorphism $X_{/Y} \cong \mathbb{P}^{2n-1}_{/\text{line}}$. By Corollary 10.6 it follows that the embeddings $Y \hookrightarrow X$ and line $\hookrightarrow \mathbb{P}^{2n-1}$ are Zariski equivalent, i.e. there exist a Zariski open subset U of $\mathbb{P}^{2n-1}$ containing a line, a Zariski open neighbourhood V of Y in X and an isomorphism $f\colon U \to V$ sending the line on Y. In particular, f yields an isomorphism $f^*(-K_V) \cong -K_U$ on anticanonical classes. Hence we get an isomorphism

$$\bigoplus_{i=0}^{\infty} H^0(U, \mathcal{O}_U(-iK_U)) \cong \bigoplus_{i=0}^{\infty} H^0(V, \mathcal{O}_V(-iK_V)).$$

Now we show that this last fact is impossible. Indeed, since $N_{Y|X}$ is ample, Y generates the homogeneous space X (otherwise Y would be contained in a fiber of a surjective morphism $f\colon X \to T$, with $\dim(T) \geq 1$). Then by Proposition 13.2 Y meets every irreducible hypersurface of X. This implies $\operatorname{codim}_X(X \setminus V) \geq 2$. Since we also have $\operatorname{codim}_{\mathbb{P}^{2n-1}}(\mathbb{P}^{2n-1} \setminus U) \geq 2)$, the above isomorphism yields an isomorphism between the anticanonical rings

$$\bigoplus_{i=0}^{\infty} H^0(X, \mathcal{O}_X(-iK_X)) \cong \bigoplus_{i=0}^{\infty} H^0(\mathbb{P}^{2n-1}, \mathcal{O}_{\mathbb{P}^{2n-1}}(2in)).$$

Since X and $\mathbb{P}^{2n-1}$ are both Fano varieties, it follows that f extends to an isomorphism between P^{2n-1} and X; but this is impossible because $\operatorname{Pic}(\mathbb{P}^{2n-1}) \cong \mathbb{Z}$ and $\operatorname{Pic}(X) \cong \mathbb{Z} \times \mathbb{Z}$. □

Remarks 13.21. i) In the case $n = 2$ Proposition 13.20 was proved by Ionescu and Voica in [85] with a completely different argument. Also H. Kurke found another proof (if $n = 2$), using the Weyl tensor (to appear). Subsequently, Ionescu and Naie found another proof of Proposition 13.20, see [84].

ii) The proof of Proposition 13.20 yields in fact the following more general assertion. Let (X_i, Y_i), $i = 1, 2$, be two pairs each consisting of a projective rational homogeneous space X_i of dimension $d \geq 2$ (char$(k) = 0$) and a quasi-line Y_i in X_i (i.e. a curve Y_i such that $Y_i \cong \mathbb{P}^1$ and $N_{Y_i|X_i} \cong (d-1)\mathcal{O}_{\mathbb{P}^1}(1)$). Assume that X_1 and X_2 are Fano varieties (in fact by a general result, every projective rational homogeneous space is always a Fano variety, see [93], page 241, Theorem 1.4). If there exists an isomorphism of formal schemes $X_{1/Y_1} \cong X_{2/Y_2}$, then there is an isomorphism $f\colon X_1 \to X_2$ such that $f(Y_1) = Y_2$.

Appendix D

Throughout this appendix the characteristic of the ground field k is assumed to be zero. The aim of this appendix is to prove, following Debarre [42] and [43], an important characterization of s-geometrically non-degenerate closed subvarieties of abelian varieties (see Definition 13.11 above). Precisely, we have:

Theorem 13.22 (Debarre [42]). *A closed irreducible subvariety Y of an abelian variety X is s-geometrically non-degenerate in X if and only if Y meets every closed irreducible subvariety Z of X such that* $\dim(Y) + \dim(Z) \geq \dim(X) + s$.

Examples of s-geometrically non-degenerate closed subvarieties of abelian varieties are given by the following:

Proposition 13.23 (Debarre [42]). *Let Y be a closed irreducible smooth subvariety of an abelian variety X such that the normal bundle $N_{Y|X}$ is s-ample for some $s \geq 0$. Then Y is s-geometrically non-degenerate in X.*

The converse in Proposition 13.23 is false in general, see [42], page 189, for an example of a 2-dimensional smooth closed subvariety Y of an abelian variety X which is geometrically non-degenerate in X, but $N_{Y|X}$ is not ample.

To prove Theorem 13.22 we need some preparation (see [43]). Let Y be a closed subvariety of an abelian variety X, of irreducible components $Y_1,\ldots,Y_m$, and let $y_0 \in Y$ be an arbitrary point of Y. According to the discussion following Definition 13.1, denote by $\langle Y\rangle$ the subgroup generated by $Y - y_0$. Then using the fact that the law group of X is commutative we easily see that $\langle Y\rangle$ is a closed subgroup of X which is independent of the choice of the point $y_0 \in Y$. In general this group is not connected if Y is not so. Clearly, if Y is irreducible (i.e. $m = 1$), $\langle Y\rangle$ is connected, i.e. $\langle Y\rangle$ is an abelian subvariety of X. If Y is not irreducible, and if $i,j \in \{1,\ldots,m\}$ are such that $Y_i \cap Y_j \neq \varnothing$, we have $Y_i - Y_j \subseteq (Y_i - Y_i) + (Y_j - Y_j)$ (if $x \in Y_i \cap Y_j$, $y_i \in Y_i$ and $y_j \in Y_j$, then $y_i - y_j = (y_i - x) + (x - y_j)$). If $Y_{i_1},\ldots,Y_{i_s}$ are irreducible components of Y such that $Y_{i_p} \cap Y_{i_{p+1}} \neq \varnothing$, $\forall p = 1,\ldots,s-1$, then by induction $Y_{i_1} - Y_{i_s} \subseteq (Y_{i_1} - Y_{i_2}) + (Y_{i_2} - Y_{i_3}) + \cdots + (Y_{i_{s-1}} - Y_{i_s}) \subseteq (Y_{i_1} - Y_{i_1}) + (Y_{i_2} - Y_{i_2}) + \cdots + (Y_{i_s} - Y_{i_s})$. Using the discussion following Definition 13.1 we infer that if Y is connected, the closed subgroup $\langle Y\rangle$ coincides with the sum of the abelian subvarieties $\langle Y_i\rangle$, $i = 1,\ldots,m$, whence $\langle Y\rangle$ is also connected. In this case we shall call $\langle Y\rangle$ *the abelian subvariety generated by* Y. In general (i.e. if Y is not necessarily connected) we shall define *the abelian subvariety of X generated by Y* as the sum of the abelian subvarieties $\langle Y_i\rangle = \langle Y_i - Y_i\rangle$, $i = 1,\ldots,m$.

Lemma 13.24. *Let Y be a closed subvariety of an abelian variety X, and denote by K the abelian subvariety of X generated by Y. Let L be a vector subspace of the tangent space T_0X of X at the origin, and assume that for a general point y in Y, the tangent space $T_0(Y-y)$ is contained in L. Then $T_0K \subseteq L$.*

Proof. There is no loss of generality in assuming that Y is irreducible and passes through the origin. Then by the discussion following Definition 13.1 there exists an $n \geq 1$ such that the morphism $f\colon Y^{2n} \to X$ defined by

$$f(y_1, \dots, y_{2n}) = y_1 - y_2 + y_3 - \cdots + y_{2n-1} - y_{2n}, \quad \forall (y_1, \dots, y_{2n}) \in Y^{2n}$$

has K as image (Y^{2n} is projective and the corestriction $Y^{2n} \to K$ is dominant). It is a general elementary fact that, fixing a general point $(y_1, \dots, y_{2n}) \in Y^{2n}$, the morphism $g : (Y - y_1) \times \cdots \times (Y - y_{2n}) \to X$ defined by

$$g(z_1, \dots, z_{2n}) = z_1 - z_2 + z_3 - \cdots + z_{2n-1} - z_{2n},$$

$\forall (z_1, \dots, z_{2n}) \in (Y - y_1) \times \cdots \times (Y - y_{2n})$, has the differential

$$dg_{(0,\dots,0)} : T_{(0,\dots,0)}((Y-y_1)\times\cdots\times(Y-y_{2n})) = T_0(Y-y_1)\times\cdots\times T_0(Y-y_{2n}) \to T_0X$$

given by

$$dg_{(0,\dots,0)}(v_1, \dots, v_{2n}) = v_1 - v_2 + v_3 - \cdots + v_{2n-1} - v_{2n},$$

$\forall (v_1, \dots, v_{2n}) \in T_0(Y-y_1)\times\cdots\times T_0(Y-y_{2n})$ (see [30], chap. I, (3.2)). In particular, since by hypothesis $T_0(Y - y_i) \subseteq L$ for every $i = 1, \dots 2n$, we get $T_0K \subseteq L$ as soon as we know that the corestriction map

$$dg_{(0,\dots 0)} : T_0(Y - y_1) \times \cdots \times T_0(Y - y_{2n}) \to T_0K$$

is surjective. This surjectivity follows from the generic smoothness in characteristic zero (see [73], Corollary 10.7, page 272), which implies that the surjective morphism $f\colon Y^{2n} \to K$ is smooth at a general point $(y_1, \dots, y_{2n}) \in Y^{2n}$. □

Now we need the following:

Definition 13.25 ([43]). Let $Y_1, \dots, Y_r$ be closed irreducible subvarieties of an abelian variety X. One says that $(Y_1, \dots, Y_r)$ *fills up* X if for every abelian subvariety K of X one has

$$\sum_{i=1}^{r} \dim(\pi(Y_i)) \geq (r-1)\dim(X/K),$$

where $\pi\colon X \to X/K$ is the canonical morphism.

Lemma 13.26. *For every closed subvariety F of an abelian variety X we have $T_0(F-F) = T_0K$, where K is the abelian subvariety of X generated by F.*

Proof. Let $F_1, \dots, F_r$ be the irreducible components of F. If for two indices i and j we have $0 \in (F_i - F_j)$ then $F_i - F_j \subseteq (F_i - F_i) + (F_j - F_j)$. In other words, we get the inclusion $(F - F) \subseteq K$ locally around 0, whence $T_0(F - F) \subseteq T_0 K$. Conversely, for every point $x \in F$ we have $T_0(F - x) \subseteq T_0(F - F)$, so that $T_0 K \subseteq T_0(F - F)$ by Lemma 13.24. □

Lemma 13.27. *Let Z be a closed irreducible subvariety of an abelian variety X. Assume that for any general point $a \in Z$, there exists an abelian subvariety K_a of X such that $T_a(a + K_a) \subseteq T_a Z$. Then $Z + K_a = Z$ for any general point $a \in Z$.*

Proof. We shall need the following rigidity result: if $\{K_a\}_{a \in Z}$ is an algebraic family of abelian subvarieties of X, then there exists a non-empty open subset Z' of Z such that the family $\{K_a\}_{a \in Z'}$ is constant. To see this, let $U \subseteq X \times Z$ be the closed subscheme giving rise to our family, i.e. if $p \colon U \to Z$ is the restriction to U of the projection $X \times Z \to Z$, then $p^{-1}(a) = K_a$, $\forall a \in Z$. Fix a general point $a_0 \in Z$, and let $f \colon U \to X/K_{a_0}$ be the morphism obtained by composing the canonical morphism $X \to X/K_{a_0}$ with the restriction to U of the other projection $X \times Z \to X$. Since $f(K_{a_0})$ is a point, from the well-known rigidity lemma (see e.g. [29], (4.1.13)) it follows that there exists an open subset V of Z containing a_0 such that $f(K_a)$ is a point for every $a \in V$, i.e. $K_a \subseteq K_{a_0}$, $\forall a \in V$. On the other hand, since a_0 was a general point of Z, $\dim(K_{a_0}) \leq \dim(K_a)$, whence $K_a = K_{a_0}$, $\forall a \in V$.

Coming back to our situation, consider the above non-empty open subset $V \subseteq Z$ such that $K_a = K$, $\forall a \in V$. Consider then the (surjective) addition morphism $g \colon Z \times K \to Z + K$. Since $\mathrm{char}(k) = 0$ for $(a, x) \in V \times K$ a general point, (a, x) is smooth on $V \times K$ (which implies that a is smooth on Z), $g(a, x) = a + x$ is smooth on $Z + K$, and so the differential

$$dg_{(a,x)} : T_{(a,x)}(Z \times K) \longrightarrow T_{a+x}(Z + K)$$

is surjective. Therefore the differential

$$dg_{(0,0)} : T_{(0,0)}((Z - a) \times K) = T_0(Z - a) \times T_0 K \longrightarrow T_0(Z - a + K)$$

is also surjective. The hypothesis says that $T_0 K \subseteq T_0(Z - a)$. Using again [30], Chap. I, (3.2) we infer that the image of $T_0 K$ under $dg_{(0,0)}$ is contained in $T_0(Z - a)$, whence $T_0(Z - a) = T_0(Z - a + K)$, i.e. $T_{a+x}(Z + x) = T_{a+x}(Z + K)$. Recalling that a and $a + x$ are smooth points on Z and $Z + K$ respectively, it follows that $\dim(Z + K) = \dim(Z)$, and since $Z \subseteq Z + K$, this implies $Z + K = Z$. □

Proposition 13.28 ([43]). *In the notation of Definition* 13.25, *$(Y_1, \dots, Y_r)$ fills up X if and only if the morphism $\varphi \colon Y_1 \times \cdots \times Y_r \to X^{r-1}$, defined by $\varphi(y_1, \dots, y_r) = (y_2 - y_1, \dots, y_r - y_1)$, is surjective.*

Proof. If $(Y_1, \dots, Y_r)$ does not fill up X, there exists an abelian subvariety K of X such that $\sum\limits_{i=1}^{r} \dim(\pi(Y_i)) < (r-1)\dim(X/K)$, with $\pi \colon X \to X/K$ the canonical morphism. Then $\dim(\pi^{r-1}(\varphi(Y_1\times\cdots\times Y_r))) < \dim((X/K)^{r-1})$, whence φ cannot be surjective.

Assume now that $(Y_1,\ldots,Y_r)$ fills up X. We shall proceed by induction on the dimension of X. Assume that φ is not surjective, i.e. $Z := \varphi(Y_1\times\cdots\times Y_r)\neq X^{r-1}$. Since $(Y_1,\ldots,Y_r)$ fills up X we have $\sum\limits_{i=1}^{r}\dim(Y_i)\geq\dim(X^{r-1})>\dim(Z)$. Therefore if $a=(a_2,\ldots,a_r)$ is a general point of Z, there exists a closed subvariety F_a of Y_1 of positive dimension such that

$$\varphi^{-1}(a)=\{(v_1,v_1+a_2,\ldots,v_1+a_r)\mid v_1\in F_a\}.$$

Then we have $(F_a+a_i)\subseteq Y_i$, $i=2,\ldots,r$. From this we get $a+u_i(F_a-F_a)\subseteq Z$, where $u_i\colon X\to X^{r-1}$ is the $(i-1)$-th canonical inclusion (for instance $u_2(x)=(x,0,\ldots,0)$, $\forall x\in X$). Indeed, for example, taking $i=2$, we have to check that $(a_2+v_1-v_1',a_3,\ldots,a_r)\in Z$, $\forall v_1,v_1'\in F_a$. Taking $z_1=v_1'$, $z_2=a_2+v_1$, $z_3=a_3+v_1',\ldots, z_r=a_r+v_1'$, we get $(z_1,\ldots,z_r)\in Y_1\times\cdots\times Y_r$ and $\varphi(z_1,\ldots,z_r)=(a_2+v_1-v_1',a_3,\ldots,a_r)$.

Now, if in Lemma 13.26 we take $F=F_a$ we get $T_a(a+u_i(K_a'))\subseteq T_aZ$ for every $i\geq 2$, where K_a' is the abelian subvariety of X generated by F_a. Thus $T_a(a+{K_a'}^{r-1})\subseteq T_aZ$. Setting $K_a:={K_a'}^{r-1}$, we may apply Lemma 13.27 to the abelian variety X^{r-1} to get $Z+K_a=Z+{K_a'}^{r-1}=Z$ for every $a\in Z$ general. Since the images of $Y_1,\ldots,Y_r$ in the quotient abelian variety X/K_a' fill up X/K_a', by the induction hypothesis we get $Z+{K_a'}^{r-1}=X^{r-1}$. Comparing the last two equalities we get $Z=X^{r-1}$, which contradicts the hypothesis that φ was not surjective. □

We can draw the following immediate consequences of Proposition 13.28.

Corollary 13.29. *Let $Y_1,\ldots,Y_r$ be closed irrducible subvarieties of an abelian variety X. If $(Y_1,\ldots,Y_r)$ fills up X then $Y_1\cap Y_2\cap\cdots\cap Y_r\neq\varnothing$.*

Proof. By Proposition 13.28, $\varphi^{-1}(\Delta)\neq\varnothing$, where Δ is the diagonal of X^{r-1}. □

Corollary 13.30. *Let Y and Z be two closed irreducible subvarieties of an abelian variety X. Then (Y,Z) fills up X if and only if $Y+Z=X$.*

Proof. Clearly, (Y,Z) fills up X if and only if $(Y,-Z)$ does. Then the conclusion follows from Proposition 13.28. □

Lemma 13.31 ([42]). *Let Y and Z be two closed irreducible subvarieties of an abelian variety X such that Y is s-geometrically non-degenerate and* $\dim(Y)+\dim(Z)\geq\dim(X)+s$. *Then the morphism $\varphi\colon Y\times Z\to X$ defined by $\varphi(y,z)=y-z$ is surjective.*

Proof. Let $\pi\colon X\to X/K$ be a quotient abelian variety of X. We claim that $\dim(\pi(Y))+\dim(\pi(Z))\geq\dim(X/K)$. Indeed, if $\pi(Y)=X/K$ this is obvious; otherwise, $\dim(\pi(Y))\geq\dim(Y)-s_0$, where $s_0:=\min(s,\dim(K))$ (indeed, if

$s_0 = \dim(K) > s$, this follows from the theorem on the dimension of fibers applied to $Y \to \pi(Y)$). From this we get:

$$\begin{aligned}\dim(\pi(Y)) + \dim(\pi(Z)) &\geq \dim(Y) - s_0 + \dim(Z) - \dim(K) \\ &\geq \dim(X) + s - s_0 - \dim(K) \geq \dim(X/K).\end{aligned}$$

This shows that (Y, Z) fills up X. Therefore by Proposition 13.28, the map φ is surjective. □

Now we are ready to prove Theorem 13.22.

Proof of Theorem 13.22. ([42]) Assume that Y meets any closed subvariety of X of dimension $\geq \operatorname{codim}_X(Y) + s$, and let $\pi\colon X \to X/K$ be an arbitrary quotient abelian variety of X. If $\pi(Y) \neq X/K$, there exists a closed subvariety Z' of X/K of dimension $\dim(X/K) - \dim(\pi(Y)) - 1$ that does not meet $\pi(Y)$. Since Y does not meet $\pi^{-1}(Z')$, $\operatorname{codim}_X(Y) + s > \dim(\pi^{-1}(Z')) = \dim(X) - \dim(\pi(Y)) - 1$, whence $\dim(\pi(Y)) \geq \dim(Y) - s$ and Y is s-geometrically non-degenerate.

Conversely, assume that Y is s-geometrically non-degenerate, and let Z be a closed irreducible subvariety of X of dimension $\geq \dim(X) - \dim(Y) + s$. From Lemma 13.31 it follows that $Y - Z = X$, whence Y meets Z. This finishes the proof of Theorem 13.22. □

It remains to prove Proposition 13.23. To this extent, let Y be a closed irreducible smooth subvariety of an abelian variety X. For every $y \in Y$ we have the natural inclusion $T_0(Y - y) \subseteq T_0X$. The proof of Proposition 13.23 will be an easy consequence of the following simple lemma:

Lemma 13.32 ([42]). *Under the above hypotheses, let S be a closed irreducible subvariety of Y, and fix an integer $s \geq 0$. Then the following two conditions are equivalent:*

(i) *The restriction $N_{Y|X}|S$ to S of the normal bundle of Y in X is s-ample.*

(ii) *For every hyperplane H of T_0X, $\dim(\{y \in S \mid T_0(Y - y) \subseteq H\}) \leq s$.*

Proof. From the normal sequence restricted to S

$$0 \to T_Y|S \to T_X|S \to N_{Y|X}|S \to 0$$

we get the canonical closed embedding $\mathbb{P}(N_{Y|X}|S) \hookrightarrow \mathbb{P}(T_X|S)$. Since T_X is isomorphic to the trivial vector bundle of rank $\dim(X)$, we get an isomorphism $\mathbb{P}(T_X|S) \cong \mathbb{P}(T_0X) \times S$. Consider the composite

$$f : \mathbb{P}(N_{Y|X}|S) \hookrightarrow \mathbb{P}(T_X|S) \cong \mathbb{P}(T_0X) \times S \to \mathbb{P}(T_0X),$$

where the last arrow is the first projection on $\mathbb{P}(T_0X)$. Then $f^*(\mathcal{O}_{\mathbb{P}(T_0X)}(1)) \cong \mathcal{O}_{\mathbb{P}(N_{Y|X}|S)}(1)$. Now by [140], Proposition 1.7, $N_{Y|X}|S$ is s-ample if and only if the fibers of f are all of dimension $\leq s$. Finally, since the restriction of the projection

$\mathbb{P}(N_{Y|X}|S) \to S$ to any fiber of f is injective, the lemma follows since the set $\{y \in S \mid T_0(Y - y) \subseteq H\}$ is just the image of $f^{-1}([H])$ under this projection, where $[H]$ is the point of $\mathbb{P}(T_0 X)$ corresponding to the hyperplane H of $T_0 X$. $\square$

Finally, we are ready to prove Proposition 13.23.

Proof of Proposition 13.23. Let $\pi\colon X \to X/K$ be a quotient abelian variety of X such that $\pi(Y) \neq X/K$. To prove that Y is s-geometrically non-degenerate it is sufficient (by the theorem on dimension of fibers) to prove that the general fiber of the morphism $\pi' := \pi|Y\colon Y \to Y' := \pi(Y)$ is of dimension $\leq s$. Since Y is smooth, the morphism π' is proper and $\mathrm{char}(k) = 0$, by generic smoothness (see [73], Corollary 10.7, page 272) we infer that ${\pi'}^{-1}(z)$ is smooth for z a general point of Y'. Let S be an irreducible component of ${\pi'}^{-1}(z)$. Then for every $y \in S$ we have the commutative diagram

$$\begin{array}{ccc} T_yY \cong T_0(Y - y) & \longrightarrow & T_0X \\ \big\downarrow & & \big\downarrow {\scriptstyle d\pi_0} \\ T_zY' \cong T_0(Y' - z) & \longrightarrow & T_0(X/K) \end{array}$$

in which the horizontal arrows are the natural inclusions. From this diagram we infer that $S = \{y \in S \mid T_0(Y - y) \subseteq (d\pi_0)^{-1}(T_0(Y' - z))\}$. Since $Y' \neq X/K$ and since z is a smooth point of Y', there exists a hyperplane H of T_0Y such that $(d\pi_0)^{-1}(T_0(Y' - z)) \subseteq H$. In other words, $S = \{y \in S \mid T_0(Y - y) \subseteq H\}$. By Lemma 13.32 the latter subset has dimension $\leq s$, since $N_{Y|X}$ is s-ample. This proves that the dimension of ${\pi'}^{-1}(z)$ is $\leq s$. $\square$

Chapter 14

Quasi-lines on Projective Manifolds

Throughout this chapter the ground field will be assumed to be the field $\mathbb{C}$ of complex numbers.

Most of the material in this chapter is taken from Bădescu–Beltrametti–Ionescu [19]. Let X be a smooth projective complex variety of dimension $n \geq 2$, and let Y be a smooth connected curve in X.

Definition 14.1. A *quasi-line* in X is a curve Y in X as above such that $Y \cong \mathbb{P}^1$, and $N_{Y|X} \cong (n-1)\mathcal{O}_{\mathbb{P}^1}(1)$, where $N_{Y|X}$ is the normal bundle of Y in X.

Clearly, the lines in $\mathbb{P}^n$ are examples of quasi-lines. A first non-trivial example of a quasi-line in dimension ≥ 3 was given in Example 9.1; that quasi-line is formally (but not Zariski) equivalent to a line in $\mathbb{P}^n$. Finally, Example 13.1 produces examples of quasi-lines in a smooth projective complex variety of dimension $2n-1$ (with $n \geq 2$), which by Proposition 13.20 are not even formally isomorphic to a line in $\mathbb{P}^{2n-1}$. On the other hand, we shall see that every quasi-line on a smooth projective surface is Zariski equivalent to a line in the projective plane (see Corollary 14.4 below, essentially due to d'Almeida [40]).

The concept of quasi-line became of interest in mathematics especially in the framework of the so-called Kählerian twistor spaces, see Hitchin [82].

First we are interested in giving an intrinsic characterization for a curve Y on a smooth projective variety X to be a quasi-line in terms of the Picard group of the first infinitesimal neighbourhood $Y(1)$ of Y in X. The consideration of the Picard group of the infinitesimal neighbourhoods of Y in X is also motivated by some work by Griffiths and Harris [64] and d'Almeida [40]. Indeed, in [64], pages 698–699, the authors reinterpreted in modern language the classical condition of Reiss concerning the existence of a plane curve of degree d in $\mathbb{P}^2$ intersecting a given line in d different prescribed points, with prescribed tangents and second-order conditions. This involves considering the Picard group of the second infinitesimal neighbourhood of the curve L in $\mathbb{P}^2$. Later on, in connection with this problem, d'Almeida studied in [40] conditions under which the canonical restriction map

$$\alpha : \operatorname{Pic}(X) \to \operatorname{Pic}(Y(1))$$

is surjective in the case when Y is a curve with positive self-intersection on a smooth projective surface X, and showed that this happens essentially when Y is a line in $\mathbb{P}^2$ (see Corollary 14.4 below).

The first result gives the desired intrinsic characterisation of quasi-lines in terms of the restriction map $\alpha : \mathrm{Pic}(X) \to \mathrm{Pic}(Y(1))$. It is a slight improvement of Theorem (2.1) of [19].

Theorem 14.2. *Let X be a smooth projective variety over $\mathbb{C}$ of dimension $n \geq 2$, and let Y be a smooth curve in X. Assume that the normal bundle $N_{Y|X}$ of Y in X is ample. Then Y is a quasi-line if and only if the canonical restriction map $\alpha : \mathrm{Pic}(X) \to \mathrm{Pic}(Y(1))$ has finite cokernel.*

Proof. Assume first that Y is a quasi-line. The truncated exponential sequence

$$0 \to N^*_{Y|X} \to \mathcal{O}^*_{Y(1)} \to \mathcal{O}^*_Y \to 0$$

yields the exact sequence of cohomology

$$H^1(Y, N^*_{Y|X}) \to \mathrm{Pic}(Y(1)) \to \mathrm{Pic}(Y) \to H^2(Y, N^*_{Y|X}).$$

The last space is zero because Y is a curve, and the first space is also zero because $Y \cong \mathbb{P}^1$ and $N_{Y|X} \cong (n-1)\mathcal{O}_{\mathbb{P}^1}(1)$. Therefore the restriction map $\mathrm{Pic}(Y(1)) \to \mathrm{Pic}(Y)$ is an isomorphism. In other words, the map α can be identified with the restriction map $\gamma : \mathrm{Pic}(X) \to \mathrm{Pic}(Y)$. Since X is projective, $Y \cong P^1$ and $\mathrm{Pic}(\mathbb{P}^1) \cong \mathbb{Z}$, $\mathrm{Coker}(\gamma)$ (and hence also $\mathrm{Coker}(\alpha)$) is a finite (cyclic) group.

Conversely, assume that the map α has finite cokernel. Then we claim that the natural map of Picard schemes $\underline{\mathrm{Pic}}^0(X) \to \underline{\mathrm{Pic}}^0(Y(1))$ is surjective.

Indeed, the commutative diagram with exact rows

$$\begin{array}{ccccccccc}
0 & \longrightarrow & \mathrm{Pic}^0(X) & \longrightarrow & \mathrm{Pic}(X) & \longrightarrow & \mathrm{NS}(X) & \longrightarrow & 0 \\
 & & \downarrow{\scriptstyle \alpha_0} & & \downarrow{\scriptstyle \alpha} & & \downarrow{\scriptstyle \beta} & & \\
0 & \longrightarrow & \mathrm{Pic}^0(Y(1)) & \longrightarrow & \mathrm{Pic}(Y(1)) & \longrightarrow & \mathrm{NS}(Y(1)) & \longrightarrow & 0
\end{array}$$

and the snake lemma yield the exact sequence

$$\mathrm{Ker}(\beta) \to \mathrm{Coker}(\alpha_0) \to \mathrm{Coker}(\alpha).$$

By Néron–Severi's theorem, $\mathrm{NS}(X)$ is a finitely generated abelian group, whence $\mathrm{Ker}(\beta)$ is also a finitely generated abelian group. On the other hand, $\mathrm{Coker}(\alpha_0)$ has a natural stucture of a connected commutative algebraic group (because α_0 is a homomorphism of connected commutative algebraic groups). Recalling that $\mathrm{Coker}(\alpha)$ is finite we infer that $\mathrm{Coker}(\alpha_0) = 0$, i.e. the homomorphism of algebraic groups $\alpha_0 : \underline{\mathrm{Pic}}^0(X) \to \underline{\mathrm{Pic}}^0(Y(1))$ is surjective.

But $\underline{\mathrm{Pic}}^0(X)$ is an abelian variety because X is a smooth projective variety over $\mathbb{C}$. Therefore the surjectivity of α_0 implies that $\underline{\mathrm{Pic}}^0(Y(1))$ is also an abelian variety.

On the other hand, the above truncated exponential sequence yields the exact sequence of cohomology

$$H^0(\mathcal{O}^*_{Y(1)}) \to H^0(\mathcal{O}^*_Y) \to H^1(Y, N^*_{Y|X}) \to \mathrm{Pic}(Y(1)) \to \mathrm{Pic}(Y) \to H^2(Y, N^*_{Y|X}),$$

in which the last space is zero. Since $H^0(\mathcal{O}^*_Y)$ consists only of non-zero constants, the first map is surjective, and therefore we get the exact sequence

$$0 \to H^1(Y, N^*_{Y|X}) \to \mathrm{Pic}(Y(1)) \to \mathrm{Pic}(Y) \to 0.$$

This yields the exact sequence of algebraic groups

$$0 \to H^1(Y, N^*_{Y|X}) \to \underline{\mathrm{Pic}}^0(Y(1)) \to \underline{\mathrm{Pic}}^0(Y) \to 0,$$

where $H^1(Y, N^*_{Y|X})$ is regarded as a product of $h^1(Y, N^*_{Y|X})$ copies of the additive group $\mathbb{G}_a$. Since we just proved that $\underline{\mathrm{Pic}}^0(Y(1))$ is an abelian variety and since $H^1(Y, N^*_{Y|X})$ is a linear algebraic group, we get

$$H^1(Y, N^*_{Y|X}) = 0. \tag{14.1}$$

Applying duality on Y, (14.1) becomes

$$H^0(Y, N_{Y|X} \otimes \omega_Y) = 0, \tag{14.2}$$

and by Riemann–Roch we get

$$\begin{aligned} h^0(Y, N_{Y|X} \otimes \omega_Y) &\geq \deg(N_{Y|X} \otimes \omega_Y) + (n-1)(1-g) \\ &= \deg(N_{Y|X}) + (g-1)(n-1), \end{aligned} \tag{14.3}$$

where g is the genus of Y. Since by hypothesis $N_{Y|X}$ is ample, $\deg(N_{Y|X}) > 0$, whence the right hand side is > 0 when $g \geq 1$. Therefore by (14.2) and (14.3) we get that $g = 0$, i.e. $Y \cong \mathbb{P}^1$.

Then by a theorem of Grothendieck [70], there are $n-1$ integers $a_1, \dots, a_{n-1}$ such that

$$N_{Y|X} \cong \mathcal{O}(a_1) \oplus \dots \oplus \mathcal{O}(a_{n-1}),$$

where $\mathcal{O}(a) := \mathcal{O}_{\mathbb{P}^1}(a)$, for each $a \in \mathbb{Z}$. Since by hypothesis $N_{Y|X}$ is ample, whence $a_i > 0$, for each $i = 1, \dots, n-1$. Recalling (14.1) and the explicit cohomology of the projective line we get $a_1 = \dots = a_{n-1} = 1$, which concludes the proof of the theorem. □

Remark 14.3. The proof of Theorem 14.2 also yields the following assertion: under the assumptions of Theorem 14.2, the map $\alpha : \mathrm{Pic}(X) \to \mathrm{Pic}(Y(1))$ is surjective if and only if Y is a quasi-line in X and the restriction map $\mathrm{Pic}(X) \to \mathrm{Pic}(Y)$ is surjective.

The ampleness assumption on the normal bundle $N_{Y|X}$ is essential for the validity of the conclusion of Theorem 14.2, as the following example shows.

Example 14.1. Fix an arbitrary smooth projective curve C of genus $g > 0$, and let L be a line bundle on C of degree $d \leq 2 - 2g$. Set $E := \mathcal{O}_C \oplus L$ and $X := \mathbb{P}(E)$, and let $\pi\colon X \to C$ denote the canonical projection. Let $i\colon Y \hookrightarrow X$ be the section of π corresponding to the canonical map $E = \mathcal{O}_C \oplus L \to L$. Then the normal bundle of Y in X is isomorphic to L (and in particular, $(Y^2) = d \leq 2 - 2g \leq 0$). Since $\pi \circ i = \mathrm{id}_C$, we get $i^* \circ \pi^* = \mathrm{id}$, and in particular, the restriction map $i^* : \mathrm{Pic}(X) \to \mathrm{Pic}(Y)$ is surjective.

On the other hand, consider the exact sequence of cohomology

$$H^1(Y, N^*_{Y|X}) = H^1(Y, L^{-1}) \to \mathrm{Pic}(Y(1)) \to \mathrm{Pic}(Y) \to H^2(Y, L^{-1}) = 0.$$

By duality $H^1(Y, L^{-1}) \cong H^0(Y, L \otimes \omega_Y)^*$, and if $L \not\cong \omega_Y^{-1}$, then this last space is zero under the above assumptions. Thus, if $d \leq 2 - 2g$ and $L \not\cong \omega_Y^{-1}$, then the map $\mathrm{Pic}(Y(1)) \to \mathrm{Pic}(Y)$ is an isomorphism.

Putting everything together it follows that under the above assumptions, the map $\alpha : \mathrm{Pic}(X) \to \mathrm{Pic}(Y(1))$ is surjective if $d \leq 2 - 2g$ and $L \not\cong \omega_Y^{-1}$.

In conclusion, under the absence of the ampleness assumption of the normal bundle, there are many examples of curves of arbitrary genus Y on a smooth projective surface X such that the map $\alpha : \mathrm{Pic}(X) \to \mathrm{Pic}(Y(1))$ is surjective.

Corollary 14.4. *Let X be a smooth projective complex surface containing a closed smooth connected curve Y such that $Y^2 > 0$. Then the following conditions are equivalent:*

(i) *The cokernel of the map $\alpha : \mathrm{Pic}(X) \to \mathrm{Pic}(Y(1))$ is finite.*

(ii) *The map $\alpha : \mathrm{Pic}(X) \to \mathrm{Pic}(Y(1))$ is surjective.*

(iii) *There is a birational morphism $f\colon X \to \mathbb{P}^2$ such that f is an isomorphism in a Zariski open neighbourhood of Y in X and $f(Y)$ is a line in $\mathbb{P}^2$.*

Proof. (i)$\Longrightarrow$(iii) By Theorem 14.2, (i) implies that $Y \cong \mathbb{P}^1$ and $Y^2 = 1$. Then it is well known (and easy to see, cf e.g. [12]) that the linear system $|Y|$ is base point free and yields a birational morphism $f\colon X \to \mathbb{P}^2$ satisfying (iii). Clearly (iii)$\Longrightarrow$(ii) because $Y \cong \mathbb{P}^1$ and $(f^*(\mathcal{O}_{\mathbb{P}^2}(1)) \cdot Y) = 1$. Finally, the implication (ii)$\Longrightarrow$(i) is trivial. □

Remark 14.5. Corollary 14.4 is a slight improvement of a result of d'Almeida [40]. Note also that the above proof is completely different from d'Almeida's proof.

Corollary 14.6. *Let X be a smooth projective complex variety of dimension $n \geq 2$ such that $\mathrm{Pic}(X) \cong \mathbb{Z}$, and let Y be a smooth connected curve in X with ample normal bundle such that the restriction map $\alpha : \mathrm{Pic}(X) \to \mathrm{Pic}(Y(1))$ is surjective. Then $X \cong \mathbb{P}^n$ and Y is a line in $\mathbb{P}^n$.*

Proof. By Theorem 14.2, the surjectivity of the map α is equivalent with the following two facts:

i) Y is a quasi-line and

ii) the map $\alpha_0 : \mathrm{Pic}(X) \to \mathrm{Pic}(Y)$ is surjective.

In particular, ii) implies that there is a line bundle $L \in \mathrm{Pic}(X)$ such that $L|Y \cong \mathcal{O}(1)$ (i.e. $(L\cdot Y) = 1$). Since $\mathrm{Pic}(X) \cong \mathbb{Z}$ and the restriction $L|Y$ is ample, we infer that L is ample on X. From i) we get $N_{Y|X} \cong (n-1)\mathcal{O}_{\mathbb{P}^1}(1)$, and in particular, $\det(N_{Y|X}) \cong \mathcal{O}_{\mathbb{P}^1}(n-1)$. Then the adjunction formula yields in our situation $-(K_X \cdot Y) = n+1$ (where K_X denotes the canonical class on X). This shows in particular that the line bundle $K_X + (n+1)L$ is not ample (because its restriction to Y is trivial). Then applying a general result from the adjunction theory (see e.g. [29], (7.2.1)), we infer that $X \cong \mathbb{P}^n$ and $L \cong \mathcal{O}_{\mathbb{P}^n}(1)$ and, since $(L \cdot Y) = 1$, Y is a line. Alternatively, an even more direct argument is to notice that, since $\mathrm{Pic}(X) \cong \mathbb{Z}$ and $(K_X + (n+1)L)|Y \cong \mathcal{O}_Y$, it follows that $K_X + (n+1)L$ is numerically trivial, and one can conclude by Kobayashi–Ochiai's characterization of the projective space (see e.g. [29], Theorem 3.1.6, page 71). □

Further remarkable examples of quasi-lines are given by the following result of Oxbury [119]. This result is particularly interesting because by Clemens and Griffiths [37] all smooth cubic hypersurfaces in $\mathbb{P}^4$ are not rational. In particular, this result will also show that the conditions (i) and (ii) stated in Corollary 14.4 need not be equivalent when $n \geq 3$.

Theorem 14.7 (Oxbury [119]). *Let X be a smooth cubic hypersurface in $\mathbb{P}^4$ over $\mathbb{C}$. Then X contains smooth conics Y which are quasi-lines on X. In particular,* $\mathrm{Coker}(\alpha) \cong \mathbb{Z}/2\mathbb{Z}$, *where $\alpha : \mathrm{Pic}(X) \to \mathrm{Pic}(Y(1))$ is the canonical restriction map.*

Proof. The proof that follows, different from Oxbury's original proof [119], is taken from [19]. Let $Z := X \cap H$, where H is a general hyperplane in $\mathbb{P}^4$. Then Z is a smooth cubic surface in $H = \mathbb{P}^3$. By the structure theorem of cubic surfaces (see e.g. [73] or [64]), Z is isomorphic to the blowing up of $\mathbb{P}^2$ in six points $P_1, \ldots, P_6$ in general position. Let C be the proper transform of a smooth conic in $\mathbb{P}^2$ passing through $P_1, \ldots, P_4$. Then C is a smooth conic on Z such that $C^2 = 0$. Thus $N_{C|Z} \cong \mathcal{O}_C$. Now consider the exact sequence of normal bundles

$$0 \to N_{C|Z} \cong \mathcal{O}_{\mathbb{P}^1} \to N_{C|X} \to N_{Z|X}|C \cong \mathcal{O}_{\mathbb{P}^1}(2) \to 0. \tag{14.4}$$

In particular, $\deg(N_{C|X}) = 2$, whence by Grothendieck's theorem [70], we get

$$N_{C|X} \cong \mathcal{O}_{\mathbb{P}^1}(a) \oplus \mathcal{O}_{\mathbb{P}^1}(2-a), \quad a \geq 0. \tag{14.5}$$

Assume first that sequence (14.4) does not split. Since $\mathcal{O}_{\mathbb{P}^1}(2)$ is ample, by claim 1 of the proof of Theorem 12.14, $N_{C|X}$ is ample, and so by (14.5) we get $N_{C|X} \cong 2\mathcal{O}_{\mathbb{P}^1}(1)$. In other words C is a smooth conic whose normal bundle in X is isomorphic to $2\mathcal{O}_{\mathbb{P}^1}(1)$, and in this case we are done.

We can therefore assume that sequence (14.4) splits, i.e.

$$N_{C|X} \cong \mathcal{O}_{\mathbb{P}^1} \oplus \mathcal{O}_{\mathbb{P}^1}(2). \tag{14.6}$$

Consider the Hilbert scheme $\mathcal{P}$ parametrizing all curves in X having the same Hilbert polynomial as C. Pick a general point $x \in C$ and consider the closed subscheme $\mathcal{P}_x$ of $\mathcal{P}$ parametrizing all curves of $\mathcal{P}$ which pass through the point x. Let us denote by $[C]$ the closed point of $\mathcal{P}_x$ corresponding to the conic C. From the theory of Hilbert schemes (see [69], exposé No. 221, or also [93], Theorem 1.7, page 95) the tangent space $T_{\mathcal{P}_x,[C]}$ of $\mathcal{P}_x$ at $[C]$ is isomorphic to $H^0(C, N_{C|X} \otimes \mathcal{J}_x)$, where $\mathcal{J}_x$ is the ideal sheaf of x in C. Moreover if

$$H^1(C, N_{C|X} \otimes \mathcal{J}_x) = 0, \tag{14.7}$$

then $\mathcal{P}_x$ is smooth at the point $[C]$ (loc. cit). In our case, by (14.6) we get $N_{C|X} \otimes \mathcal{J}_x \cong \mathcal{O}_{\mathbb{P}^1}(-1) \oplus \mathcal{O}_{\mathbb{P}^1}(1)$, so that (14.7) holds true. Thus we conclude that $[C]$ is a smooth point of $\mathcal{P}_x$ and $\dim_{[C]}(\mathcal{P}_x) = \dim T_{\mathcal{P}_x,[C]} = 2$.

In particular, $[C]$ belongs to a unique irreducible component $\mathcal{P}'$ of $\mathcal{P}_x$. If $\mathcal{T} \subset \mathcal{P} \times X$ is the universal family of the Hilbert scheme $\mathcal{P}$, denote it by $\mathcal{T}' := \mathcal{T}|\mathcal{P}' \times X$. Denote also by $p : \mathcal{T}' \to \mathcal{P}'$ the restriction to $\mathcal{T}'$ of the first projection of $\mathcal{P}' \times X$, and by $q : \mathcal{T}' \to X$ the restriction to $\mathcal{T}'$ of the second projection of $\mathcal{P}' \times X$. Finally, set $S := q(\mathcal{T}')$. Thus S is the locus of all curves in X parametrized by $\mathcal{P}'$ (all of them are conics in X).

Since $\dim(\mathcal{P}') = 2$, we have $\dim(\mathcal{T}') = 3$. Assume that q is surjective, i.e.

$$\dim(S) = 3. \tag{14.8}$$

Then, by [93], Corollary 3.10.1, page 117, C can be deformed to a smooth rational curve C' such that $T_X|C'$ is ample (more precisely, C' is the image via q of a general fiber of p). Since $N_{C'|X}$ is a quotient of $T_X|C'$, $N_{C'|X}$ is also ample. (Note that there is a perfect analogue of Corollary 3.10.1 of [93] in the context of Hilbert schemes, which allows one to get directly the ampleness of $N_{C'|X}$.) Now, by construction, C' is also a conic, whence $\deg(N_{C'|X}) = 2$. Using (14.5) for C' and the ampleness of $N_{C'|X}$ we get $N_{C'|X} \cong 2\mathcal{O}_{\mathbb{P}^1}(1)$. In other words, if (14.8) holds, we can take $Y = C'$ the conic we were looking for.

It remains therefore to prove (14.8). Assume to the contrary. Then S is a surface. Observe that $q^{-1}(x) \cong \mathcal{P}'$ via the morphism p, so $\dim(q^{-1}(x)) = 2$. Furthermore, for a general point $y \in S$, the theorem of dimension of fibers applied to q yields $\dim(q^{-1}(y)) = 1$. In other words, there is a 2-dimensional family of (smooth) conics passing through x and contained in S. Moreover, there is a 1-dimensional family of conics passing through x and a fixed general point y, and which are contained in S.

Assume that S is not a plane. Then we claim that the surface S spans a $\mathbb{P}^3$. To see this, consider the projection π from the point x. Then the image $S' := \pi(S)$ is a surface which satisfies the condition that for each general point $y' \in S'$ there is a 1-dimensional family of lines through y'. This implies that S' is a plane. This yields the fact that S spans a linear space L of dimension 3. Observe also that S is not a cone with vertex x because otherwise every conic passing through x would be degenerate.

Now consider the linear system $\mathcal{L}_x \cong (\mathbb{P}^2)^*$ of all planes in L passing through x. Since S is not a cone, the linear system cut out on S by $\mathcal{L}_x$ is not composed with a pencil. Then by Bertini, the intersection of S with a general plane $P \in \mathcal{L}_x$ is an irreducible reduced curve Γ_P.

Consider also the algebraic variety $\mathcal{C}_x$ (which is isomorphic to $\mathcal{P}'$) consisting of all conics in S passing through x. Then we have the natural rational map $\varphi_x : \mathcal{C}_x \dashrightarrow \mathcal{L}_x$ which sends a conic $D \in \mathcal{C}_x$ (which is not a double line) into the plane generated by D. Since two conics $D, D' \in \mathcal{C}_x$ have the same image P if and only if they are contained in $P \cap S$, the general fiber of φ_x is finite. This, together with the equalities $\dim(\mathcal{C}_x) = \dim(\mathcal{L}_x) = 2$, implies that every general plane $P \in \mathcal{L}_x$ is of the form $P = \varphi(C_P)$, with C_P a conic in S. It follows that $\Gamma_P = C_P$, i.e. the intersection of S with a general plane P of L through x, is a conic. This last fact clearly implies that S is a quadric surface in L. So, assuming that S is a surface, we have proved it is either a plane, or a quadric.

On the other hand, by Lefschetz' theorem, $\mathrm{Pic}(X) = \mathbb{Z}[\mathcal{O}_X(1)]$. Therefore there is a positive integer b such that $\mathcal{O}_X(S) \cong \mathcal{O}_X(b)$. It follows that $\deg(S)$ is a multiple of $\deg(X) = 3$, a contradiction. □

Remark 14.8. The above proof of Theorem 14.7 applies almost word by word to every Fano 3-fold of index 2 whose Picard group is $\mathbb{Z}$, see [19], Theorem (3.2). According to Fano–Iskovskih classification [86], apart from smooth cubic hypersurfaces in $\mathbb{P}^4$, these are: the smooth complete intersections of two hyperquadrics in $\mathbb{P}^5$ and the intersection of the image of the Grassmann variety of lines in $\mathbb{P}^4$ (via the Plücker embedding) with three general hyperplanes of $\mathbb{P}^9$.

The Hilbert scheme of a quasi-line

Our next objective is to study the geometry of smooth projective complex varieties containing quasi-lines. Let Y be a quasi-line on a projective n-fold X, and let $[Y]$ be a point of the Hilbert scheme $\mathrm{Hilb}(X)$ which corresponds to the curve Y. Since $H^1(Y, N_{Y|X}) = H^1(\mathbb{P}^1, (n-1)\mathcal{O}_{\mathbb{P}^1}(1))$ we conclude that $[Y]$ is a smooth point of $\mathrm{Hilb}(X)$ (see [69], exposé 221, or [113], or also [93], page 95). Therefore $[Y]$ belongs to a unique irreducible component, $\mathcal{H}$, of $\mathrm{Hilb}(X)$.

Consider the universal flat family in $\mathrm{Hilb}(X) \times X$

$$\begin{array}{ccccc} \mathcal{F} & \xrightarrow{\;i\;} & \mathcal{H} \times X & \xrightarrow{\;p_2\;} & X \\ \Big\downarrow p & & \Big\downarrow p_1 & & \\ \mathcal{H} & \xrightarrow{\;\mathrm{id}\;} & \mathcal{H} & & \end{array}$$

Denote $q := p_2 \circ i : \mathcal{F} \to X$. Note that for each $t \in \mathcal{H}$, $Y_t := q(p^{-1}(t))$ is the curve on X given by t and the restriction map

$$q|p^{-1}(t) : p^{-1}(t) \to Y_t$$

is an isomorphism.

For any given points $x, x' \in X$, set $\mathcal{H}_x := p(q^{-1}(x))$ and $\mathcal{H}_{x,x'} := \mathcal{H}_x \cap \mathcal{H}_{x'}$. Then $\mathcal{H}_x$ (respectively $\mathcal{H}_{x,x'}$) is the closed subscheme of $\mathcal{H}$ consisting of those points $t \in \mathcal{H}$ such that the curve Y_t contains x (respectively x, x').

Lemma 14.9. *Let X be a smooth projective complex variety of dimension $n \geq 2$. Let $Y_0 := q(p^{-1}(t_0))$ be a quasi-line in X, $t_0 \in \mathcal{H}$. Then there exists an open neighbourhood $\mathcal{U}$ of t_0 in $\mathcal{H}$ such that for each closed point $t \in \mathcal{U}$, the curve $Y_t := q(p^{-1}(t))$ is a quasi-line in X.*

Proof. Since the morphism $p : \mathcal{F} \to \mathcal{H}$ is flat and $p^{-1}(t_0) \cong \mathbb{P}^1$, there exists an open neighbourhood $\mathcal{U}$ of t_0 in $\mathcal{H}$ such that $Y_t \cong \mathbb{P}^1$ for every $t \in \mathcal{U}$. By the base-change theorems (see e.g. [73], III, §9) the functions $f(t) := \dim H^0(Y_t, N_{Y_t|X})$ and $g(t) := \dim H^1(Y_t, N_{Y_t|X})$ are upper semi-continuous. Thus, shrinking $\mathcal{U}$ if necessary, $g(t) = 0$ for every $t \in \mathcal{U}$ (because $H^1(Y_{t_0}, N_{Y_{t_0}|X}) = 0$). Hence $f(t) = f(t) - g(t) = \chi(Y_t, N_{Y_t|X})$ is constant in $\mathcal{U}$ by base-change theorems again. Moreover, since ampleness is an open condition, shrinking $\mathcal{U}$ once again, we can assume that

$$N_{Y_t|X} \cong \mathcal{O}_{\mathbb{P}^1}(a_1) \oplus \cdots \oplus \mathcal{O}_{\mathbb{P}^1}(a_{n-1}), \quad \text{for every } \; t \in \mathcal{U},$$

where $a_i = a_i(t)$ are positive integers depending on t, $\forall i = 1, \ldots, n-1$. Since

$$\dim H^0(Y_t, N_{Y_t|X}) = \sum_{i=1}^{n-1} (a_i + 1) = n - 1 + \sum_{i=1}^{n-1} a_i,$$

and

$$\dim H^0(Y_0, N_{Y_0|X}) = \dim H^0(\mathbb{P}^1, (n-1)\mathcal{O}_{\mathbb{P}^1}(1)) = 2n - 2,$$

we conclude that $\sum_{i=1}^{n-1} a_i = n - 1$ (because the function f is constant in $\mathcal{U}$), or else, $a_i = 1$, for every $i = 1, \ldots, n-1$ (because $a_i > 0$, $\forall i = 1, \ldots, n-1$). $\square$

Fix now a quasi-line Y and two distinct points $x, x' \in Y$. From general deformation theory (see e.g. [93], Theorem 1.7, page 95) we know that the tangent space to $\mathcal{H}_x$ at $[Y]$ is isomorphic to $H^0(Y, N_{Y|X} \otimes \mathcal{J}_x)$ and the tangent space to $\mathcal{H}_{x,x'}$ at $[Y]$ is isomorphic to $H^0(Y, N_{Y|X} \otimes \mathcal{J}_{x,x'})$, where $\mathcal{J}_x$, $\mathcal{J}_{x,x'}$ denote the ideal sheaves in Y of the subsets $\{x\}$, $\{x, x'\}$ with reduced structures. Note that, since $N_{Y|X} \cong (n-1)\mathcal{O}(1)$, one has

$$N_{Y|X} \otimes \mathcal{J}_x \cong N_{Y|X} \otimes \mathcal{O}_Y(-x) \cong N_{Y|X} \otimes \mathcal{O}_{\mathbb{P}^1}(-1) \cong (n-1)\mathcal{O}_{\mathbb{P}^1}.$$

Similarly we get

$$N_{Y|X} \otimes \mathcal{J}_{x,x'} \cong (n-1)\mathcal{O}_{\mathbb{P}^1}(-1).$$

We conclude that $H^1(Y, N_{Y|X} \otimes \mathcal{J}_x) = H^1(Y, N_{Y|X} \otimes \mathcal{J}_{x,x'}) = 0$. By loc.cit. we infer that $\mathcal{H}_x$ and $\mathcal{H}_{x,x'}$ are smooth at the point $[Y]$. Thus there exists a unique irreducible component, $\mathcal{H}_{x,[Y]}$, of $\mathcal{H}_x$ containing $[Y]$. From Lemma 14.9 we know that there exists an open subset, $\mathcal{U}_x \subset \mathcal{H}_x$, such that for each $t \in \mathcal{U}_x$, $q(p^{-1}(t))$ is a quasi-line containing x. Set $\mathcal{H}_x^\circ := \mathcal{H}_{x,[Y]} \cap \mathcal{U}_x$ and $\mathcal{H}_{x,x'}^\circ := \mathcal{H}_x^\circ \cap \mathcal{H}_{x,x'}$.

The following result shows that there are only finitely many quasi-lines passing through two distinct points, x, x', of a given quasi-line Y. In particular, $[Y]$ will be an isolated point of $\mathcal{H}_{x,x'}^0$. It will also follow that there are only finitely many quasi-lines passing through any two different points of X.

Lemma 14.10. *Let X be a smooth projective complex variety of dimension $n \geq 2$. Let Y be a quasi-line in X, and let $x, x' \in Y$. With the notation and assumptions as above we have:*

(i) $\dim_{[Y]}(\mathcal{H}_x^\circ) = n-1$;

(ii) $\dim_{[Y]}(\mathcal{H}_{x,x'}^\circ) = 0$.

Proof. The tangent space to $\mathcal{H}_x^0$ at $[Y]$ is isomorphic to $H^0(Y, N_{Y|X} \otimes \mathcal{J}_x)$ and the tangent space to $\mathcal{H}_{x,x'}^0$ at $[Y]$ is isomorphic to $H^0(Y, N_{Y|X} \otimes \mathcal{J}_{x,x'})$. Moreover, $\mathcal{H}_x^0$ and $\mathcal{H}_{x,x'}^0$ are both smooth at $[Y]$. Thus we get

$$\dim_{[Y]} \mathcal{H}_x^0 = \dim H^0(Y, N_{Y|X} \otimes \mathcal{J}_x) = \dim H^0(Y, (n-1)\mathcal{O}_{\mathbb{P}^1}) = n-1, \quad \text{and}$$

$$\dim_{[Y]} \mathcal{H}_{x,x'}^0 = \dim H^0(Y, N_{Y|X} \otimes \mathcal{J}_{x,x'}) = \dim H^0(Y, (n-1)\mathcal{O}_{\mathbb{P}^1}(-1)) = 0. \quad \square$$

The following result shows in particular that the family of quasi-lines in X fills up a dense open subset of X.

Theorem 14.11 ([19]). *Let X be a smooth projective complex variety of dimension $n \geq 2$. Let Y be a quasi-line in X, and let x be a fixed point of Y. With the notation and assumptions as above, we have:*

(i) *There exists a non-empty open subset U_x of X such that for each point $y \in U_x$, there exists a quasi-line passing through x and y;*

(ii) *The restriction map $q_x := q|p^{-1}(\mathcal{H}_{x,[Y]}) : p^{-1}(\mathcal{H}_{x,[Y]}) \to X$ is surjective;*

(iii) *For every point $y \in X$ there exists a connected curve with rational (but possibly singular) irreducible components passing through x and y.*

Proof. (i) Since $\mathcal{H}_x^0$ is an irreducible open subset of $\mathcal{H}_{x,[Y]}$ and the fiber of p over an arbitrary closed point of $\mathcal{H}_x^0$ is isomorphic to $\mathbb{P}^1$ (and in particular, all the fibers of $p^{-1}(\mathcal{H}_x^0) \to \mathcal{H}_x^0$ are irreducible of dimension 1), we infer that $p^{-1}(\mathcal{H}_x^0)$ is irreducible of dimension n (because by Lemma 14.10, $\dim \mathcal{H}_x^0 = n-1$). We claim that the morphism $q_x^\circ := q|p^{-1}(\mathcal{H}_x^\circ) : p^{-1}(\mathcal{H}_x^\circ) \to X$ is dominant. Indeed, using the theorem of dimension of the fibers of a morphism between two irreducible schemes, it will be sufficient to show that there is a fiber of q_x° with an isolated point. To this extent, pick an arbitrary point $y \in Y$, $y \neq x$. Since there is a one-to-one correspondence between $(q_x^\circ)^{-1}(y)$ and all the curves parametrized by $\mathcal{H}_{x,y}^\circ$, and since $[Y] \in \mathcal{H}_{x,y}^\circ$, Lemma 14.10, (ii) implies that $[Y]$ corresponds to an isolated point of $(q_x^\circ)^{-1}(y)$.

Now, by a theorem of Chevalley, there exists a non-empty open subset U_x in X such that $U_x \subseteq q_x^\circ(p^{-1}(\mathcal{H}_x^\circ))$. This means that for every point $y \in U_x$ there is a quasi-line passing through x and y. This proves (i).

Part (ii) follows from (i) because $p^{-1}(\mathcal{H}_{x,[Y]})$ is a projective scheme containing $p^{-1}(\mathcal{H}_x^\circ)$ as an open subset.

Part (iii) is a consequence of the following simple well-known lemma (applied to the proper flat morphism $p|p^{-1}(\mathcal{H}_{x,[Y]}) : p^{-1}(\mathcal{H}_{x,[Y]}) \to \mathcal{H}_{x,[Y]}$). □

Lemma 14.12. *Let $f\colon U \to V$ be a proper flat morphism of projective algebraic schemes over $\mathbb{C}$ such that V is irreducible and the general fiber of f is isomorphic to $\mathbb{P}^1$. Then every fiber of f is a connected curve with rational (but possibly singular) irreducible components.*

Proof. Since the general fiber of f is $\mathbb{P}^1$, it is geometrically irreducible. Then the flatness of f together with this imply that U is irreducible as well. Let $x \in V$ be a general point and $y \in V$ be an arbitrary point. Since V is irreducible, by a well-known elementary result we can find a closed irreducible curve C in V passing through x and y. (Hint: blow up V at x and at y to obtain the projective variety $\tilde{V}$ and the morphism $\pi : \tilde{V} \to V$; then choose a projective embedding $\tilde{V} \subset \mathbb{P}^N$; by Bertini one can find an irreducible hyperplane section $\tilde{W}$ in $\tilde{V}$ which necessarily meets $\pi^{-1}(x)$ and $\pi^{-1}(y)$; taking $W := \pi(\tilde{W})$ one finds an irreducible hypersurface of V passing through x and through y; finally, use induction on $\dim(V)$.) Let $g\colon C' \to V$ be composite of the inclusion $C \subset V$ with the normalization morphism $C' \to C$, and set $f'\colon U' := C' \times_V U$. Then we may replace V by C', U by U' and f by f'. In other words, there is no loss of generality in assuming that V is a smooth curve. Then U is an irreducible surface. Replacing U by a resolution of singularities $u : U_1 \to U$ and f by $f \circ u$, we may assume U smooth as well (the morphism $f \circ u$ remains flat since V is a non-singular curve). Then by the elementary theory of surfaces any fiber of a proper morphism $f\colon U \to V$, with V a smooth curve, U a smooth surface and the general fiber of f isomorphic to $\mathbb{P}^1$, is a connected curve whose irreducible components are all smooth rational curves.

Indeed, in this case $D := f^{-1}(y)$ is an effective divisor on U. Since the function $x \to h^2(\pi^{-1}(x), \mathcal{O}_{\pi^{-1}(x)})$ is constantly zero, the function $x \to h^1(\pi^{-1}(x), \mathcal{O}_{\pi^{-1}(x)})$ is constant by the base-change theorems (see e.g. [73]). Since for $f^{-1}(x) \cong \mathbb{P}^1$ for $x \in V$ general, we get $H^1(D, \mathcal{O}_D) = 0$. Let E be any irreducible component of D. Then the surjection $\mathcal{O}_D \to \mathcal{O}_E \to 0$ yields a surjection $H^1(D, \mathcal{O}_D) \to H^1(E, \mathcal{O}_E) \to 0$, whence $E \cong \mathbb{P}^1$. To prove the connectedness of D, we use Zariski's connectedness theorem, see [66] III, Théorème (4.3.1), or also [73], Corollary 11.3, page 279, according to which the morphism f factors as $U \to V' \to V$, where the first morphism has connected fibers and the second is finite (the Stein factorization of f). Since the general fiber of f is connected ($\cong \mathbb{P}^1$), the second morphism is birational, whence an isomorphism because V is a smooth curve. It follows that every fiber of f is connected. □

Remark 14.13. It is worth noting that, if X is a smooth projective surface, and if Y is a quasi-line in X, then through any of two distinct points $x, y \in Y$ there is no quasi-line in X passing through x and y, other than Y. To see this, first we have $Y \cong \mathbb{P}^1$ and $Y^2 = 1$. Then, by a standard elementary fact in the theory of surfaces (see e.g. [26] or [12]) the linear system $|Y|$ is base point free and defines a birational morphism $f\colon X \to \mathbb{P}^2$, such that $f^*\mathcal{O}_{\mathbb{P}^2}(1) \cong \mathcal{O}_X(Y)$. Therefore Y is the pullback of a line, ℓ, of $\mathbb{P}^2$. Since Y is irreducible we thus conclude that Y does not meet the locus of X where f is not biregular. Then $Y \cong \ell$ and, since there is a unique line in $\mathbb{P}^2$ passing through two distinct points, we are done.

Theorem 14.14 ([19]). *Let X be a smooth projective complex variety of dimension $n \geq 2$. Then the following conditions are equivalent.*

(i) *X contains a quasi-line Y.*

(ii) *The locus of all quasi-lines passing through a general point of X is dense in X.*

(iii) *There exists a non-empty open subset V of $X \times X$ such that for every point $(x, y) \in V$, there is a quasi-line joining x and y.*

If furthermore $n \geq 3$, the above conditions are also equivalent to any of the following ones:

(iv) *There is a morphism $f : \mathbb{P}^1 \to X$ such that $f^*(T_X) \cong \mathcal{O}_{\mathbb{P}^1}(2) \oplus (n-1)\mathcal{O}_{\mathbb{P}^1}(1)$.*

(v) *There is a morphism $f : \mathbb{P}^1 \to X$ such that $\deg(f^*(T_X)) = n + 1$ and the locus of all deformations of f sending $0 \in \mathbb{P}^1$ to a given general point of X (see [93]) is dense in X.*

Proof. The equivalence (i)$\Longleftrightarrow$(ii) is a direct consequence of Theorem 14.11. Clearly (iii)$\Longrightarrow$(i). Assume therefore that (i) holds. Consider the open dense subset $\mathcal{H}'$ of $\mathcal{H}$ which parametrizes the set of quasi-lines of $\mathcal{H}$, and define the incidence variety

$$\Gamma := \{([C], x, x') \in \mathcal{H}' \times X \times X \mid x, x' \in C\}.$$

Denote by $\rho : \Gamma \to X \times X$ the restriction to Γ of the projection onto $X \times X$. By Lemma 14.10, (ii), there is a point $(x, x') \in X \times X$, with $x \neq x'$, such that $\rho^{-1}(x, x')$ is finite. Thus, by the theorem of dimension of the fibers, the morphism ρ is dominant. Then, as above, there is a non-empty open subset V of $X \times X$ such that $V \subseteq \rho(\Gamma)$. This yields the required open subset of (iii), i.e. (i)$\Longrightarrow$(iii).

Using the normal exact sequence of Y in X we easily get (i)$\Longrightarrow$(iv). Noting that the condition (iv) can be restated as $\deg(f^*(T_X)) = n+1$ and $f^*(T_X)$ is ample, the equivalence (iv)$\Longleftrightarrow$(v) follows from [106], Lemma (2.6), page 101. Finally, (iv)$\Longrightarrow$(i) follows from [93], page 120, Theorem 3.14, (iii) (with $B = \varnothing$), since $n \geq 3$. $\square$

Now we need the following:

Definition 14.15 ([94]). Let X be a smooth projective variety over $\mathbb{C}$ of dimension $n \geq 2$. X is said to be *rationally connected* if one of the following equivalent conditions holds:

i) Given arbitrary points $x_1, \dots, x_m \in X$ there is an irreducible rational curve which passes through $x_1, \dots, x_m$.

ii) Given two arbitrary points $x_1, x_2 \in X$ there is an irreducible rational curve which passes through x_1 and x_2.

iii) For sufficiently general $(x_1, x_2) \in X \times X$ there is an irreducible rational curve which passes through x_1 and x_2.

iv) Given two arbitrary points $x_1, x_2 \in X$ there is a connected curve which passes through x_1 and x_2 and whose irreducible components are all rational curves.

v) There is a morphism $f : \mathbb{P}^1 \to X$ such that $f^*(T_X)$ is ample.

For the proof of the fact that these conditions are equivalent we refer the reader to [94]. Moreover, by [93], Theorem 3.14, page 120, if $n \geq 3$ the condition v) above is also equivalent to the following one:

vi) There is a smooth rational curve Y in X such that $N_{Y|X}$ is ample.

Remarks 14.16. i) The property iii) in Definition 14.15 is obviously a birationally invariant property. Therefore the concept of rationally connectedness is birationally invariant.

ii) A smooth projective surface X is rationally connected if and only if X is rational. Indeed, by i) we may assume that X is minimal. Then the property v) of Definition 14.15 implies that the curve $C := f(\mathbb{P}^1)$ satisfies $(C \cdot K) < 0$ and $(C^2) > 0$, where K is the canonical class of X. By the classification of surfaces (see e.g. [12], Proposition 13.1, page 195), $p_{12} = 0$, whence by Enriques' criterion (see [12], Theorem 13.2, page 195), X is ruled. If X were irregular, then the Albanese morphism $p : X \to B$ would yield a fibration onto a smooth projective curve B of genus $q = \dim H^1(X, \mathcal{O}_X) > 0$. But this is impossible because otherwise the rational curve C would dominate the curve B of positive genus (because $(C^2) > 0$).

Thus X is a surface with $p_{12} = q = 0$, whence X is rational by Castelnuovo's criterion of rationality (see [12], Theorem 13.3, page 195). Conversely, if X is a rational surface, then X is birationally equivalent to $\mathbb{P}^2$ which is obviously rationally connected.

iii) Any unirational manifold is rationally connected. Moreover, any Fano manifold is rationally connected (see [34], [94]). One of the important properties of rationally connected manifolds is that they are simply connected; moreover, if X is a rationally connected manifold then $H^0(X, \otimes^m \Omega^1_X) = 0$ for all $m > 0$, and $H^i(X, \mathcal{O}_X) = 0$ for all $i > 0$ (see [94], Proposition (2.5)). The rational connectedness is expected to be the "correct analogue of rationality" in dimension $n \geq 3$.

Definition 14.15 and the above remarks motivate the following:

Definition 14.17 ([19]). Let X be a smooth projective complex variety (or a projective manifold) of dimension ≥ 2. We say that X is *quasi-linearly connected* if X contains a quasi-line. Theorem 14.14 gives various characterizations of such manifolds.

Remarks 14.18. i) If X is a smooth projective surface, by Corollary 14.4 X is quasi-linearly connected if and only if X birationally dominates $\mathbb{P}^2$. For instance, the quadric surface $\mathbb{F}_0 = \mathbb{P}^1 \times \mathbb{P}^1$ does not contain any quasi-line, i.e. is not quasi-linearly connected. However, if we blow up any point of $\mathbb{F}_0$ we get a surface which birationally dominates $\mathbb{P}^2$ (and hence is quasi-linearly connected). In particular, the quasi-linearly connectedness is not birationally invariant. We also observe that any rational surface X is birationally dominated by a quasi-linearly connected surface X'. Indeed, let $\varphi : X \dashrightarrow \mathbb{P}^2$ be a birational map. By [138], Theorem 3, page 254, there exists a birational morphism $f : X' \to X$, with X' a smooth projective surface, such that $\varphi \circ f : X' \to \mathbb{P}^2$ is a (birational) morphism. Then, by what we have said above, X' is quasi-linearly connected.

ii) Let X be a smooth projective variety. If X is quasi-linearly connected then X is clearly rationally connected in the sense of Definition 14.15. Theorem 14.14 may be considered, in the case of quasi-lines, a strengthened form of Theorem (2.1) in [94].

iii) It is well known that any smooth projective deformation of a rationally connected manifold is again rationally connected (see [106], page 107, Proposition (2.13)). This statement is no longer true for quasi-linear connectedness. For example the Hirzebruch surface $\mathbb{F}_1$ (the projective plane blows up at a point), which is obviously quasi-linearly connected, can degenerate into a Hirzebruch surface $\mathbb{F}_{2e+1}$ with $e \geq 1$, which is not quasi-linearly connected. However the following holds:

Proposition 14.19 ([19]). *Any small (smooth) projective deformation of a quasi-linearly connected manifold X is again quasi-linearly connected.*

Proof. Let $f : \mathcal{X} \to T$ be a smooth projective morphism such that there is a point $t_0 \in T$ with the property that $f^{-1}(t_0) \cong X$. By taking an appropriate base change we may assume that T is a smooth curve. Let Y be a quasi-line in X. We may view Y as a curve in $\mathcal{X}$. The proof of the openness of the deformations of rationally connected manifolds works in our situation as well (see the first part of the proof of Proposition (2.13) of [106], page 107). In fact in our case it becomes even simpler, working with the Hilbert scheme (instead of the Hom-scheme). In fact, the canonical exact sequence

$$0 \to N_{Y|X} \to N_{Y|\mathcal{X}} \to N_{X|\mathcal{X}}|Y \cong \mathcal{O}_Y \to 0$$

splits to get $N_{Y|\mathcal{X}} \cong N_{Y|X} \oplus \mathcal{O}_Y$. Therefore

$$H^0(Y, N_{Y|\mathcal{X}}) \cong H^0(Y, N_{Y|X}) \oplus \mathbb{C}, \text{ , and } H^1(Y, N_{Y|\mathcal{X}}) = 0.$$

Hence there exists a one parameter family of curves $\{Y_s\}_{s\in D}$ (parametrized by the unit disk D) such that $Y_0 = Y$ and $Y_s \not\subseteq X_{t_0} = X$ for $s \neq 0$. Since Y_0 is contained in the fiber $f^{-1}(t_0)$, Y_s is contained in some fiber $X_{t_s} = f^{-1}(t_s)$, $\forall s$, and the morphism $s \to t_s$ is unramified near 0. Furthermore, ampleness of the normal bundle is an open condition on s. Therefore $N_{Y_s|X_{t_s}}$ is also ample for s near 0. But since $N_{Y|X} \cong (n-1)\mathcal{O}_{\mathbb{P}^1}(1)$, as in the proof of Lemma 14.9 we deduce that $N_{Y_s|X_{t_s}} \cong (n-1)\mathcal{O}_{\mathbb{P}^1}(1)$ for s near 0. Thus, since T is a smooth curve, there is an open neighbourhood of t_0 in T over which X_t contains a quasi-line, i.e. X_t is quasi-linearly connected. □

We remarked above already that the property of containing a quasi-line is not birationally invariant. However, we have the following result:

Proposition 14.20. *Let $\varphi\colon X \to X'$ be a birational morphism between smooth projective complex varieties of dimension $n \geq 2$. Assume that X' contains a quasi-line. Then X also contains a quasi-line.*

Proof. Let E be the locus where φ is not biregular. Then the image $\varphi(E)$ is a closed subset of X' of dimension $\leq n-2$. Fix a general point x on X'. Then there is a quasi-line through x which does not meet $\varphi(E)$. Indeed, by Lemma 14.10, (ii), there is only a finite number of quasi-lines through x and some point y of $\varphi(E)$. Thus the locus of quasi-lines through x which meet $\varphi(E)$ is of dimension $\leq \dim(\varphi(E)) + 1 \leq n-1$. But from Theorem 14.11 we know that the quasi-lines through x fill up a dense open subset of X'. Thus we conclude that there exists a quasi-line, C, on X' such that $C \cap \varphi(E) = \varnothing$. Therefore $Y := \varphi^{-1}(C)$ is a quasi-line on X. □

Remark 14.21. Let us point out that the converse does not hold in Proposition 14.20: if $\varphi\colon X \to X'$ is as in Proposition 14.20 and X contains a quasi-line, it does not follow in general that X' does. For example, take $X' = \mathbb{F}_0 = \mathbb{P}^1 \times \mathbb{P}^1$, and as $\varphi\colon X \to X'$ the blow up morphism of X' at some of its points. Clearly, X' does not contain any quasi-line. On the other hand, X contains quasi-lines because X dominates the surface $\mathbb{F}_1$ (and hence X dominates $\mathbb{P}^2$).

From now on X will denote a complex smooth projective complex variety of dimension $n \geq 2$.

Definition 14.22 ([19]). We say that X is *strongly rational* if there exist non-empty open sets $U \subseteq X$, $V \subseteq \mathbb{P}^n$ such that U is isomorphic to V and $\text{codim}_{\mathbb{P}^n}(\mathbb{P}^n \setminus V) \geq 2$.

Remarks 14.23. i) Any (smooth) hyperquadric of dimension $n \geq 2$ is certainly rational, but it is not strongly rational. Indeed, any strongly rational manifold contains a quasi-line. But we noticed above that the hyperquadrics do not contain quasi-lines.

ii) If there is a birational morphism $\psi\colon X \to \mathbb{P}^n$, X is strongly rational. Indeed, if E is the locus of X where ψ is not biregular, then $\text{codim}_{\mathbb{P}^n}(\psi(E)) \geq 2$, whence X contains an open subset isomorphic to $\mathbb{P}^n \setminus \psi(E)$. It is easy to see that any strongly rational surface is of this type, i.e. it dominates $\mathbb{P}^2$. This no longer holds if $n \geq 3$ as we shall see later.

Now, if X' is any (smooth) rational variety and $\beta\colon X' \dashrightarrow \mathbb{P}^n$ is a fixed birational map, it follows from Hironaka's results [80] that we can find a composite of blow ups with smooth centers, $\varphi\colon X \to X'$, and a birational morphism $\psi\colon X \to \mathbb{P}^n$ such that $\psi = \beta \circ \varphi$. Thus we see that any rational variety is dominated by a strongly rational one.

We need the following general elementary fact.

Lemma 14.24. *Let Y be a quasi-line on X. Then there exist only finitely many prime divisors H on X such that $(H \cdot Y) = 0$.*

Proof. Let H be a prime divisor on X such that $(H \cdot Y) = 0$. Then we claim that $H \cap Y = \varnothing$. Indeed, if $x \in H \cap Y$ is a point, by Theorem 14.11, (i), there exists a quasi-line Y' passing through x, not contained in H, and which is numerically equivalent to Y. In particular, $(H \cdot Y) = 0$ implies $(H \cdot Y') = 0$. Since Y' is not contained in H, this last equality implies $H \cap Y' = \varnothing$, contradicting the fact that $x \in H \cap Y'$. This shows that H does not meet the locus of quasi-lines which are in the same family with Y, and hence H is contained in a proper closed subset of X. In particular, there are only finitely many prime divisors of X that are orthogonal to Y. □

We can now prove the following result.

Theorem 14.25 ([19]). *Let X be a smooth projective complex variety of dimension $n \geq 2$. The following conditions are equivalent.*

(i) *X is strongly rational;*

(ii) *X contains a quasi-line Y such that the pair (X, Y) is Zariski equivalent to $(\mathbb{P}^n$, line$)$;*

(iii) *X contains a quasi-line Y and there exists a divisor D on X such that $(D \cdot Y) = 1$ and $\dim |D| \geq n$.*

Proof. The implications (i)$\Longleftrightarrow$(ii) and (ii)$\Longrightarrow$(iii) are easy consequences of the definitions. To show that (iii)$\Longrightarrow$(i) we shall proceed by induction on n. Let us first prove this in the case $n = 2$. Consider the exact sequence

$$0 \to \mathcal{O}_X(D - Y) \to \mathcal{O}_X(D) \to \mathcal{O}_Y(D) \cong \mathcal{O}_{\mathbb{P}^1}(1) \to 0.$$

Assume $\dim H^0(X, \mathcal{O}_X(D)) \geq 4$. Then $\dim |D - Y| \geq 1$, so that the divisor $D - Y$ moves. Since $((D - Y) \cdot Y) = 0$, this contradicts Lemma 14.24. Therefore $\dim H^0(X, \mathcal{O}_X(D)) = \dim H^0(X, \mathcal{O}_X(Y)) = 3$ and $\dim |D - Y| = 0$. This implies that $E := D - Y$ is an effective divisor which must be the fixed part of $|D|$. Thus the rational maps associated to $|D|$ and $|Y|$ coincide. But, as is well known, the linear system $|Y|$ has no base points and yields a birational morphism $X \to \mathbb{P}^2$ such that Y is the pull-back of a line (see e.g. [12]).

Assume now $n \geq 3$. Write $D = E + M$, where E and M are the fixed and the moving part of the linear system D respectively. Since Y moves, we have $(E \cdot Y) \geq 0$ and $(M \cdot Y) \geq 0$. Therefore, by Lemma 14.24, the assumption $(D \cdot Y) = 1$ yields $(E \cdot Y) = 0$ and $(M \cdot Y) = 1$. Since $\dim |M| = \dim |D| \geq n$ we conclude that the divisor M satisfies the same assumptions as the divisor D. Thus we may assume that the linear system D is free from fixed components. Then from Hironaka's desingularization theory [80], we infer that there exists a morphism $\sigma \colon X' \to X$, which is a composite of blow ups along smooth centers, such that $\sigma^*(|D|) = E' + |D'|$, with $|D'|$ a base point free linear system on X', $E' \geq 0$ the fixed part of $|E' + D'|$, and $\dim(\sigma(E')) \leq n - 2$. In particular, $(\sigma^{-1}(Y) \cdot D') = (Y \cdot D) = 1$. Therefore by the previous argument $|D'|$ is not composed with a pencil. Since $|D'|$ is base point free and not composed with a pencil, by Bertini a general member, Δ', of $|D'|$ is smooth and connected. Fix such a Δ' and set $\Delta := \sigma(\Delta')$.

Now take two general points $x, y \in X'$. Since $\dim |D| \geq n \geq 2$ we may assume that $x, y \in \Delta'$. From Theorem 14.14, (iii) it follows that there exists a quasi-line Y on X passing through $\sigma(x)$, $\sigma(y)$. If Z denotes the locus of points where σ^{-1} is not defined, we have $\operatorname{codim}_X(Z) \geq 2$. The proof of Proposition 14.20 then shows that we may further assume that Y does not meet Z. So $Y' := \sigma^{-1}(Y)$ is a quasi-line on X' passing through x and y. Therefore $\{x, y\} \subset \Delta' \cap Y'$ and the assumption that $(\Delta \cdot Y) = 1$ implies $(\Delta' \cdot Y') = 1$. Since $\{x, y\} \subset \Delta' \cap Y'$ this last equality forces $Y' \subset \Delta'$.

Then we claim that Y' is a quasi-line on Δ'. To see this, consider the exact sequence

$$0 \to N_{Y'|\Delta'} \to N_{Y'|X'} \cong (n-1)\mathcal{O}_{\mathbb{P}^1}(1) \to N_{\Delta'|X'}|Y' \cong \mathcal{O}_{\mathbb{P}^1}(1) \to 0.$$

By tensoring with $\mathcal{O}_{\mathbb{P}^1}(-1)$ and dualizing we get the exact sequence

$$0 \to \mathcal{O}_{\mathbb{P}^1} \to (n-1)\mathcal{O}_{\mathbb{P}^1} \to N^*_{Y'|\Delta'}(1) \to 0$$

on $\mathbb{P}^1$. From this we see that $N^*_{Y'|\Delta'}(1)$ is globally generated and of degree zero. Hence it is trivial. This is equivalent to $N_{Y'|\Delta'} \cong (n-2)\mathcal{O}_{\mathbb{P}^1}(1)$, i.e. Y' is a quasi-line on Δ'.

Now consider the exact sequence

$$0 \to \mathcal{O}_{X'} \to \mathcal{O}_{X'}(\Delta') \to \mathcal{O}_{\Delta'}(\Delta') \to 0. \tag{14.9}$$

Since $\dim|\Delta'| \geq n$, we get $h^0(\Delta', \mathcal{O}_{\Delta'}(\Delta')) \geq n$. Now apply induction on Δ'.

Then the rational map given by $H^0(\Delta', \mathcal{O}_{\Delta'}(\Delta'))$ maps a neighbourhood of a fixed quasi-line, $Y' \subset \Delta'$, isomorphically to some open subset of $\mathbb{P}^{n-1}$, such that Y' corresponds to a line. Since $H^1(X', \mathcal{O}_{X'}) = 0$ (see the remark following the proof of Theorem 14.14) we see from the cohomology sequence associated to the exact sequence (14.9) that the rational map given by $H^0(X', \mathcal{O}_{X'}(\Delta'))$ maps a neighbourhood of $Y' \subset X'$ isomorphically to some open subset of $\mathbb{P}^n$, such that Y' corresponds to a line. We obtain the same conclusion for the pair (X, Y) via the birational morphism σ. This completes the proof of our theorem. □

Before passing to applications we need the following simple lemma:

Lemma 14.26. *Let X be a smooth projective complex variety of dimension $n \geq 1$, and let D be a divisor on X such that the linear system $|D|$ has no fixed components and yields a birational map $\varphi\colon X \dashrightarrow \mathbb{P}^n$ which is an isomorphism between an open subset U of X and an open subset V of $\mathbb{P}^n$ whose complement is of codimension ≥ 2 in $\mathbb{P}^n$. Then for every $m \geq 0$, φ induces isomorphisms $H^0(X, \mathcal{O}_X(mD)) \cong H^0(\mathbb{P}^n, \mathcal{O}_{\mathbb{P}^n}(m))$. In particular, if $|mD|$ is base point free for some $m > 0$, then $|D|$ is also base point free.*

Proof. Let U' be the open subset of X on which φ is defined. Since $|D|$ has no fixed components, $\operatorname{codim}_X(X \setminus U') \geq 2$, the restriction map $H^0(X, \mathcal{O}_X(mD)) \to H^0(U', \mathcal{O}_X(mD)|U')$ is an isomorphism. By hypothesis $\operatorname{codim}_{\mathbb{P}^n}(\mathbb{P}^n \setminus V) \geq 2$, whence the restriction map $\psi\colon H^0(\mathbb{P}^n, \mathcal{O}_{\mathbb{P}^n}(m)) \to H^0(V, \mathcal{O}_{\mathbb{P}^n}(m)|V)$ is also an isomorphism. Now the lemma follows easily from the commutative diagram

$$\begin{array}{ccc} H^0(\mathbb{P}^n, \mathcal{O}_{\mathbb{P}^n}(m)) & \xrightarrow{\varphi^*} & H^0(U', \mathcal{O}_X(mD)|U') \\ \Big\downarrow{\scriptstyle \psi} & & \Big\downarrow{\scriptstyle \cap} \\ H^0(V, \mathcal{O}_{\mathbb{P}^n}(m)|V) & \xrightarrow{\cong} & H^0(U, \mathcal{O}_X(mD)|U) \end{array}$$

in which the second vertical (restriction) map is injective and the bottom horizontal map is an isomorphism.

For the last assertion, observe that φ is a morphism, being the composite of the inverse of the isomorphism $v_m : \mathbb{P}^n \to v_m(\mathbb{P}^n)$ given by the m-th Veronese embedding of $\mathbb{P}^n$, and the morphism associated to $|mD|$. □

A first consequence of Theorem 14.25 is the following "birational version" of Kobayashi–Ochiai's result (see e.g. [29], (3.1.6)):

Corollary 14.27 ([19]). *Assume that X contains a quasi-line Y. Let H be a nef and big divisor on X such that $(H \cdot Y) = 1$. Let $n := \dim X$. Then we have:*

(i) *The pair (X, Y) is Zariski equivalent to $(\mathbb{P}^n, \text{line})$;*

(ii) *Assume moreover that there is no effective divisor orthogonal to Y. Then $(X, Y) \cong (\mathbb{P}^n, \text{line})$.*

Proof. (i) Let $d := (H^n) > 0$. Consider the Hilbert polynomial $p(t) := \chi(\mathcal{O}_X(tH))$ (of degree n). By duality and the Kawamata–Viehweg vanishing theorem we have (for $i = 1, \ldots, n$)

$$p(-i) = \chi(\mathcal{O}_X(-iH)) = (-1)^n \chi(\mathcal{O}_X(K_X + iH)) = (-1)^n h^0(\mathcal{O}_X(K_X + iH)).$$

Since $(K_X \cdot Y) = -n-1$, we get $((K_X + iH) \cdot Y) < 0$, and hence $h^0(\mathcal{O}_X(K_X + iH)) = 0$, for $i = 1, \ldots, n$. Therefore

$$p(t) = \frac{d}{n!}(t+1) \cdots (t+n).$$

On the other hand,

$$p(-(n+2)) = (-1)^n d(n+1) = (-1)^n h^0(\mathcal{O}_X(K_X + (n+2)H)).$$

Thus $h^0(\mathcal{O}_X(K_X + (n+2)H)) = d(n+1) \geq n+1$, so that there exists an effective divisor $D \in |K_X + (n+2)H|$ such that $\dim |D| \geq n$. Note that $(D \cdot Y) = 1$. Then Theorem 14.25 applies to give the result.

(ii) With the same notation, arguing as in the proof of part (i), we have

$$p(-(n+1)) = (-1)^n d = (-1)^n h^0(\mathcal{O}_X(K_X + (n+1)H)) \neq 0.$$

Thus $K_X + (n+1)H$ is an effective divisor. Since $(K_X + (n+1)H) \cdot Y = 0$ we conclude that $K_X + (n+1)H$ is trivial and hence H is linearly equivalent to $D := K_X + (n+2)H$. Since H is nef and big, we can apply the Kawamata-Reid-Shokurov base point free theorem (see e.g. [29], (1.5.1)) to deduce that $|mD|$ is base point free for some $m \geq 1$. From Lemma 14.26 it follows that $|D|$ is base point free. So $|D|$ defines a birational morphism to $\mathbb{P}^n$ which is an isomorphism by Zariski's Main Theorem and our hypothesis. □

As a further application of Theorem 14.25, we give an example of strongly rational manifolds in any dimension $n \geq 3$ which does not dominate birationally $\mathbb{P}^n$.

Example 14.2. ([19]). Let $n \geq 3$ and let $\sigma \colon Z \to \mathbb{P}^{n-1}$ be the blow up a point. Denote by E' the exceptional locus of σ and by H' the pullback of a hyperplane in $\mathbb{P}^{n-1}$. Moreover, let $L' := H' - E'$. Denote by X the projective bundle $\mathbb{P}(\mathcal{O}_Z \oplus \mathcal{O}_Z(L'))$ and let $\pi \colon X \to Z$ be the canonical projection. Denote by S the section of π given by the natural surjection $\mathcal{O}_Z \oplus \mathcal{O}_Z(L') \to \mathcal{O}_Z(L')$. Set $E := \pi^* E'$, $H := \pi^* H'$ and $L := \pi^* L'$.

We first claim that there is no birational morphism from X to $\mathbb{P}^n$. To prove the claim it is enough to see that for any nef divisor, say A, on X, we have $A^n \neq 1$. To this end we shall compute intersection numbers with respect to the basis of $\mathrm{Pic}(X)$ given by the classes of L, H and S. Note that the normal bundle of S in X is identified via π to $\mathcal{O}_Z(L')$. We get the following: $(L^i \cdot H^{n-i}) = (L^i \cdot S^{n-i}) = 0$ for any $0 \leq i \leq n$, and $(S^i \cdot L^j \cdot H^{n-i-j}) = 1$ if $i+j \leq n-1$ and $1 \leq i \leq n-1$, $0 \leq j \leq n-2$. Now write $A = aL + bH + cS$, where a, b, c are integers. Assuming A nef and noting that H and L are also nef, we get $(A \cdot S^{n-1}) = b \geq 0$, $(A \cdot H^{n-1}) = c \geq 0$. Assume moreover that $A^n = 1$. Computing A^n by the preceeding formulae we first see that A^n is divisible by b and c, so we get $b = c = 1$; thus, the expression for A^n becomes:

$$A^n = \sum_{i=1}^{n-1} \sum_{j=0}^{i-1} \binom{n}{i} \binom{i}{j} a^j.$$

Next we get $(A^2 \cdot H^{n-2}) = 3 + 2a \geq 0$, giving $a \geq -1$ since a is an integer. Now the above equality gives $A^n > 1$ if $a \geq 0$, while for $a = -1$ we get $A^n = 2$ if n is even and $A^n = 0$ if n is odd. In conclusion, we proved that $A^n \neq 1$ if A is nef, which shows that there is no birational morphism from X to $\mathbb{P}^n$.

Secondly, consider the linear system $|D|$, where $D := S + E$. As is easily seen, $H^1(X, \mathcal{O}_X(E)) = 0$, $H^0(X, \mathcal{O}_X(E)) \cong \mathbb{C}$ and $H^0(X, \mathcal{O}_X(D)) \cong \mathbb{C}^{n+1}$. Let $Y \subset S$ be the curve corresponding via π to the pullback of a general line in $\mathbb{P}^{n-1}$ via σ. Clearly Y is a quasi-line on S; as we also have $(Y \cdot S) = (Y \cdot L) = 1$, from the normal bundles sequence it follows that Y is in fact a quasi-line of X. Moreover we have $(D \cdot Y) = 1$ and $\dim |D| = n$. So, Theorem 14.25 implies that X is strongly rational. □

We close this chapter by mentioning the following result of Ionescu and Naie [84] which explains the role of the quasi-lines in the framework of rationally connected varieties:

Theorem 14.28 ([84]). *Let X be an arbitrary rationally connected smooth projective variety X of dimension $n \geq 2$. Then there exists a birational morphism $f : X' \to X$, which is a composite of blow ups of smooth 2-codimensional centers, such that X' is quasi-linearly connected.*

Idea of proof. Note that the case $n = 2$ was already settled in Remark 14.18, i). So we may assume $n \geq 3$. Then by the condition vi) of Definition 14.15 there exists a curve $Y \cong \mathbb{P}^1$ in X such that $N_{Y|X}$ is ample. Therefore, by Grothendieck's theorem (see [70]) we have

$$N_{Y|X} \cong \mathcal{O}_{\mathbb{P}^1}(a_1) \oplus \cdots \oplus \mathcal{O}_{\mathbb{P}^1}(a_{n-1}), \quad \text{with } a_1 \geq a_2 \geq \cdots \geq a_{n-1} > 0. \tag{14.10}$$

If $a_1 = 1$ then Y is a quasi-line and we have nothing to prove. Assuming $a_1 \geq 2$, then the proof of the theorem follows by repeatedly applying the following elementary (but crucial) lemma:

Lemma 14.29 ([84], Lemma 2.2). *Let X be a smooth projective variety of dimension $n \geq 2$ and let $Y \cong \mathbb{P}^1$ be a smooth rational curve on X such that the normal bundle $N_{Y|X}$ satisfies* (14.10)*. Let Z be any 2-codimensional smooth closed subvariety of X such that Z intersects Y transversely in just one point (given any point $y \in Y$ by Bertini there always exists such a Z intersecting Y only in y, tranversely). Let $\sigma : X' \to X$ be the blow up morphism of X of center Z, and let Y' be the proper transform of Y via σ. Then*

$$N_{Y'|X'} \cong \mathcal{O}_{\mathbb{P}^1}(a_1 - 1) \oplus \mathcal{O}_{\mathbb{P}^1}(a_2) \oplus \cdots \oplus \mathcal{O}_{\mathbb{P}^1}(a_{n-1}).$$

Remark 14.30. In case $n = 2$ Lemma 14.29 follows from [138], Corollary 3, page 253. If $n \geq 3$ the idea is to use generalized elementary transformations, see [84] for details.

Bibliography

[1] A. Altman and S. Kleiman, *Introduction to Grothendieck Duality Theory*, Lecture Notes in Math. Vol. **146**, Springer-Verlag, Berlin-Heidelberg-New York, 1970.

[2] A. Andreotti, *Théorèmes de dépendence algébrique sur les espaces pseudo-concaves*, Bull. Soc. Math. France **91** (1963), 1–38.

[3] E. Arbarello, M. Cornalba, Ph. Griffiths and J. Harris, *Geometry of Algebraic Curves I.*, Grundlehren, **267** Springer-Verlag, 1985.

[4] M. Artin, *Some numerical criteria of contractibility of curves on algebraic surfaces*, Amer. J. Math. **84** (1962), 485–496.

[5] M. Artin, *Deformations of Singularities*, Tata Inst. Fund. Research, Bombay, 1976.

[6] M. Artin, *Algebraic approximation of structures over complete local rings*, Publ. Math. IHES **36** (1969), 23–58.

[7] M. F. Atiyah and I. G. Macdonald, *Introduction to Commutative Algebra*, Adison-Wesley Publ. Company, Reading, Mass., 1969.

[8] A. Babakarian and H. Hironaka, *Formal functions over Grassmannians*, Illinois J. Math. **26** (1982), 201–211.

[9] L. Bădescu, *On a result of Zak-L'vovsky*, in Lect. Notes in Pure and Appl. Math, Vol. **166**, Marcel Dekker, Inc. New York-Basel-Hong Kong, (1994), 57–73.

[10] L. Bădescu, *Infinitesimal deformations of negative weights and hyperplane sections*, in Lect. Notes in Math. Vol. **1417**, Springer-Verlag, (1990), 1–22.

[11] L. Bădescu, *Algebraic Barth-Lefschetz theorems*, Nagoya Math. J. **142** (1996), 17–38.

[12] L. Bădescu, *Algebraic Surfaces*, Universitext, Springer, New York-Berlin-Heidelberg-Barcelona-Hong Kong-London-Milan-Paris-Tokyo, 2001.

[13] L. Bădescu, *Special chapters of projective geometry*, Rendiconti del Seminario Matematico e Fisico di Milano **69** (1999-2000), 239–326.

[14] L. Bădescu, *Formal functions in algebraic geometry*, Mathematical Reports (formerly Stud. Cerc. Mat.) **4 (54)** (2002), 1–51.

[15] L. Bădescu, *Semi-affiness and formal functions*, Mathematical Reports (formerly Stud. Cerc. Mat.) **3 (53)** (2001), 133–138.

[16] L. Bădescu and E. Ballico, *Formal-rational functions along zero loci of sections of ample vector bundles* Rev. Roumaine Math. Pures Appl. **38** (1993), 609–630.

[17] L. Bădescu and M. Schneider, *A criterion for extending meromorphic functions*, Math. Annalen **305** (1996), 393–402.

[18] L. Bădescu and M. Schneider, *Formal functions, connectivity and homogeneous spaces*, in "Algebraic Geometry - A Volume in Memory of Paolo Francia" (M.C. Beltrametti, F. Catanese, C. Ciliberto, A. Lanteri, C. Pedrini, editors), Walter de Gruyter, Berlin-New York, 2002, pp. 1–21.

[19] L. Bădescu, M. C. Beltrametti and P. Ionescu, *Almost-lines and quasi-lines on projective manifolds*, in "Complex Analysis and Algebraic Geometry - A Volume in Memory of Michael Schneider" (Th. Peternell, F-O. Schreyer, editors), Walter de Gruyter, Berlin-New York, 2000, pp. 1–27.

[20] C. Bănică and O. Stănăşilă, *Algebraic Methods in the Global Theory of Complex Spaces*, John Wiley, New York, 1976.

[21] W. Barth, *Fortsetzung meromorpher Funktionen in Tori und komplexprojektiven Räumen*, Invent. Math. **5** (1968), 42–62.

[22] W. Barth, *Submanifolds of low codimension in projective space*, Proc. ICM Vancouver (1975), 409–413.

[23] W. Barth, *Verallgemeinerung des Bertinisches Theorems in Abelschen Mannigfaltigkeiten*, Ann. Sc. Norm. Sup. Pisa **23** (1969), 317–330.

[24] W. Barth, M. E. Larsen, *On the homotopy groups of complex projective algebraic manifolds* Math. Scand. **30** (1972), 88–94.

[25] W. Barth, C. Peters and A. Van de Ven, *Compact Complex Surfaces*, Springer-Verlag, 1984.

[26] A. Beauville, *Surfaces Algébriques Complexes*, Astérisque 54, Soc. Math. France, Paris, 1978.

[27] A. Beauville and J.-Y. Mérindol, *Sections hyperplanes des surfaces K3*, Duke Math. J. **55** (1987), 873–878.

[28] M. Beltrametti and L. Robbiano, *Introduction to the theory of weighted projective spaces*, Expo. Math. **4** (1986), 111–162.

[29] M. C. Beltrametti and A. J. Sommese, *The Adjunction Theory of Complex Projective Varieties*, Expositions in Math. **16**, Walter de Gruyter, Berlin-New York, 1995.

[30] A. Borel, *Linear Algebraic Groups*, W.A. Benjamin, New York-Amsterdam, 1969.

[31] N. Bourbaki, *Algèbre: Polynômes et Corps. (Chapitres IV et V)*, Hermann, Paris, 1960.

[32] R. Braun, *On the normal bundle of Cartier divisors on projective varieties*, Arch. Math. **59** (1992), 403–411.

[33] R. Braun, *Über das Normalenbündel von projektiven Varietäten*, Disertazion, Universität Bayreuth, 1989.

[34] F. Campana, *Connexité rationnelle des variétés de Fano*, Ann. Sci. Éc. Norm. Sup. **25** (1992), 539–545.

[35] W. L. Chow, *On meromorphic maps of algebraic varieties*, Annals of Math. **89** (1969), 391–403.

[36] W. L. Chow, *Formal functions on homogeneous spaces*, Invent. Math. **86** (1986), 115–130.

[37] H. Clemens and Ph. A. Griffiths, *The intermediate Jacobian of the cubic threefold*, Annals of Math. **95** (1972), 281–356.

[38] C. Ciliberto, J. Harris and R. Miranda, *On the surjectivity of the Wahl map*, Duke Math. J. **57** (1988), 829–858.

[39] C. Ciliberto and R. Miranda, *The Gaussian map for canonical curves of low genus*, Duke Math. J. **61** (1990), 417–443.

[40] J. d'Almeida, *Une caractérisation du plan projectif complexe*, Enseign. Math. **41** (1995), 135–139.

[41] O. Debarre, *Sur un théorème de connexité de Mumford pour les espaces homogènes*, Manuscripta Math. **89** (1996), 407–425.

[42] O. Debarre, *Fulton-Hansen and Barth-Lefschetz theorems for subvarieties of abelian varieties*, J. reine angew. Math. **467** (1995), 187–197.

[43] O. Debarre, *Théorèmes de connexité et variétés abéliennes*, Amer. J. Math. **117** (1995), 787–805.

[44] P. Deligne, *Le groupe fondamental du complément d'une courbe plane n'ayant que des points double ordinaires*, Séminaire Bourbaki, No. **543**, Paris, November, 1979.

[45] I. Dolgachev, *Weighted projective varieties*, Lect. Notes in Math. Vol. **956**, Springer-Verlag, Berlin-Heidelberg-New York, 1982, pp. 34–71.

[46] G. Ellingsrud, L. Gruson, C. Peskine and S. A. Strømme, *On the normal bundle of curves on smooth projective surfaces*, Invent. Math. **80** (1985), 181–184.

[47] G. Faltings, *Formale Geometrie und homogene Räume*, Invent. Math. **64** (1981), 123–165.

[48] G. Faltings, *A contribution to the theory of formal meromorphic functions*, Nagoya Math. J. **77** (1980), 99–106.

[49] G. Faltings, *Algebraisation of some formal vector bundles*, Annals of Math. **110** (1979), 501–514.

[50] T. Fujita, *Rational retractions onto ample divisors*, Sci. Papers Coll. Art. Sci. Univ. Tokyo **33** (1983), 33–39.

[51] W. Fulton, *Intersection Theory*, Ergebnisse der Math. und ihres Grenzgebiete, 3. Series, Band **2**, Springer-Verlag, Berlin-Heidelberg-New York-Tokyo, 1984.

[52] W. Fulton and J. Hansen, *A connectedness theorem for proper varieties with applications to intersections and singularities*, Annals of Math. **110** (1979), 159–166.

[53] W. Fulton and R. Lazarsfeld, *Connectivity and its applications in algebraic geometry*, Lect. Notes in Mathematics Vol. **862**, Springer-Verlag, 1981, pp. 26–92.

[54] W. Fulton and R. Lazarsfeld, *Positivity and excess intersection*, Enumerative Geometry and Classical Algebraic Geometry, Progress in Math. Vol. **24**, Birkhäuser, Boston-Basel-Stuttgart, 1982, pp. 97–105.

[55] W. Fulton and R. Lazarsfeld, *On the connectedness of degeneracy loci and special divisors*, Acta Math. **146** (1981), 271–283.

[56] D. Gallarati, *Ricerche sul contatto di superficie algebriche lungo curve*, Mémoires de l'Académie Royale de Belgique **32 (3)** (1960), 1–78.

[57] D. Gallarati, *Problemi di completa interferenza in geometria algebrica*, Rend. Sem. Fac. Sc. Univ. Cagliari (RSFSAK 203–245) **46 (3-4)** (1976), 242–270.

[58] D. Gieseker, *On two theorems of Griffiths about embeddings with ample normal bundle*, Amer. J. Math. **99** (1977), 1137–1150.

[59] D. Gieseker, *P-ample bundles and their Chern classes*, Nagoya Math. J. **43** (1971), 91–116.

[60] N. Goldstein, *Ampleness and connectedness in complex G/P*, Trans. Amer. Math. Soc. **274** (1982), 361–373.

[61] J. E. Goodman and A. Landman, *Varieties proper over affine schemes*, Invent. Math. **20** (1973), 267–312.

[62] M. Green, *Koszul cohomology and the geometry of projective varieties*, J. Diff. Geometry **19** (1984), 125–171.

[63] Ph. A. Griffiths, *The extension problem in complex analysis II: Embeddings with positive normal bundle*, Amer. J. Math. **88** (1965), 366–446.

[64] Ph. Griffiths and J. Harris, *Principles of Algebraic Geometry*, Wiley, Interscience, New York, 1978.

[65] A. Grothendieck, *Local Cohomology*, Lect. Notes in Math. Vol. **41**, Springer-Verlag, 1966.

[66] A. Grothendieck and J. Dieudonné, *Éléments de Géométrie Algébrique*, Publ. Math. IHES, Vol. **4, 8, 11, 17, 20, 24, 28, 32**, Bures-sur-Yvette, 1960-1967.

[67] A. Grothendieck, *Revêtements Étales et Groupe Fondamental*, Lecture Notes in Math. Vol. **224**, Springer-Verlag, Berlin-Heidelberg-New York, 1971.

[68] A. Grothendieck, *Cohomologie Locale des Faisceaux Cohérents et Théorèmes de Lefschetz Locaux et Globaux*, North-Holland, Amsterdam, 1968.

[69] A. Grothendieck, *Fondements de la Géométrie Algébrique*, Extraits du Séminaire Bourbaki (Éxposés No. **221, 232, 236**), Paris, 1957–1962.

[70] A. Grothendieck, *Sur la classification des fibrés vectoriels sur la sphère de Riemann*, Amer. J. Math. **79** (1957), 121–138.

[71] A. Grothendieck, *Dix Éxposés sur la Cohomologie des Schémas*, North Holland, Amsterdam, 1968.

[72] J. Hansen, *A connectedness theorem for flagmanifolds and Grassmannians*, Amer. J. Math. **105** (1983), 633–639.

[73] R. Hartshorne, *Algebraic Geometry*, Graduate Texts in Math., Springer-Verlag, 1977.

[74] R. Hartshorne, *Ample Subvarieties of Algebraic Varieties*, Lect. Notes in Math. Vol. **156**, Springer-Verlag, 1970.

[75] R. Hartshorne, *Cohomological dimension of algebraic varieties*, Annals of Math. **88** (1968), 403–450.

[76] R. Hartshorne, *Ample vector bundles on curves*, Nagoya Math. J. **43** (1971), 73–89.

[77] R. Hartshorne, *Curves with high self-intersection on an algebraic surface*, Publ. Math. IHES Vol. **36** (1969), 111–126.

[78] H. Hironaka, *On some formal embeddings*, Illinois J. Math. **12** (1968), 587–602.

[79] H. Hironaka, *Smoothing algebraic cycles of small dimensions*, Amer. J. Math. **90** (1968), 1–54.

[80] H. Hironaka, *Resolution of singularities of an algebraic variety over a field of characteristic zero*, Annals of Math. **79** (1964), 109–326.

[81] H. Hironaka and H. Matsumura, *Formal functions and formal embeddings*, J. Math. Soc. Japan **20** (1968), 52–82.

[82] N. J. Hitchin, *Kählerian twistor spaces*, Proc. London Math. Soc. **43** (1981), 133–150.

[83] C. Huneke and G. Lyubeznik, *On the vanishing of local cohomology modules*, Invent. Math. **102** (1990), 73–93.

[84] P. Ionescu and D. Naie, *Rationality properties of manifolds containing quasi-lines*, Intern. J. Math. **14** (2003), 1053–1080.

[85] P. Ionescu and C. Voica, *Models of rationally connected manifolds*, J. Math. Soc. Japan **55** (2003), 143–164.

[86] V. A. Iskovskih, *Fano 3-folds. I*, Math. USSR Izvestija **11** (1977), 485–527.

[87] J.-P. Jouanolou, *Théorèmes de Bertini et Applications*, Birkhäuser, Boston-Basel-Stuttgart, 1983.

[88] J. Kajiwara and E. Sakai, *Generalization of a theorem of Oka-Levi on meromorphic functions*, Nagoya Math. J. **29** (1967), 75–84.

[89] S. Kleiman, *Toward a numerical theory of ampleness*, Annals of Math. **84** (1966), 293–344.

[90] S. Kleiman, *On the vanishing of $H^n(X, F)$ for an n-dimensional variety*, Proc. Amer. Math. Soc. **18** (1967), 940–944.

[91] S. L. Kleiman, *The transversality of a general translate*, Compositio Math. **28** (1974), 287–297.

[92] S. L. Kleiman and J. Landolfi, *Geometry and deformation of special Schubert varieties*, Comp. Math. **23** (1971), 407–430.

[93] J. Kollár, *Rational Curves on Algebraic Varieties*, Ergebnisse der Mathematik und ihre Grenzgebiete, Springer, 1996.

[94] J. Kollár, Y. Miyaoka and S. Mori, *Rationally connected varieties*, J. Algebraic Geometry **1** (1992), 429–448.

[95] D. Laksov, *Indecomposability of the restricted tangent bundles*, Young tableaux and Schur functors in algebra and geometry, Astérisque **87-88**, Soc. Math. France, 1981, 207–219.

[96] A. Lanteri, *Ample vector bundle with sections vanishing on surfaces of Kodaira dimension zero*, Matematiche (Catania) **51** (1996), 115–125.

[97] A. Lanteri and H. Maeda, *Geometrically ruled surfaces as zero loci of ample vector bundles*, Forum Math. **9** (1997), 1–15.

[98] R. Lazarsfeld, *A sampling of vector bundle techniques in the study of linear series*, Lectures on Riemann surfaces (M. Cornalba et al., eds), World Scientific Press, 1989, pp. 500–559.

[99] R. Lazarsfeld, *Some applications of the theory of positive vector bundles*, in *Complete Intersections* (Acireale, 1983), Lecture Notes in Math. Vol. **1092**, Springer (1984), 21–61.

[100] R. Lazarsfeld, *Positivity in Algebraic Geometry*, Vols. I., II., Springer, Berlin Heidelberg New York Honk Kong London Milan Paris Tokyo, 2004 (to appear).

[101] J. Le Potier, *Annulation de la cohomologie à valeurs dans un fibré vectoriel holomorphe positif de rang quelconque*, Math. Ann. **218** (1975), 35–53.

[102] M. Lübke, *Beweis einer Vermutung von Hartshorne für den Fall homogener Mannigfaltigkeiten*, J. Reine Angew. Math. **316** (1980), 215–220.

[103] S. M. L'vovsky, *Extension of projective varieties I.*, Michigan Math. J. **39** (1992), 41–51.

[104] G. Lyubeznik, *Étale cohomological dimension and the topology of algebraic varieties*, Annals of Math. **137** (1993), 71–128.

[105] J. Milnor, *Morse Theory*, Annals of Math. Studies **51** Princeton Univ. Press. 1963.

[106] Y. Miyaoka and T. Peternell, *Geometry of Higher Dimensional Algebraic Varieties*, DMV Seminar, Birkhäuser Vol. 26, 1997.

[107] S. Mori and S. Mukai, *The uniruledness of the moduli space of curves of genus* 11, in Lecture Notes in Math. Vol. **1016**, Springer-Verlag, 1983, pp. 334–353.

[108] S. Mori and H. Sumihiro , *On Hartshorne's conjecture*, J. Math. Kyoto Univ. **18** (1978), 523–533.

[109] D. Mumford, *A remark on a paper of M. Schlessinger*, Rice Univ. Studies (Complex Analysis, 1972, Vol. I: Geometry of Singularities) **51** (1973), 113–117.

[110] D. Mumford, *Pathologies: III*, Amer. J. Math. **89** (1967), 94–104.

[111] D. Mumford, *Varieties defined by quadratic equations*, in "Questions on Algebraic Varieties", Centro Internazionale Matematico Estivo, Varenna, 1969 pp. 31–100.

[112] D. Mumford, *Introduction to Algebraic Geometry*, Harvard Univ., 1968.

[113] D. Mumford, *Lectures on Curves on Algebraic Surfaces*, Annals of Math. Studies **59**, Princeton, 1966.

[114] D. Mumford, *Abelian Varieties*, Lecture in TATA Institute of Fundamental Research, Bombay, 1968.

[115] M. Mustaţă and M. Popa, *A new proof of a theorem of Van de Ven*, Bull. Math. Soc. Sc. Math. Roumanie **39 (87)**, 1996, 243–249. (See also the corrected typographical version in the next volume of 1997.)

[116] M. Nagata, *Local Rings*, Wiley, New York, 1962.

[117] A. Ogus, *Local cohomological dimension of algebraic varieties*, Annals of Math. **98** (1973), 327–365.

[118] C. Okonek, M. Schneider and H. Spindler, *Vector Bundles on Complex Projective Spaces*, Progress in Math. **3**, Birkhäuser, Boston, Basel, Stuttgart, 1980.

[119] W. M. Oxbury, *Twistor spaces and Fano threefolds*, Quart. J. Math. Oxford **45** (1994), 343–366.

[120] H. Pinkham, *Deformations of Algebraic Varieties with $\mathbb{G}_m$-action*, Astérisque Vol. **20**, Soc. Math. France, 1974.

[121] N. Radu, *Local Rings (in Romanian)*, Vols. I–II, Edit. Acad. Române, Bucharest, 1968, 1970.

[122] Z. Ran, *On subvarieties of abelian varieties*, Invent. Math. **62** (1981), 459–479.

[123] R. Remmert and A. Van de Ven, *Zur Funktionentheorie homogener komplexer Mannigfaltigkeiten*, Topology **2** (1963), 137–157.

[124] H. Rossi, *Continuation of subvarieties of projective varieties*, Amer. J. Math. **91** (1969), 565–575.

[125] F. Russo, *On the complement of a nef and big divisor on an algebraic variety*, Math. Proc. Camb. Phil. Soc. **120** (1996), 411–422.

[126] F. Russo, *Tangents and Secants of Algebraic Varieties (Notes of a Course)*, **24**^{o} Colóquio Brasileiro de Matemática, Publicações Matemáticas IMPA, Rio de Janeiro, 2003.

[127] M. Schlessinger, *Infinitesimal Deformations of Singularities*, Ph.D. Thesis, Harvard, 1964.

[128] M. Schlessinger, *Rigidity of quotient singularities*, Invent. Math **14** (1971), 17–26.

[129] M. Schlessinger, *On rigid singularities*, Rice Univ. Studies **59** (1973), 147–162.

[130] M. Schneider, *Ein einfacher Beweis des Verschwindungssatzes für positive holomorphe Vektorraumbundel*, Manuscr. Math. **11** (1974), 95–101.

[131] G. Scorza, *Sopra una certa classe di varietà razionali*, Rend. Circ. Matem. Palermo **28** (1909), 400–401.

[132] G. Scorza, *Sulle varietà di Segre*, Opere scelte, Vol. **I**, pp.376–386.

[133] J.-P. Serre, *Algèbre Locale. Multiplicités*, Lecture Notes in Math. Vol. **11** Springer-Verlag, Berlin-Heidelberg-New York, 1965.

[134] J.-P. Serre, *Faisceaux algébriques cohérents*, Annals of Math. **61** (1955), 197–278.

[135] J.-P. Serre, *Géométrie Algébrique et Géométrie Analytique*, Ann. Inst. Fourier **6** (1956), 1–42.

[136] J.-P. Serre, *Sur la cohomologie des variétés algébriques*, Ann. Inst. Fourier **6** (1957), 1–16.

[137] F. Severi, *Alcune proprietà fondamentali dell'insieme dei punti singolari di una funzione analitica di più variabili*, Mem. Accad. Ital. **3** (1932).

[138] I. R. Shafarevich, *Basic Algebraic Geometry* **1**: *Varieties in Projective Space* (Second, Revised and Expanded Edition), Springer-Verlag, 1994.

[139] I. R. Shafarevich, *Basic Algebraic Geometry* **2**: *Schemes and Complex Manifolds* (Second, Revised and Expanded Edition), Springer-Verlag, 1997.

[140] A. J. Sommese, *Submanifolds of abelian varieties*, Math. Ann. **233** (1978), 229–256.

[141] R. Speiser, *Cohomological dimension and abelian varieties*, Amer. J. Math. **95** (1973), 1–34.

[142] R. Speiser, *Projective varieties of low codimension in characteristic* $p > 0$, Trans. Amer. Math. Soc. **240** (1978), 329–343.

[143] F. Steffen, *Connectedness theorems for determinantal schemes*, J. Algebraic Geometry **8** (1978), 169–179.

[144] A. Van de Ven, *A property of algebraic varieties in complex projective spaces, in "Colloque de Géométrie Différentielle Globale"*, Bruxelles, 1958, 151–152.

[145] L. Verdi, *Le curve razionali normali come intersezioni complete insiemistiche*, Boll. Unione Mat. Ital. **16-A (5)** (1979), 385–390.

[146] J. Wahl, *A cohomological characterization of* $\mathbb{P}^n$, Invent. Math. **72** (1983), 315–322.

[147] J. Wahl, *The Jacobian algebra of a graded Gorenstein singularity*, Duke Math. J. **55** (1987), 105–128.

[148] J. Wahl, *Introduction to Gaussian maps on an algebraic curve, in "Complex Projective Geometry"*, London Math. Soc. Lect. Notes Ser. **179**, Cambridge Univ. Press, 1992, 304–323.

[149] F. L. Zak, *Some properties of dual varieties and their applications in projective geometry*, Lect. Notes in Math. Vol. **1479**, Springer-Verlag, 1991, 274–280.

[150] F. L. Zak, *Tangents and Secants of Algebraic Varieties*, AMS Transl. Math. Monographs Vol. **127**, Providence, Rhode Island, 1993.

[151] O. Zariski, *Theory and Applications of Holomorphic Functions on Algebraic Varieties over Arbitrary Ground Fields*, Memoirs of the American Math. Soc., Vol. **5**, 1951.

[152] O. Zariski, *The theorem of Riemann-Roch for high multiples of an effective divisor on an algebraic surface*, Annals of Math., **76** (1962), 560–615.

[153] O. Zariski, *Interprétation algébro-géométrique du quatorzième problème de Hilbert*, Bull. Sci. Math. **78** (1954), 155–164.

[154] O. Zariski and P. Samuel, *Commutative Algebra*, Vol. **I, II**, D. Van Nostrand Company, Inc., Princeton, New Jersey, 1958, 1960.

Glossary of Notation

$\mathcal{O}_X$	structural sheaf of a variety X, 3
F^*	dual of a coherent sheaf F, 3
T_X	tangent sheaf of a variety X, 3
$N_{Y\|X}$	normal sheaf of a closed subvariety Y of a variety X, 3
$H^0(X,F)$	space of global sections of a coherent sheaf F, 4
$\mathrm{codim}_X(Y)$	codimension of a closed subvariety Y in X, 5
$\mathrm{Reg}(X)$	smooth locus of a variety X, 5
$\mathrm{Sing}(X)$	singular locus of a variety X, 5
$H^i_Y(X,F)$	i-th local cohomology of a sheaf F with supports in Y, 8
I-depth(M)	I-depth of an A-module M, 8
$\mathrm{Spec}(A)$	sprectrum of a commutative ring A, 8
$Z(s)$	zero locus of a global section S of a coherent sheaf F, 10
$\mathrm{char}(k)$	characteristic of a field k, 10
$\mathrm{Proj}(S)$	projective spectrum of a graded k-algebra S, 15
$\mathrm{Spec}(R)$	affine spectrum of a k-algebra R, 15
$D_+(s)$	principal open subset on Proj, 16
$\mathrm{Der}_k(R)$	R-module of k-derivations, 16
$\Omega^1_{X\|Y}$	sheaf of 1-differential forms relative to a morphism $X \to Y$, 3, 17
$\varphi_{L,M}$	Gaussian map associated two line bundles L and M, 31
w_L	Wahl's map associated to a line bundle L, 33
$T^1_{Y,y}$	space of first order infinitesimal deformations of (Y,y), 39
$\mathrm{Der}_k(A,M)$	A-module of k-derivations from A to M, 40
$T^1(A\|k,M)$	deformation module of $A\|k$ by M, 40
$\mathrm{Ex}^1(A\|k,M)$	set of classes of extensions of $A\|k$ by M, 40
$Y(1)$	first infinitesimal neighbourhood of Y in an ambient variety, 49
$\mathrm{NS}(X)$	$= \mathrm{Pic}(X)/\mathrm{Pic}^0(X)$: Néron–Severi group of a variety X, 51, 73
$\mathrm{Pic}(X)$	Picard group of a variety X, 52
$\mathrm{Coh}(X)$	category of coherent sheaves on X, 55
$\mathrm{cd}(X)$	cohomological dimension of a variety X, 55
$\mathrm{Supp}(F)$	support of a coherent sheaf F, 56
$K(X)$	field of rational functions of an irreducible variety X, 60, 82
$\mathbb{P}^n(e)$	weighted projective space of weights $e = (e_0, \ldots, e_n)$, 61
Δ_X	diagonal of $X \times X$, 64

T_y	projective tangent space at y of a subvariety $Y \subseteq \mathbb{P}^n$, 65
$(\mathbb{P}^n)^*$	projective space dual to $\mathbb{P}^n$, 66
$\mathrm{Pic}^0(X)$	group of algebraically trivial line bundles on X, 73
$\mathrm{Pic}^\tau(X)$	group of numerically trivial line bundles on X, 73
$\mathrm{Num}(X)$	the group $\mathrm{Pic}(X)/\mathrm{Pic}^\tau(X)$, 73
$H^i(X,\mathbb{Z})$	i-th singular cohomology with coefficients in $\mathbb{Z}$, 76
F^{an}	analytic sheaf associated to a coherent sheaf F, 76
$X_{/Y}$	formal completion of X along a closed subvariety Y, 82
$Y(n)$	n-th infinitesimal neighbourhood of Y in an ambient variety X, 82
$[A]_0$	total ring of fractions of a commutative ring A, 82
$K(X_{/Y})$	ring of formal-rational functions of X along Y, 82
$\alpha_{X,Y}$	canonical map $K(X) \to K(X_{/Y})$, 84
$\mathcal{M}_{X_{/Y}}$	sheaf of formal-rational functions of X along a closed subvariety Y,85
G_i in X	Y G_i in X $(i = 2,3)$, with Y a closed subvariety in X, 95 Y analytically G_i in X $(i = 2,3)$, 118
$\mathrm{Lef}(X,Y)$	Grothendieck–Lefschetz condition, 111
$\mathrm{Leff}(X,Y)$	effective Grothendieck–Lefschetz condition, 111
$\mathrm{Tors}(E)$	torsion subgroup of an abelian group E, 116
$D_s(\sigma)$	s-th degeneracy locus of a map $\sigma : E \to F$ of vector bundles, 132
$\mathrm{Alb}(X)$	Albanese variety of a variety X, 148
$\mathrm{amp}(X)$	ampleness of a homogeneous space X, 165
$\mathrm{ca}(X)$	coampleness of a homogeneous space X, 165
$\underline{\mathrm{Pic}}^0(X)$	Picard scheme of an algebraic scheme X, 176
$\mathrm{Hilb}(X)$	Hilbert scheme of a projective variety X, 181

Index

Monografie Matematyczne

[1] S. Banach, *Théorie des opérations linéaires*, 1932
[2] S. Saks, *Théorie de l'integrale*, 1933
[3] C. Kuratowski, *Topologie* I, 1933
[4] W. Sierpiński, *Hypothèse de continu*, 1934
[5] A. Zygmund, *Trigonometrical Series*, 1935
[6] S. Kaczmarz, H. Steinhaus, *Theorie der Orthogonalreihen*, 1935
[7] S. Saks, *Theory of the integral*, 1937
[8] S. Banach, *Mechanika*, T. I, 1947
[9] S. Banach, *Mechanika*, T. II, 1947
[10] S. Saks, A. Zygmund, *Funkcje analityczne*, 1948
[11] W. Sierpiński, *Zasady algebry wyższej*, 1946
[12] K. Borsuk, *Geometria analityczna w n wymiarach*, 1950
[13] W. Sierpiński, *Działania nieskończone*, 1948
[14] W. Sierpiński, *Rachunek różniczkowy poprzedzony badaniem funkcji elementarnych*, 1947
[15] K. Kuratowski, *Wykłady rachunku różniczkowego i całkowego*, T. I, 1948
[16] E. Otto, *Geometria wykreślna*, 1950
[17] S. Banach, *Wstęp do teorii funkcji rzeczywistych*, 1951
[18] A. Mostowski, *Logika matematyczna*, 1948
[19] W. Sierpiński, *Teoria liczb*, 1950
[20] C. Kuratowski, *Topologie* I, 1948
[21] C. Kuratowski, *Topologie* II, 1950
[22] W. Rubinowicz, *Wektory i tensory*, 1950
[23] W. Sierpiński, *Algèbre des ensembles*, 1951
[24] S. Banach, *Mechanics*, 1951
[25] W. Nikliborc, *Równania różniczkowe*, Cz. I, 1951
[26] M. Stark, *Geometria analityczna*, 1951
[27] K. Kuratowski, A. Mostowski, *Teoria mnogości*, 1952
[28] S. Saks, A. Zygmund, *Analytic functions*, 1952
[29] F. Leja, *Funkcje analityczne i harmoniczne*, Cz. I, 1952
[30] J. Mikusiński, *Rachunek operatorów*, 1953
* [31] W. Ślebodziński, *Formes extérieures et leurs applications*, 1954
[32] S. Mazurkiewicz, *Podstawy rachunku prawdopodobieństwa*, 1956
[33] A. Walfisz, *Gitterpunkte in mehrdimensionalen Kugeln*, 1957
[34] W. Sierpiński, *Cardinal and ordinal numbers*, 1965
[35] R. Sikorski, *Funkcje rzeczywiste*, 1958
[36] K. Maurin, *Metody przestrzeni Hilberta*, 1959
[37] R. Sikorski, *Funkcje rzeczywiste*, T. II, 1959
[38] W. Sierpiński, *Teoria liczb* II, 1959

* [39] J. Aczél, S. Gołąb, *Funktionalgleichungen der Theorie der geometrischen Objekte*, 1960
[40] W. Ślebodziński, *Formes extérieures et leurs applications*, II, 1963
[41] H. Rasiowa, R. Sikorski, *The mathematics of metamathematics*, 1963
[42] W. Sierpiński, *Elementary theory of numbers*, 1964
* [43] J. Szarski, *Differential inequalities*, 1965
[44] K. Borsuk, *Theory of retracts*, 1967
[45] K. Maurin, *Methods of Hilbert spaces*, 1967
[46] M. Kuczma, *Functional equations in a single variable*, 1967
[47] D. Przeworska-Rolewicz, S. Rolewicz, *Equations in linear spaces*, 1968
[48] K. Maurin, *General eigenfunction expansions and unitary representations of topological groups*, 1968
[49] A. Alexiewicz, *Analiza funkcjonalna*, 1969
* [50] K. Borsuk, *Multidimensional analytic geometry*, 1969
* [51] R. Sikorski, *Advanced calculus. Functions of several variables*, 1969
[52] W. Ślebodziński, *Exterior forms and their applications*, 1971
[53] M. Krzyżański, *Partial differential equations of second order*, vol. I, 1971
[54] M. Krzyżański, *Partial differential equations of second order*, vol. II, 1971
[55] Z. Semadeni, *Banach spaces of continuous functions*, 1971
[56] S. Rolewicz, *Metric linear spaces*, 1972
[57] W. Narkiewicz, *Elementary and analytic theory of algebraic numbers*, 1974
[58] Cz. Bessaga, A. Pełczyński, *Selected topics in infinite dimensional topology*, 1975
* [59] K. Borsuk, *Theory of shape*, 1975
[60] R. Engelking, *General topology*, 1977
[61] J. Dugundji, A. Granas, *Fixed point theory*, 1982
* [62] W. Narkiewicz, *Classical problems in number theory*, 1986

The volumes marked with * are available at the exchange department of the library of the Institute of Mathematics, Polish Academy of Sciences.

MIX
Papier aus verantwortungsvollen Quellen
Paper from responsible sources
FSC® C105338

If you have any concerns about our products,
you can contact us on
ProductSafety@springernature.com

In case Publisher is established outside the EU,
the EU authorized representative is:
Springer Nature Customer Service Center GmbH
Europaplatz 3, 69115 Heidelberg, Germany

Printed by Libri Plureos GmbH
in Hamburg, Germany